24 Springer Series in Solid-State Sciences
Edited by Peter Fulde and Hans-Joachim Queisser

Springer Series in Solid-State Sciences

Editors: M. Cardona P. Fulde H.-J. Queisser

1 **Principles of Magnetic Resonance**
2nd Edition   2nd Printing
By C. P. Slichter

2 **Introduction to Solid-State Theory**
By O. Madelung

3 **Dynamical Scattering of X-Rays in Crystals** By Z. G. Pinsker

4 **Inelastic Electron Tunneling Spectroscopy**
Editor: T. Wolfram

5 **Fundamentals of Crystal Growth I**
Macroscopic Equilibrium and Transport Concepts. By F. Rosenberger

6 **Magnetic Flux Structures in Superconductors** By R. P. Huebener

7 **Green's Functions in Quantum Physics**
By E. N. Economou

8 **Solitons and Condensed Matter Physics**
Editors: A. R. Bishop and T. Schneider

9 **Photoferroelectrics**
By. V. M. Fridkin

10 **Phonon Dispersion Relations in Insulators** By H. Bilz and W. Kress

11 **Electron Transport in Compound Semiconductors** By B. R. Nag

12 **The Physics of Elementary Excitations**
By S. Nakajima, Y. Toyozawa, and R. Abe

13 **The Physics of Selenium and Tellurium**
Editors: E. Gerlach and P. Grosse

14 **Magnetic Bubble Technology**
By A. H. Eschenfelder

15 **Modern Crystallography I**
Symmetry of Crystals,
Methods of Structural Crystallography
By B. K. Vainshtein

16 **Organic Molecular Crystals**
Their Electronic States
By E. Silinsh

17 **The Theory of Magnetism I**
Ground State and Elementary Excitations
By D. Mattis

18 **Relaxation of Elementary Excitations**
Editors: R. Kubo and E. Hanamura

19 **Solitons,** Mathematical Methods for Physicists
By G. Eilenberger

20 **Theory of Nonlinear Lattices**
By M. Toda

21 **Modern Crystallography II**
Structure of Crystals
By B. K. Vainshtein, A. A. Chernov, and L. A. Shuvalov

22 **Point Defects in Semiconductors I**
M. Lannoo and J. Bourgoin

23 **Physics in One Dimension**
Editors: J. Bernasconi, T. Schneider

24 **Physics in High Magnetic Fields**
Editors: S. Chikazumi and N. Miura

Physics in High Magnetic Fields

Proceedings of the Oji International Seminar
Hakone, Japan, September 10–13, 1980

Editors:
S. Chikazumi and M. Miura

With 257 Figures

Springer-Verlag Berlin Heidelberg New York 1981

Professor Dr. *Sōshin Chikazumi*
Professor Dr. *Noború Miura*

Institute for Solid-State Physics, University of Tokyo,
Roppongi, Minato-ku, Tokyo, Japan

Series Editors:

Professor Dr. Manuel Cardona
Professor Dr. Peter Fulde
Professor Dr. Hans-Joachim Queisser

Max-Planck-Institut für Festkörperforschung, Heisenbergstrasse 1
D-7000 Stuttgart 80, Fed. Rep. of Germany

Executive Committee:
S. Chikazumi   N. Miura   S. Tanuma

Organizing Committee:

S. Chikazumi	H. Fukuyama	S. Mase	Y. Nakagawa	S. Tanaka
(Chairman)	C. Hamaguchi	N. Miura	S. Narita	S. Tanuma
M. Date	S. Kawaji	*(Secretary)*	Y. Nishina	

International Advisory Committee:
L. Esaki (Yorktown Heights)   G. Landwehr (Grenoble)   R. A. Stradling (St. Andrews)
S. Foner (Cambridge)   R. Pauthenet (Grenoble)

Sponsors:
The Japan Society for the Promotion of Science, and
The Fujihara Foundation of Science

ISBN 3-540-10587-5 Springer-Verlag Berlin Heidelberg New York
ISBN 0-387-10587-5 Springer-Verlag New York Heidelberg Berlin

This work is subject to copyright. All rights are reserved, whether the whole or part of the material is concerned, specifically those of translation, reprinting, reuse of illustrations, broadcasting, reproduction by photocopying machine or similar means, and storage in data banks. Under § 54 of the German Copyright Law, where copies are made for other than private use, a fee is payable to 'Verwertungsgesellschaft Wort', Munich.

© by Springer-Verlag Berlin Heidelberg 1981
Printed in Germany

The use of registered names, trademarks, etc. in this publication does not imply, even in the absence of a specific statement, that such names are exempt from the relevant protective laws and regulations and therefore free for general use.

Offset printing: Beltz Offsetdruck, 6944 Hemsbach/Bergstr. Bookbinding: J. Schäffer oHG, 6718 Grünstadt.
2153/3130-543210

Preface

This volume represents the Proceedings of the Oji International Seminar on the Application of High Magnetic Fields in the Physics of Semiconductors and Magnetic Materials, which was held at the Hakone Kanko Hotel, Hakone, Japan, from 10 to 13 September 1980. The Seminar was organized as a related meeting to the 15th International Conference on the Physics of Semiconductors which was held in Kyoto between 1 and 5 September 1980. From 12 countries, 77 delegates participated in the Seminar. This Seminar was originally planned to be a formal series of International Conferences on the Application of High Magnetic Fields in the Physics of Semiconductors, which was first started by Professor G. Landwehr in 1972 in Würzburg as a satellite conference to the 11th Semiconductor Conference in Warsaw. The Conference in Würzburg was conducted in an informal atmosphere which was followed by three conferences, in Würzburg in 1974 and 1976, and in Oxford in 1978. At the current Seminar the physics of magnetic materials was added to the scope of the Seminar, because high-field magnetism is also an important research area in the physics of high magnetic fields and is also one of the most active fields in physics in Japan.

In the last decade, considerable effort has been devoted to develop the techniques for generating the high magnetic fields in many high-field laboratories in the world. As a result, the progress of the magnet technology for both steady and the pulsed high magnetic fields has enabled high precision measurements in much higher fields than ever possible before. For example, steady fields higher than 20 T are now practically available for solid-state experiments in some laboratories, and pulsed fields in the megagauss range (i.e., higher than 100 T) are also utilized for the actual experiments on solid-state physics. Owing to such rapid progress in measuring techniques, a great deal of new findings have been accumulated in both semiconductor physics and magnetism under high magnetic fields.From this point of view, the scope of the Seminar was mainly focused on new developments in semiconductor physics and magnetism in high magnetic fields, and on techniques for generating high fields in continuous and pulsed forms. Scientific interaction between scientists working on semiconductors and magnetism was another aim. Along this line, properties of magnetic semiconductors and the transport phenomena related to magnetic ions were discussed as one of the main topics.

In the last two Würzburg meetings, the main topic was the physics in the space charge layers in MOS devices. This was also one of the main streams in this Seminar, but at present our knowledge on this system seems to have been deepened to a great extent. In addition to MOS, another type of two-dimensional electronic system, i.e., superlattice arose much interest. Layered materials and intercalation compounds were also discussed extensively at the Seminar. Various new aspects were reported on infrared spectroscopy, impurity states and excitons, magneto-optics, magneto-transport phenomena, narrow-gap semiconductors, and so on.

In the first two chapters of this volume, new developments in the steady high-field systems and the pulsed high magnetic fields are described. We believe that they give a good overview of the present state of affairs.

Fortunately, the Seminar was conducted successfully in an informal atmosphere full with stimulating discussion. We think that this Seminar with a small number of participants was very useful to the participants for exchanging ideas and establishing basis for future collaboration. Although there were no Proceedings published from the Würzburg and Oxford meetings except for the Conference Booklets, we decided to publish the Proceedings of this Seminar, because we hope that it will be helpful to general readers who are interested in this subject, as well as to the Seminar participants. A photograph of the participants is on page 356. They can also be identified by the number given in italics in the list of contributors on page 358.

The Oji International Seminar was totally supported by Fujihara Foundation of Science financially and was operated under the auspices of the Japan Society for the Promotion of Science. Taking this opportunity, we would like to express our sincere gratitude to them.

Tokyo, November 1980　　　　　　　　　　　　　　*S. Chikazumi and N. Miura*

Contents

Part I. *Recent Progress in the Steady High Field System*

Recent Progress in the Generation of Continuous High Magnetic Fields
By G. Landwehr 2

Recent Developments in High Field Superconductors
By K. Tachikawa 12

Research at the Nijmegen High Field Magnet Laboratory
By H.W. Myron, C.J.M. Aarts, A.R. de Vroomen and P. Wyder 24

Part II. *Physics in Pulsed High Magnetic Fields*

Strong and Ultrastrong Magnetic Fields: Development, State of the
Art and Applications. By F. Herlach 34

Recent Topics on the Generation and Application of Pulsed High
Magnetic Fields at Osaka. By M. Date, M. Motokawa, K. Okuda,
H. Hori and T. Sakakibara 44

Explosive Generation of High Magnetic Fields in Large Volumes and
Solid State Applications. By C.M. Fowler, R.S. Caird, D.J. Erickson,
B.L. Freeman and W.B. Garn 54

Generation of Megagauss Fields by Electromagnetic Flux Compression
By N. Miura, G. Kido, H. Miyajima, K. Nakao and S. Chikazumi 64

Application of Megagauss Fields to Studies of Semiconductors and
Magnetic Materials. By G. Kido, N. Miura, M. Akihiro, H. Katayama
and S. Chikazumi 72

Part III. *Cyclotron Resonance and Laser Spectroscopy of Semiconductors*

Far-Infrared Spectroscopy of Semiconductors in High Magnetic Fields
By A.M. Davidson, P. Knowles, P. Makado, R.A. Stradling, S. Porowski
and Z. Wasilewski 84

Submillimeter-Magneto Spectroscopy of Semiconductors
By M. von Ortenberg and U. Steigenberger 94

Four-Wave Spectroscopy of Shallow Donors in Germanium
By R.L. Aggarwal 105

New Magneto-Optical Transitions in n-InSb: Mid-Gap Deep Defect Level
and Three-Phonon Assisted Processes. By D.G. Seiler, M.W. Goodwin
and W. Zawadzki 112

An Infrared Bolometer for Use in High Magnetic Fields
By J.P. Kotthaus, C. Gaus and P. Stallhofer 116

Part IV. *Impurity States in High Magnetic Fields*

D⁻ Centers in Semiconductors in High Magnetic Fields
By D.M. Larsen 120

High Magnetic Field Zeeman Splitting and Anisotropy of Cr-Related
Photo-Luminescence in Semi-Insulating GaAs and GaP. By L. Eaves,
T. Englert and C. Uihlein 130

Impurity States of Tellurium in High Magnetic Fields
By K. von Klitzing and J. Tuchendler 139

Part V. *Magneto-Transport Phenomena*

Anomalous Anisotropies in Rare-Earth Magnetic Compounds and Their
Behavior under High Magnetic Field. By T. Kasuya, K. Takegahara,
M. Kasaya, Y. Ishikawa, H. Takahashi, T. Sakakibara and M. Date .. 150

Anisotropic Scattering by Rare-Earth Impurities: Effect of a High
Magnetic Field. By J.C. Ousset, S. Askenazy and A. Fert 161

Magnetophonon Effect of Hot Electrons in n-InSb and n-GaAs
By C. Hamaguchi, K. Shimomae and J. Takayama 169

Transport Equations Treating Phonon Drag Effect in a Strong Magnetic
Field. By Y. Ono 174

Part VI. *Excitons and Magneto-Optics*

Exciton and Shallow Impurity States of Semiconductors in an Arbitrary
Magnetic Field. By N.O. Lipari and M. Altarelli 180

Magnetic Field Induced LT Mixing of the Multicomponent Polaritons in
CdTe. By K. Cho, S. Suga and W. Dreybrodt 190

Magneto-Optical Effects in III-VI Layer Compounds
By Y. Sasaki, N. Kuroda, Y. Nishina, H. Hori, M. Shinoda
and M. Date 195

Interband Faraday and Kerr Effects in Semiconductors: An Analysis by
Means of Equivalent Modulated Magneto-Optical Spectra of Ge and Si
By J. Metzdorf 199

Magnetoplasma Modes at the Interface between a Semiconductor and a
    Metallic Screen in the Faraday Geometry. By P. Halevi 203

Part VII. *Electron-Hole Drops and Semimetals*

Electron-Hole Liquid in Ge in High Magnetic Field
    By M.S. Skolnick 208

Far-Infrared Magneto-Plasma Absorptions of Electron-Hole Drops in
    Germanium. By S. Narita, K. Muro and M. Yamanaka 216

Electron Interactions in Bismuth
    By H.D. Drew and S. Baldwin 224

Part VIII. *Narrow Gap Semiconductors*

Free and Bound Magneto-Polarons in Narrow Gap Semiconductors
    By W. Zawadzki, L. Swierkowski and J. Wlasak 234

Investigation of Strain Effects in Epitaxial Semiconductor Films by
    Interband Magneto-Optical Transitions: IV-VI Compounds
    By H. Pascher, E.J. Fantner, G. Bauer and A. Lopez-Otero 244

Study of Electronic and Lattice Properties in (Pb, Sn, Ge) Te by
    Intra-Band Magneto-Optics. By T. Ichiguchi and K. Murase 249

Magnetic Freeze-Out and High Pressure Induced Metal-Nonmetal
    Transition in Low Concentration n-Type InSb. By R.L. Aulombard,
    A. Raymond, L. Konczewicz, J.L. Robert and S. Porowski 253

A Sensitive HgCdTe Bolometer for the Detection of Millimeter Wave
    Radiation. By B. Schlicht and G. Nimtz 257

Part IX. *Space Charge Layer and Superlattice*

Electric Subbands in a Magnetic Field
    By F. Koch 262

Electron Transport in Silicon Inversion Layers at High Magnetic
    Fields. By T. Englert 274

Temperature Dependence of Transverse and Hall Conductivities of
    Silicon MOS Inversion Layers under Strong Magnetic Fields
    By S. Kawaji and J. Wakabayashi 284

Two-Dimensional Charge Density Wave State in a Strong Magnetic Field
    By D. Yoshioka and H. Fukuyama 288

InAs-GaSb Superlattices in High Magnetic Fields
    By M. Voos and L. Esaki 292

Magnetic Quantization and Transport in a Semiconductor Superlattice
    By T. Ando 301

Part X. *Layered Materials and Intercalation*

Magnetoreflection and Shubnikov-de Haas Experiments on Graphite Intercalation Compounds. By M.S. Dresselhaus, G. Dresselhaus, M. Shayegan and T.C. Chieu 306

Electrical Properties of Layered Materials at High Magnetic Fields By S. Tanuma, Y. Ōnuki, R. Inada, A. Furukawa, O. Takahashi and Y. Iye 316

Possibility of the Magnetic Breakdown for $2H-MX_2$ in High Field By M. Naito, S. Tanaka and N. Miura 320

Part XI. *Magnetism and Magnetic Semiconductors*

High Field Magnetism
 By R. Pauthenet 326

Spectroscopy of Magnetic Insulating Transition Metal Dihalides in High Magnetic Fields. By J. Tuchendler 336

Semimagnetic Semiconductors in Magnetic Fields
 By T. Dietl 344

Photograph of the Participants of the Seminar 356
List of Persons in the Photograph 357
Index of Contributors 358

Part I

Recent Progress in the Steady High Field System

Recent Progress in the Generation of Continuous High Magnetic Fields

G. Landwehr

Max-Planck-Institut für Festkörperforschung
Hochfeldmagnetlabor Grenoble 166 X
38042 Grenoble-Cedex, France

High Magnetic Fields have been a valuable tool in solid state physics for a long time [1]. The highest fields, which can be produced nowadays, are in the 100 Tesla range. They are, however, of short duration, typically of the order of microseconds. Although many interesting results have been achieved with pulsed fields [2], it is true that continuous fields of a smaller magnitude are usually more advantageous. The reason for this is that for steady field measurements very sensitive and sophisticated experimental methods can be used, which for instance allow to reduce the noise level substantially. Moreover continuous fields in general offer larger volume which makes complicated experimental set-ups possible. For all these reasons the efforts to improve the methods for the generation of higher dc magnetic fields have never ended.

It is well known that a break-through in the generation of high magnetic fields occurred when the first superconducting coils were built [3]. In the meantime it has become possible to generate magnetic fields of 10 Tesla in a bore of a few centimeters quite inexpensively with NbTi coils. With somewhat higher cost 15 Tesla can be achieved with Nb_3Sn tape. In the recent past Nb_3Sn multifilament conductors have been produced commercially and the first 15 Tesla magnets with a few centimeters inner diameter, which can be energized quickly, are under construction. However, this paper will not be concerned with a review of the present status of our capabilities to generate the highest possible fields with superconductors. Instead emphasis will be on the recent development of high field water-cooled magnets and on the developments of hybrid magnets. The term hybrid magnet has been coined for a combination of a water-cooled resistive magnet and a large bore superconducting coil. By combining a 10 MW resistive coil with an 8 Tesla superconducting magnet, a dc field of 30 Tesla has been achieved in a bore of 32 mm [4].

Resistive Magnets

Because fields up to 15 Tesla can be produced economically with superconducting coils and because demand for higher magnetic fields has steadily increased, much effort has been devoted towards the generation of 20 Tesla and above in bores between 30 and 50 mm, employing 10 MW of power. It is well-known that high field magnets of this kind require a careful design. Most constructions are based on the well-known Bitter-concept. From slotted copper (or copper alloy) discs a multi-turn coil is built by interleaving adjacent discs. Electrical insulation is achieved by inserting non-conducting sheets with high durability. Bitter coils are usually cooled by water flowing through axial channels, which are generated by punching holes into each plate. For the innermost coil of a high field magnet sometimes radially oriented cooling slots are employed.

Recently it has been demonstrated at the MIT Francis Bitter National Magnet Laboratory that there is still room for the improvement of Bitter coils. By a combination of various measures the maximum field of a 10 MW, 3.2 cm bore magnet has been raised by about 1 T to 23.5 T [5]. Part of this increase was achieved by improving the current distribution in the magnet. It is well-known that a current distribution $j(r,z) \propto r/(r^2+z^2)^{3/2}$, ($r$ = radial distance from the magnet center, z = axial distance from the magnet center) as proposed by Kelvin gives the highest central field per Watt. Although such a distribution is hard to realize technically in a Bitter magnet, it can be approximated by increasing the current density near the midplane of the coil by inserting thinner discs. The efficiency has also been improved by diverting power from the outer part of the magnet to the inner one. This, however, required better cooling of the inner coil, which

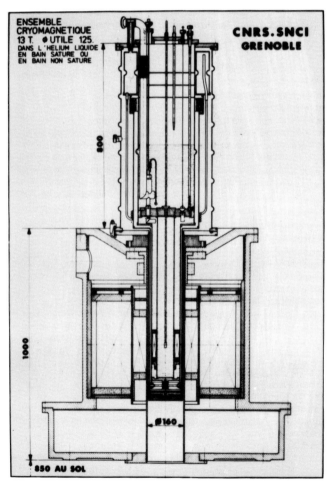

Fig.1 Schematic drawing of a 10 MW axially cooled, two stack Bitter magnet generating 14 Tesla in a 16 cm room temperature bore. With 8.5 MW 13 Tesla are produced. The helium cryostat shown has an inner diameter of 12.5 cm and allows the testing of experimental multifilament Nb_3Sn coils in the temperature range from 1.8 to 4.2 K.

was obtained by an enlargement of the cooling surface. In addition, the space factor could be enhanced by employing a thinner insulator. It was also possible, to run the magnet about 20 % cooler than before.

In the Grenoble high magnetic field facility, which is operated jointly by the Centre National de la Recherche Scientifique (CNRS) and the Max-Planck-Institut für Festkörperforschung, part of the effort was devoted towards increasing the longevity and reliability of the existing 10 MW magnets. There are 2 two-stage, axially cooled Bitter magnets, which allow to generate 20.1 Tesla in a 5 cm bore with 8.7 MW of power without problems. One of them has frequently been diverted into a 16 cm bore, 13 Tesla magnet requiring 8.5 MW of power. By tolerating a somewhat shorter lifetime, 14 Tesla can be achieved at the full power of 10 MW. This magnet is shown schematically in Fig. 1 together with an insert cryostat of 12.5 cm inner bore, which allows testing small superconducting coils made of multifilament Nb_3Sn. Because conductors of this kind are very sensitive to mechanical stress it is often necessary, to determine their critical current as a function of the magnetic field experimentally [6].

The crucial problems in building high field magnets are the mechanical stresses and the cooling. The tangential stress σ in a thin coil with radius r and current density j in a magnetic field B is $j \cdot r \cdot B$. In high field magnets the stresses at the inside of the coils - where B is large - easily can exceed the yield strength of ordinary copper. Therefore hardened copper or copper-alloys with better mechanical strength are usually employed for the construction of magnets generating high fields. However, increased tensile strength is accompanied by a larger resistivity, which results in higher power dissipation, which in turn requires enhanced cooling. The 1/r current distribution of Bitter magnets is in principle advantageous for their efficiency, but the power concentration close to the bore requires a large density of cooling holes in this area, which weakens the mechanical strength. The heat transfer can be improved by radial cooling channels, but in order to relief the stresses in magnets designed for the highest attainable fields, it is necessary to split the magnet into several subunits.

An alternative approach to generate continuous high magnetic fields has been pursued in Grenoble with the so-called polyhelix concept. A polyhelix magnet consists of a series of concentric, single layer helical coils. The principle can be visualized in Fig. 2, where a schematic presentation of the concept is given. The turns of the coils are glued together by an epoxy resin, which has sufficient strength at elevated temperatures and which does not degrade substantially when exposed to water. The cooling water flows axially through the annular space between adjacent helices. It is evident that in a polyhelix magnet the cooling area is considerably enhanced with respect to axially cooled Bitter magnets. In addition, the surface of the cooling area is well defined. A definite advantage is the possibility of reducing the current through each helix to such a level that the mechanical stress dois not exceed a tolerable level. This is possible, because each helix has a separate connection with the power supply. Moreover the tangential stress is reduced in comparison to Bitter magnets. A polyhelix magnet can be calculated rather precisely and optimized with respect to the existing boundary conditions. It turns out that the space factor is very advantageous in polyhelix magnets. Due to the compactness of the construction a power dissipation of about 2 kW/cm^3 at a heat flux of about 5 W/mm^2 can be achieved. A 5 MW polyhelix coil with an inner diameter of 57 mm and an outer diameter of 160 mm generated 14.5 Tesla in Grenoble. A 5 MW Bitter magnet, producing the same field, has about twice

the diameter and an average power dissipation of about 0.5 kW/cm^3. When
the 5 MW polyhelix coil was surrounded by a single 5 MW Bitter coil (which
was not optimized, but produced from existing Bitter discs) a field of
22.3 Tesla was generated in a 5 cm bore [8]. This duplex-magnet has been
used to investigate the electronic structure of silicon field-effect
transistors [9]. Recently the outer stack of the 10 MW duplex-magnet has
been replaced by two optimized Bitter coils, which allow producing 10.3
Tesla in 18 cm bore. With the 5 MW polyhelix insert a field of 23.4 Tesla
was achieved in a room temperature bore
of 50 mm at a power of 8.7 MW. Details
will be published elsewhere [10]. A
few modifications should allow to
generate soon close to 25 Tesla in a
50 mm bore. These results demonstrate
that the polyhelix concept is very
promising. Some early problems arising from Lorentz forces on the current
leads seem to be under control. Although the concept of a polyhelix
magnet is quite simple, its production
is relatively demanding and time consuming. In principle it is possible
to build a 10 MW magnet entirely of
helical coils. It turns out, however,
that the manufacturing of the outer
helices requires very high precision
because of the small annular space
between two adjacent helices. Optimization calculations [11] show,
however, that a 10 MW magnet consisting
of an inner 5 MW polyhelix and 2 outer
2.5 MW Bitter coils yields practically
the same field as a magnet consisting
entirely of helices. The reason for the suitability of Bitter stacks in
the outer part of the magnet is, that the cooling requirements and the
stress limitations are not serious. Consequently the more economical
duplex construction is advisable.

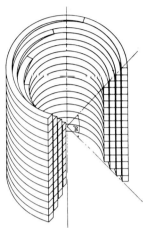

Fig. 2 Schematic drawing of
a polyhelix magnet coil after
ref. [8].

It should be possible to increase the maximum field of a 10 MW duplex magnet by reducing the bore of the magnet. Calculations have shown that a
magnet with a useful inner diameter of 30 mm should generate approximately
27 Tesla [11].

Hybrid Magnets

In principle it is possible to generate magnetic fields exceeding 25 Tesla
by employing larger power supplies. This straightforward method is, however, rather inefficient because the stress limitations require that the
additional power is dissipated in the outside region of the magnet. It is
by far more economical to generate the additional field with a large superconducting coil put around a resistive magnet. This hybrid concept, which
was proposed by Montgomery [12], has been successfully applied in several
laboratories. The largest steady field produced with a hybrid magnet was
generated at the Francis Bitter National Magnet Laboratory when an 8.5 Tesla
superconducting coil of 40 cm bore was combined with a 10 MW Bitter magnet
of 32 mm inner diameter. The maximum field achieved was 30 Tesla. The
superconducting coil was shipped afterwards to the University of Nijmegen

where it generates together with a 6 MW resisitive coil a dc field of
26 Tesla [1]. A 25 Tesla 5.6 MW hybrid coil was built at the Kurchatov
Institute in Moscow and a 2 MW, 16 Tesla system is operated at the Clarendon Laboratory of the University of Oxford [1].

Recently the CNRS and the MPG agreed to build a hybrid system in Grenoble.
A study was made taking into account several options. The already existing
hybrid systems employ NbTi superconducting coils operating at 4.2 K. The
achieved fields vary between 7 and 8.5 Tesla, depending on the amount and
kind of conductor used. A possibility, to enhance the maximum field in
NbTi coils, is to operate them below the helium λ point. This has not only
the advantage, to increase both the critical current and the critical magnetic field, but also results in a better stability due to improved heat
transfer by superfluid helium. The He II-cooling has been widely practised
for small magnets, but hardly for large ones. The reason for this is that
a substantial pumping capacity is required and that it is not easy to keep
the helium bath under reduced pressure for a long time due to the refilling-problem. In addition the electrical breakdown voltage in the gaseous phase
is decreased and the evaporation rate can become a problem. A method to
surmount these difficulties is, to employ superfluid helium at atmospheric
pressure for the cooling of a magnet [13]. This is possible by immersing
a heat exchanger through which superfluid helium is circulated into the
cooling bath. Recently it has been demonstrated that the concept is
technically feasible, a large cryostat housing a 10 Tesla NbTi magnet with
a 30 cm bore has been successfully tested [14]. The principle [15] of
such a cryostat is shown in Fig. 3. A large helium reservoir of He I is
located in the upper part of the cryostat. It accommodates most of the
heat losses from the top of the vessel and the losses arising from the
current leads. The He I bath is separated from He II reservoir by a lid of
insulating material containing a safety valve. The cooling of the lower
helium bath, which is always under atmospheric pressure, is achieved by a
heat exchanger of relatively modest dimensions. The helium for the primary
cooling circuit is drawn from the upper He I reservoir and the necessary
pumping capacity is relatively low, two 100 m^3/h pumps are sufficient to
handle a heat load of 1.5 W. Certainly the refilling of the cryostat is
easily done.

Calculations have shown that the application of the 1.8 K, atmospheric
pressure helium cooling concept to the booster coil of a hybrid system
weighing about 3.5 tons is feasible. An advantage of the cooling by superfluid helium is that in principle no axial cooling channels are necessary.
The cooling power can be provided entirely by radial cooling slots [14].

Another possibility is to employ multifilament Nb_3Sn as conductor for the
booster coil of a hybrid system. The maximum field, which can be produced
with the present day technology without unduly high cost, is in the vicinity of 13 Tesla. In order to handle the stresses and in order to protect
the coil against damage after possible quenching, it is necessary to provide
the conductor with a relatively large amount of copper and to reinforce it,
for instance with stainless steel. This results in a rather large weight
of the coil.

From the beginning it seemed advisable to use a 10 MW polyhelix or duplex
magnet as an insert of the projected hybrid magnet. In order to find out
which maximum fields could be obtained with the various options of the
superconducting booster coils, optimization calculations were performed
for inner polyhelix coils [16]. Such calculations seemed mandatory because

not only the stress limitations of the available conductor materials are important, but also the tolerable heat flux. In the calculations made the polyhelix magnets were optimized with respect to the stress, heat flux, materials and dimensions for each individual helix for a given external magnetic field. Due to the many variables involved such calculations are relatively difficult. The strategy has been to limit the stress level in the helices to a chosen, tolerable value. Simultaneously the heat flux was limited to a level which is technically feasible. Moreover it was necessary, to find out which kind of conductor can be used most advantageously, because it is by no means obvious whether it would be better, for instance to employ a copper-silver alloy with 98 % of the conductivity of copper and a yield strength of 400 N/mm^2 or copper containing dispersed Al_2O_3 with a yield strength of 550 N/mm^2 at a reduced conductivity of 87 % relative to copper.

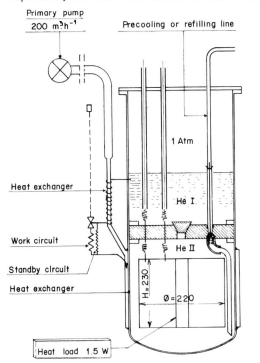

Fig.3 Schematic concept of a cryostat in which a superconducting magnet is cooled by superfluid helium at atmospheric pressures (after ref. [15])

Details of the calculations have been published elsewhere [16]. The first calculations were made for an external field of 13 Tesla generated by a Nb$_3$Sn coil of 40 cm bore. The maximum central field for a tolerated heat flux of 5 W/mm^2 has been plotted as a function of the input power of the polyhelix magnet for Cu-Al and Cu(Al$_2$O$_3$) in Fig. 4. It can be seen, that in a room temperature bore of 50 mm a field well in excess of 30 Tesla can be generated. It turns out that for power inputs above 6 MW, Cu(Al$_2$O$_3$) is the better material. An examination of the stress and the heat flux of the single helices indicates that at 10 MW a magnet made of Cu-Ag is entirely stress limited for the given tolerated stress level of 300 N/mm^2. That means that the power dissipation is below the value given by the upper heat flux limit of 5 W/mm^2. On the other hand it turns out that the Al$_2$O$_3$-magnet is entirely heat flux limited. The stress in the helices never reaches the predetermined, admitted values (which correspond to a strain of 0.2 %). At input powers below 6 MW with Cu-Ag a higher field can be generated than with Cu(Al$_2$O$_3$). A reduced number of helices is stress limited and the superior conductivity of Cu-Ag partially compensates for the shifting of power to the outside region of the magnet. The dotted lines in the figure indicate which central fields could be obtained if the magnets

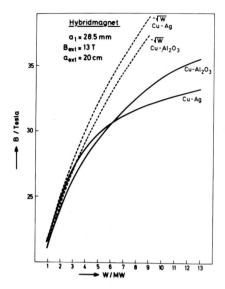
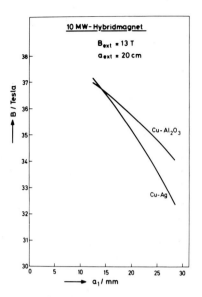

Fig. 4 Central magnetic field vs. input power of a hybrid magnet consisting of an inner polyhelix coil (bore 50 mm) and an outer 13 Tesla superconducting coil with 40 cm room temperature bore, for two resistive insert materials, Cu-Ag and Cu(Al$_2$O$_3$). The maximum heat flux is 5 W/mm^2. The dotted line would apply if no heat flux and stress limits would exist (after ref. [16]).

Fig. 5 Maximum central field of a 10 MW polyhelix magnet in a 13 Tesla external field as a function of the radius of the inner bore for two different materials (after ref. [16]).

were not limited with respect to stress and heat flux. In Fig. 5 the dependence of the central field on the radius of the innermost helix coil has been plotted for Cu/Ag and Cu(Al$_2$O$_3$). A reduction of the radius to 18 mm corresponding to a 30 mm room temperature bore results in a central field in the vicinity of 36 Tesla for Cu(Al$_2$O$_3$), the maximum fields obtained with Cu-Ag are always below those obtained with Cu(Al$_2$O$_3$), again for a heat flux limit of 5 W/mm^2. Because a heat flux limit of 5 W/mm^2 is rather conservative - as the experience with polyhelix magnets in Grenoble has shown - calculations of the maximum center field in an external field of 13 Tesla have been performed for radii a_1 of the inner bore of 13, 18, 23 and 28 mm for maximum heat flux values between 4 and 8 W/mm^2. The results shown in Table 1 indicate that in a 30 mm room temperature bore a field around 39 Tesla could be obtainable. A further increase seems possible by grading the polyhelix coil, that means by manufacturing the inner part out of Cu(Al$_2$O$_3$) and the outer one out of Cu-Ag. In Fig. 6 the maximum calculated central field of a 10 MW (B_{ex} = 13 T) hybrid magnet with 5 cm room temperature bore has been plotted for a polyhelix insert consisting of N_1 inner coils manufactured of Cu(Al$_2$O$_3$) and of (15-N_1) outer coils made of Cu-Ag.

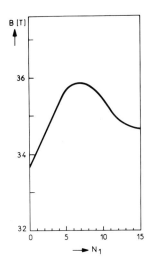

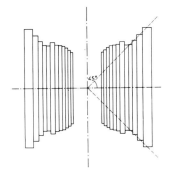

Fig. 7 Cross section through an optimized polyhelix magnet. The shape results from the imposed stress and heat flux limitations in conjunction with the radial dependence of the magnetic field (after ref. [16]).

Fig.6 Maximum central field of a 10 MW, $B_{ext} = 13$ Tesla hybrid magnet with 5 cm room temperature bore with a polyhelix magnet made of $Cu(Al_2O_3)$ for the N_1 inner coils and of Cu-Ag for the remaining helices. The number N of the helices is 15 (after ref. [16]).

Table 1 Central field for 10 MW, $B_{ext} = 13$ Tesla, hybrid magnets with $Cu(Al_2O_3)$ insert for various radii and heat flux limits

Maximum heat flux [W/mm²]	Radius a_1 of the inner bore [mm]				
	13	18	23	28	
4	36.5	35.5	34.6	33.6	[Tesla]
5	37.9	36.8	35.6	34.4	[Tesla]
6	39.1	37.7	36.3	35.0	[Tesla]
7	39.9	38.4	36.7	35.1	[Tesla]
8	40.6	38.8	36.9	35.3	[Tesla]

It can be recognized that the central field is roughly enhanced by 1 Tesla by the grading. In Fig. 7 a cross section through a polyhelix insert consisting of 12 helices is shown. The last 4 helices give a contour of conical shape indicating that no stress and heat flux limitations are present. The inner 8 helices do not differ very much in height. Calculations [11] show that the maximum field does not vary substantially if they are made of equal height, which is desirable for structural reasons. Because the outer sections of the coil are not heat flux or stress limited, it is possible to replace them by two Bitter coils of different height without substantial loss in the central field. It was already mentioned, that Bitter coils are easier and faster to manufacture than polyhelix magnets. Consequently the duplex construction seems to be most suitable for the insert coil.

Optimization calculations were also made for external fields of 10 and 8 Tesla, both for duplex and polyhelix inserts. The maximum tolerated temperature is 80° C. Results for the dependence of the central field for a 10 MW duplex-insert coil made of $Cu(Al_2O_3)$ (B_{ext} = 10 T) on the radius of the inner helix are shown in Fig. 8. One can recognize that for a heat flux of 6 W/mm^2 in a room temperature bore of 5 cm about 31.4 Tesla should be obtainable and about 34 Tesla in a 3 cm bore. An increase in the heat flux to 8 W/mm^2 will raise the maximum field by about 1 Tesla. This can be seen from Tab. 2. The calculation was made for a polyhelix insert made of $Cu(Al_2O_3)$ and a maximum stress of 450 N/mm^2. It turns out, however, that a reduction of the stress to 400 N/mm^2 does not decrease the central field in a 30 mm bore more than 0.3 Tesla. With a tolerated stress of 350 N/mm^2 and a heat flux of 8 W/mm^2 still a field of 34.4 Tesla is obtained. It turns out that for this σ the coil is stress limited already at a heat flux of 6 W/mm^2.

Table 2 Central field for 10 MW, B_{ext} = 10 Tesla hybrid magnets with polyhelix $Cu(Al_2O_3)$ insert for various radii and heat flux limits.

Maximum heat flux [W/mm^2]	Radius a_1 of the inner bore [mm]				
	13	18	23	28	
4	33.3	32.8	32.0	31	[Tesla]
5	34.9	34.0	32.8	31.6	[Tesla]
6	36.0	34.8	33.4	32.0	[Tesla]
7	36.6	35.1	33.9	32.3	[Tesla]
8	37.3	35.8	34.0	32.3	[Tesla]

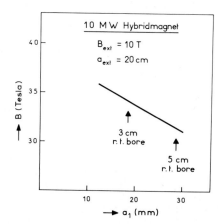

Fig. 8 Maximum central field of a 10 MW $Cu(Al_2O_3)$ duplex insert in a 10 T external field as a function of the inner radius a_1.

After it was found that with an external field of 10 Tesla it seems possible to generate a central field above 30 Tesla in a 5 cm bore it was decided to build the first Grenoble hybrid magnet with a 10 Tesla NbTi coil, cooled with superfluid helium of 1.8 K at atmospheric pressure. Although it seems also feasible to build a 13 Tesla multifilament Nb_3Sn booster coil at present, it seemed advisible to postpone the construction of such a magnet somewhat, until more experience with large bore 13 Tesla Nb_3Sn multifilament coils is available. It is clear that the specifications for a hybrid magnet booster coil are especially tough, because quenching of the coil should by all means be avoided, even in the case of an emergency shut down of the resistive part.

Acknowledgements

The author would like to acknowledge interesting discussions with R. Pauthenet, J.C. Vallier, P. Rub, J.C. Picoche, H.J. Schneider-Muntau and K. Hackbarth. Special thanks are deserved by H. Weber, who performed the hybrid magnet optimization calculations reported here.

References

1 See, e.g., Proc.Int.Conf.on Solids and Plasmas in High Magnetic Fields, Cambridge 1978, R.L. Aggarwal, A.H. Freeman and B.B. Swartz Eds., J. Magnetism Magn. Mat. 11 (1979)
2 N. Miura, G. Kido, M. Akihiro and S. Chikazumi, ref. [1], p. 275
3 J.E. Kunzler, E. Buehler, F.S.L. Hsu and J.J. Weinick, Phys. Rev. Lett. 6, 89 (1961)
4 M.J. Leupold, R.J. Weggel and Y. Iwasa, Proc. 6th Int.Conf. on Magnet Technology, Bratislava 1977, p. 400
5 R.J. Weggel and M.J. Leupold, Annual Report 1977, Francis Bitter National Magnet Laboratory, MIT, Cambridge, USA
6 P. Rub, M. Lombardi and H.J. Schneider-Muntau, Proc. 6th Int.Conf.on Magnet Technology, Bratislava 1977, p. 406; J.C. Vallier, personal comm.
7 H.J. Schneider-Muntau, Lecture Notes Intern.Conf.on Applications of High Magnetic Fields in Semiconductor Physics (Phys.Inst.Univ.Würzburg, 1974), p. 120
8 G. Landwehr, Journ.Magnetism Magn.Materials 13, 13 (1979)
9 Th.Englert, D.C. Tsui and G. Landwehr, Solid State Comm. 33, 1167 (1980)
10 H.J. Schneider-Muntau, to be published
11 H. Weber, unpublished
12 D.B. Montgomery, Solenoid Magnet Design, Wiley Interscience, 1969, p. 107
13 P. Roubeau, C.R. Acad.Sci Paris 273, 581 (1971)
14 J.C. Vallier, personal communication
15 G. Bon Mardion, G. Claudet and J.C. Vallier, Proc. 6th Int.Cryogenic Eng. Conf. Denver (1976) p. 159
16 H. Weber, H.J. Schneider-Muntau and G. Landwehr, Applied Physics 20, 163 (1979)

Recent Developments in High Field Superconductors

K. Tachikawa

National Research Institute for Metals, 1-2-1, Sengen, Sakura, Niiharigun
Ibaraki 305, Japan

1. Introduction

High-field superconducting magnets capable of generating magnetic fields over 10 T are much useful for research works in solid state physics. The development of compound-type superconductors is indispensable for generating magnetic fields over 10 T. Up to date, a variety of processes, such as the surface diffusion process, the composite process, the insitu process, have been developed to fabricate mechanically brittle A15 crystal-type Nb_3Sn and V_3Ga compounds.

The Nb_3Sn tape produced by the surface diffusion process first generated magnetic fields over 10 T in 1966, and is being widely used for 15 T-class superconducting magnets. Then, the V_3Ga tape with better current-carrying capacity than the Nb_3Sn tape in high magnetic fields was developed, and a 17.5 T superconducting magnet was successfully constructed in 1975. Figure 1 schematically shows the chronological increase of magnetic field generated by superconducting magnets.

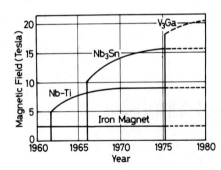

Fig.1 Magnetic field generated by superconducting magnet at 4.2 K versus year

2. Developments of Nb_3Sn and V_3Ga Tapes by the Surface Diffusion Process

The A15 crystal-type Nb_3Sn tape was first developed by the surface diffusion process (SDP) [1]. In the SDP, tin diffuses from the surface of a niobium tape passing continuously through a molten tin bath held about 900°C. In the subsequent heat treatment at about 950°C, Nb_3Sn layers are formed on the both sides of the tape. Finally the tape is sandwiched by foils of a low resistivity copper for the stabilization and, if necessary, by stainless-steel foils for the reinforcement. The SDP Nb_3Sn tape has been widely used for the construction of small or medium size high-field superconducting magnets.

The critical current density versus magnetic field, J_c-H of V_3Ga is convex upward in field above 10 T [2], whereas that of Nb$_3$Sn is convex downward. Thus, V$_3$Ga shows a considerably higher J$_c$ at fields above 13 T than Nb$_3$Sn, although both compounds have nearly the same upper critical field, H$_{c2}$(4.2K), of about 22 T. The SDP to produce V$_3$Ga tape consists of two stages [3,4]. In the first stage, a vanadium substrate tape, about 50 μm in thickness, is passed continuously through a molten gallium bath held about 500°C and then through a tube electric furnace held about 800°C. After the first stage, V$_3$Ga$_2$ and VGa$_2$ compound layers, totaling about 10 μm in thickness, are formed on both sides of the substrate. In the second stage, the tape is electroplated with copper layers, and then heat treated at about 650°C to form V$_3$Ga layers. The copper rapidly diffuses into the V$_3$Ga$_2$ and VGa$_2$ compound layers to form a ternary Cu-Ga-V alloy which has a much lower melting point than those compounds. The copper changes the diffusion mode from a grain boundary one to a bulk one, and significantly enhances the formation of V$_3$Ga [5]. The copper does not dissolve into the V$_3$Ga, and hence does not degrade the intrinsic superconducting properties of V$_3$Ga. Instead of electroplating, the copper may be added to the molten gallium bath. On both sides of the diffusion tape, copper foils of 25 μm in thickness are soldered for stabilization. The resulting V$_3$Ga tape is shown in Fig. 2.

Fig.2 The surface diffusion processed V$_3$Ga tape which carries 180 A at 17.5 T (10 mm wide and 0.13 mm thick)

Fig.3 The 17.5 T superconducting magnet system. The magnet has a clear bore of 31 mm, an outer diameter of 400 mm and a height of 600 mm [6]

In 1975, a 17.5 T superconducting magnet with an outer Nb_3Sn and an inner V_3Ga section has been constructed and successfully operated in NRIM [6]. The outer Nb_3Sn section generates 13.5 T in a 160 mm bore, and the inner V_3Ga section generates an incremental 4.0 T in a 31 mm clear bore. The magnet is cooled down from room temperature to about 10 K by the parallel operation of two hellium refrigerators with a total refrigeration power of 746 W at 20 K. It takes about 6 hr to cool the magnet from room temperature to 10 K, and subsequently about 30 l of liquid helium are required to cool the magnet to 4.2 K. The 17.5 T superconducting magnet system including the cryogenic system is shown in Fig.3. The SDP A15 tapes, however, are not stable under a rapid field change, and the magnet wound by the SDP tape can not be energized rapidly.

3. The Developments of Multifilamentary V_3Ga and Nb_3Sn Wires by the Composite Diffusion Process

The multifilamentary A15 compound wires were expected to be more stable than tapes and to be useful for applications in time varying fields. The composite diffusion process (CDP) to fabricate multifilamentary V_3Ga wires has been developed as an extention of the SDP [7]. In the CDP, a composite of a Cu-Ga solid solution alloy matrix containing 15-20 at% Ga and vanadium cores is fabricated into a thin wire, and then heat treated at a temperature between 600-650°C. In the heat treatment, gallium in the Cu-Ga alloy matrix selectively diffuses with vanadium, and only V_3Ga layers are formed around the vanadium cores. The copper acts as a mother metal of gallium, and scarcely dissolves into the V_3Ga layer [8]. When the gallium concentration in the Cu-Ga matrix exceeds 40 %, compounds richer in gallium appear. The CDP V_3Ga shows about the same T_c, but a 0.5-1.0 T lower $H_{c2}(4.2K)$, as compared to those of the SDP V_3Ga [9].

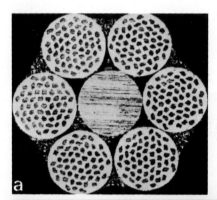

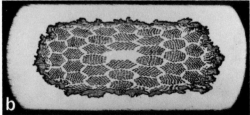

Fig.4 Photomicrographs of the cross section of the multifilamentary V_3Ga wires; (a) 6 stranded wire. Overall diameter: ~10μm, core number: 55 × 6 = 330 [11]. (b) Multifilamentary V/Cu-Ga/Nb/Cu composite tape (before final rolling). Final dimension: 0.16 5.0 mm, core diameter: 6μm, core number: 2574

The CDP Nb_3Sn wire has been also fabricated by a similar process in which a composite of niobium cores and a Cu-Sn matrix is drawn into a thin wire and then heat treated at about 700°C [10]. The typical tin concentration in the Cu-Sn matrix is 6-8 at%.

The multifilamentary V_3Ga wire with excellent stability and high-field performance has been commercially produced by the CDP. A cross section of stranded V_3Ga wire is shown in Fig.4a. This wire shows an overall J_c of 3.3×10^4 A/cm^2 at 15 T [11]. The critical current I_c of small coils wound with the multifilamentary V_3Ga wire has been measured under pulsed current excitation. The result shows that the multifilamentary V_3Ga wire is stable even under a rapid field change of over 20 T/sec [12]. A stable and compact 10 T magnet has been wound with this wire in 1975 [13]. Recently, a 13 T magnet has been constructed using the multifilamentary V_3Ga tape with larger current capacity, of which cross section is shown in Fig.4b. The outerview of the 13 T magnet is shown in Fig.5. The magnet is very stable and convenient to operate; it can be energized very quickly and does not quench even if the power source is turned off at 13 T.

The multifilamentary Nb_3Sn wires produced by the CDP are also being commercially developed for various applications. Especially for fusion reactor magnets, CDP Nb_3Sn conductors with large current capacities, such as 5 kA at 12 T, are being developed using a large scale fabrication techniques [14].

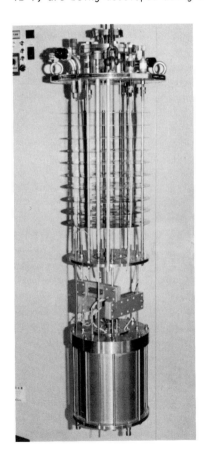

Fig.5 The 13 T superconducting magnet with high stability. Outer diameter 250 mm, clear bore 30 mm, height 250 mm.

4. Improvements in High-Field Performances of the Composite Diffusion Processed V_3Ga and Nb_3Sn

Recently several studies have been made on the effects of additional elements on the superconducting properties of the CDP A15 conductors. Here a few improvements in high-field properties of the CDP V_3Ga and Nb_3Sn conductors achieved in NRIM are descrived. The addition of a small amount of magnesium to the Cu-Ga matrix has been found to increase the formation rate and J_c of the CDP V_3Ga [15]. The scanning electron microscopy (SEM) has revealed that the magnesium addition to the matrix decreases the grain size of V_3Ga. This grain refinement may attribute to the increase of formation rate through the enhancement of the grain boundary diffusion. The decrease of the grain size is also effective for the increase of J_c, since grain boundaries are the most predominant defect which accounts for the flux pinning. The T_c and H_{c2} of the CDP V_3Ga are slightly increaseed by the gallium addition to the vanadium core. According to the X-ray microanalysis (XMA), the addition of gallium to the core increases the gallium concentration in the V_3Ga from 22.5 at% to 23.5 at%. This increase in the gallium concentration, which shifts the composition to the stoichiometry, may account for the increases in T_c and H_{c2}. The simultaneous addition of gallium to the core and magnesium to the matrix increases H_{c2} by about 2 T and significantly increases the critical current I_c of the CDP V_3Ga in high magnetic field. The I_c and J_c versus H curves for the specimens heat treated at different temperatures are shown in Fig. 6. The compositions of the specimens are listed in the caption. In the 3GM and 6GM specimens, J_c of over 1×10^5 A/cm^2 is obtained at 20 T even for thick V_3Ga layers of 5-10 μm. This may produce a marked improvement in overall J_c of the multifilamentary V_3Ga wire in high magnetic fields. The addition of small amount

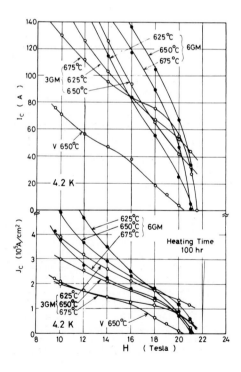

Fig.6 I_c and J_c versus magnetic field curves for the improved CDP V_3Ga specimens heat treated at quoted temperatures for 100 hr. The thicknesses of V_3Ga layers in the 3GM and 6GM specimens heat treated at 625°C and at 650°C are about 6 μm and 9 μm, respectively [15].

Specimen	Composition (at%)	
	Core	Matrix
V	pure V	Cu-19Ga
3GM	V-3Ga	Cu-19Ga-0.5Mg
6GM	V-6Ga	Cu-19Ga-0.5Mg

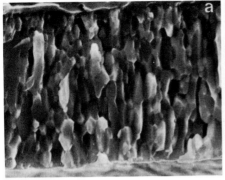

Fig.7 Scanning electron micrographs of the CDP Nb_3Sn grain structure heat treated at 800°C for 100 hr [16]. (a) Nb/Cu-6at%Sn composite. (b) Nb/Cu-6at%Sn-0.5at%Mg composite.

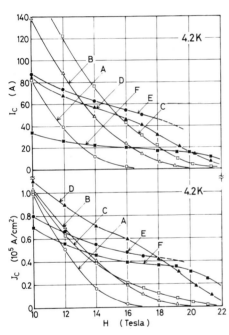

Fig.8 I_c versus magnetic field curves for the improved CDP Nb_3Sn specimens heat treated at 800°C for 100 hr [17].

Specimen	Composition (at%)		Nb_3Sn Layer Thickness (μm)
	Core	Matrix	
A	pure Nb	Cu-7Sn	18
B	Nb-2Hf	Cu-7Sn	32
C	Nb-5Hf	Cu-7Sn	47
D	Nb-2Hf	Cu-5Sn-4Ga	18
E	Nb-5Hf	Cu-5Sn-4Ga	27
F	Nb-5Hf	Cu-3Sn-9Ga	19

of magnesium to the Cu-Sn matrix also produces much finer Nb_3Sn grains in the CDP Nb_3Sn as shown in Fig.7 [16].

The CDP Nb_3Sn with pure niobium cores and a Cu-Sn matrix is difficult to stand practical use in fields higher than 12 T, because its J_c falls rapidly. The addition of hafnium to the niobium core significantly increases the Nb_3Sn formation rate, H_{c2} and J_c in high magnetic fields of the CDP Nb_3Sn [17]. The substitution of gallium for tin in the Cu-Sn matrix also increases H_{c2} and J_c in high fields. The growth rate of the Nb_3Sn layer in the specimens with a niobium core containing 2 to 5 at% hafnium is 2-3 times as large as that of the specimen with a pure niobium core and a matrix of the same composition. The substitution of gallium for tin in the matrix decreases the formation rate of Nb_3Sn. However, the decrease is compensated by the hafnium addition to the niobium core. Figure 8 shows I_c and J_c versus magnetic field curves at 4.2 K for the specimens, of which core and matrix compositions are listed in the caption [17]. The hafnium addition to the niobium core shifts the J_c-H curve to higher field without changing the shape, while the gallium addition to the matrix makes the J_c-H curve of Nb_3Sn convex upward in fields above 12 T like that of V_3Ga. The results in Fig.8 indicate that the simultaneous addition of hafnium to the core and gallium to the matrix is most effective for increasing high-field current-carrying capacity of the CDP Nb_3Sn. The $H_{c2}(4.2K)$ of over 24 T has been obtained in the specimens E and F. In the 19 core Nb-5Hf/Cu-5Sn-4Ga and Nb-5Hf/Cu-3Sn-9Ga wire specimens heat treated at 700-750°C, of which Nb_3Sn layers are 5-10 μm in thickness, J_c's of over 1×10^5 A/cm^2 have been obtained at 18 T [18].

These improved CDP V_3Ga and Nb_3Sn conductors are being developed and a 20 T superconducting magnet made by the stable multifilamentary A15 wires may be realized in the near future.

5. Superconducting and Mechanical Properties of the in situ Processed A15 Conductors

In 1973, Tsuei first applied the so-called insitu process to fabricate filamentary A15 composites using Cu-Nb-Sn ternary alloys richer in copper [19]. The characteristic of this process is to melt the constituent elements into an ingot, followed by cold working and appropriate heat treatment. External diffusion of tin after the cold working was found to raise the overall J_c to the value comparable to the commercial multifilamentary Nb_3Sn wires [20,21]. The advantages of the in situ process are relatively easy fabricating process compared to other techniques and insensitivity of superconducting properties to the applied stress or strain. In NRIM, studies on the metallurgical aspects and superconducting properties of the insitu processed V_3Ga have been carried out [22].

An optical micrograph of the as-cast ternary alloy with a composition of Cu-20at%V-12at%Ga is shown in Fig.9(a). The dendritic particles composed mostly of vanadium are precipitated in the Cu-Ga matrices. Figure 9(b) shows transverse and longitudinal cross sections of the same alloy after the cold drawing to the final diameter of 0.5 mm. The V-base dendrites shown in Fig.9 (a) are elongated into fine filaments. The T_c of V_3Ga formed in the ternary insitu Cu-V-Ga alloy is strongly dependent on the alloy composition. The maximum T_c of 14.8 K is obtained at the composition of Cu-13at%V-13at%Ga after the heat treatment at about 500°C. J_c of V_3Ga formed in the ternary in situ Cu-V-Ga alloy tends to increase with increasing vanadium concentration, because the sample containing more vanadium has a larger number of vanadium filaments. On the contrary, J_c at high magnetic fields depends on the gallium concentration in the alloy. Namely, the sample with lower gallium concentration exhibits lower T_c which leads to a rapid degradation of J_c in high magnetic fields. In the in situ processed V_3Ga using ternary Cu-V-Ga alloys, the overall J_c of 1.5×10^4 A/cm^2 is obtained at 10 T for the alloy composition of Cu-20at%V-12at%Ga. J_c of the in situ processed V_3Ga is significantly improved by the external diffusion of gallium after the fabrication, increasing the gallium concentration of the alloy. The overall J_c's of 6.5×10^4 A/cm^2 and 1.5×10^4 A/cm^2 are obtained at 10 and 15 T, respectively, for the external diffusion specimen with the composition of Cu-18at%V-20at%Ga. These overall J_c values are comparably as large as those of the commercial CDP V_3Ga conductor. The optimum heat treatment for obtaining large J_c is 500-600°C. The in situ processed V_3Ga shows appreciably better high-field performances as compared to the in situ processed Nb_3Sn.

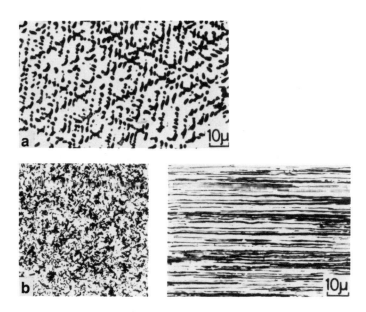

Fig.9 (a) Optical micrograph of the as cast Cu-20at%V-12at%Ga alloy. (b) Transverse (left) and longitudinal (right) cross section of the same alloy in (a) after the cold drowing.

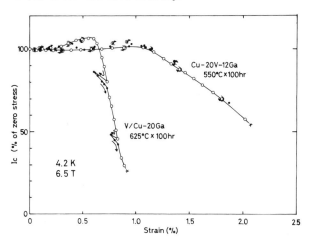

Fig.10 I_c versus tensile strain relationship for the in situ processed V_3Ga tape (Cu-20V-12Ga) and the single-core CDP V_3Ga tape (V/Cu-20Ga) [22].

Figure 10 shows the relationship between I_c (normalized by I_c at zero strain) and tensile strain for the insitu processed V_3Ga tape and that for the single-core CDP V_3Ga tape measured at a magnetic field of 6.5 T [22]. Closed circles show the I_c values obtained on the way back to unload. The I_c of the CDP V_3Ga gradually increases with strain up to a small maximum and then rapidly decreases. For the in situ V_3Ga, however, the I_c maximum does not appear appreciably. I_c degradation of the CDP V_3Ga begins at about 0.5 % strain, while that of the in situ V_3Ga begins at appreciably larger strain of about 1.0 %. Moreover, I_c of the in situ V_3Ga decreases much more slowly than that of the CDP V_3Ga. When the load is removed, the insitu V_3Ga shows complete recovery of I_c, namely the I_c-strain curve is reversible, whereas the CDP V_3Ga shows irreversible degradation of I_c. Scanning electron microscopy observations after the breaking of the sample indicated that cracks in the V_3Ga layer are dispersed rather uniformly in the insitu V_3Ga, while in the CDP V_3Ga they are concentrated near the fractured surface. Such differences in the fracture behavior seem to be responsible for the difference in the I_c-strain curve and the unloaded behavior. The high resistivity of I_c against strain of the insitu processed A15 composite makes them technologically attractive for a variety of applications in high-field superconducting magnets. The large scale production techniques of the in situ processed Nb_3Sn and V_3Ga superconductors are being developed.

6. Developments of New High-Field Superconductors

The formation of A15 compounds with T_c and H_{c2} higher than those of Nb_3Sn, e.g. Nb_3Al [23], Nb_3Ga [24] and $Nb_3(Al,Ge)$ [25], has been studied using the diffusion methods. However, there is a difficulty to get high enough T_c and J_c together, in these compounds; for getting high T_c requires high diffusion reaction temperature which leads to low J_c due to the grain growth of the A15 compounds. For the $Nb_3(Al,Ge)$, the so-called infiltration process has been applied [25]. In this process, a niobium rod with a controlled porosity is prepared by pressing and sintering niobium powders. Then the niobium rod is immersed in a liquid Al-Ge alloy bath. The Al-Ge infiltrated niobium rod is fabricated into a thin wire and then heat treated at 1250-1700°C. By this process, the fine filamentary $Nb_3(Al,Ge)$ phase is dispersed in the niobium matrix. T_c of about 19 K and J_c of about 1×10^5 A/cm^2 at 5 T have been obtained. The T_c and H_{c2} of Nb_3Al, Nb_3Ga and $Nb_3(Al,Ge)$ are appreciably increased by the secondary heat treatment at 700-800°C after the diffusion reaction.

The stoichiometric A15 Nb_3Ge shows the highest T_c of about 23 K in the existing superconductors. The A15 Nb_3Ge prepared by an ordinary arc-melting forms with much excess niobium, germanium concentration being 17.5 at%. Such an off-stoichiometric Nb_3Ge shows a low T_c of about 7 K. A nearly stoichiometric Nb_3Ge has been synthesized by sputtering [26], co-evaporation [27] and chemical vapour deposition [28,29]. The Nb_3Ge prepared by these deposition processes has not only high T_c but also large J_c due to its relatively small grain size. Moreover, the Nb_3Ge has high $H_{c2}(4.2K)$ of 32-37 T, which makes this compound attractive as a high-field superconductor.

The CVD process is considered to be most promising for the mass production of Nb_3Ge, since its deposition rate is much larger than that of sputtering and co-evaporation. Figure 11 shows the principle of the CVD process to prepare the Nb_3Ge tape [28]. The $GeCl_4$ is kept at 0°C and the temperature of $NbCl_5$ is adjusted to maintain an adequate vapour pressure. The hydrogen is used as both the carrier and the reduction gas. The Nb_3Ge is synthesized via the following chemical reaction.

$$3NbCl_5 + GeCl_4 + \frac{19}{2} H_2 \rightarrow Nb_3Ge + 19HCl$$

The reaction proceeds on the joule-heated hastelloy substrate tape. The appropriate deposition temperature is 850-900°C, and the deposition rate is about 5 μm/min. The J_c of Nb_3Ge is much increased by the precipitation of σ-Nb_5Ge_3 phase and is as high as 1×10^5 A/cm^2 at 18 T [30]. The continuous CVD process to produce long-length Nb_3Ge tape is being developed.

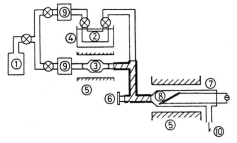

Fig.11 The schematic diagram of the CVD synthesis apparatus. 1: Hydrogen Gas, 2: $GeCl_4$, 3: $NbCl_5$, 4: Ice-Water Bath, 5: Electric Furnace, 6: Optical Window, 7: Reaction Tube, 8: Substrate, 9: Flow Meter, 10: Exhaust

The ternary V_2Hf-base C15 type Laves phase compounds discovered in NRIM have high $H_{c2}(4.2K)$ of over 20 T, and are not as brittle as A15 compounds [31]. Moreover, V_2Hf is reported to be highly resistive to the heavy neutron irradiation. The large electronic specific heat coefficient and the large spin-orbit scattering due to the heavy hafnium atoms seem to enhance H_{c2} in this material. The addition of zirconium, niobium, tantalum or chromium increases significantly both T_c and H_{c2} of the V_2Hf [32,33]. Among the V_2Hf-base C15 compounds, $V_2(Hf,Zr)$ and $V_2(Hf,Nb)$ are most promising for practical use since they can be fabricated by the CDP or by the direct plastic deformation process, respectively.

In the CDP, a composite of the vanadium sheath and the Hf_xZr_{1-x} alloy core is fabricated and then heat treated at 900-1050°C. The $V/Hf_{0.4}Zr_{0.6}$ composite shows the highest values of T_c and $H_{c2}(4.2K)$ which are 10.2 K and 20.8 T, respectively. J_c of the Laves phase layer in the $V/Hf_{0.4}Zr_{0.6}$ composite at 4.2 K is about the same as that of CDP Nb_3Sn in fields higher than 10 T. In Fig.12 are shown J_c versus magnetic field curves for the $V/Hf_{0.4}Zr_{0.6}$ compo-

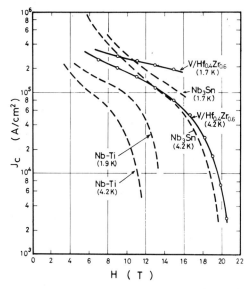

Fig.12 J_c versus magnetic field curves for the $V/Hf_{0.4}Zr_{0.6}$ composite, for the CDP Nb_3Sn and for the Nb-Ti wire measured at quoted temperatures.

site measured at different temperatures, together with those for the Nb_3Sn and Nb-Ti wires. Both H_{c2} and J_c of the $V/Hf_{0.4}Zr_{0.6}$ tape rapidly increase with decreasing temperature. At 1.7 K the $V/Hf_{0.4}Zr_{0.6}$ tape shows H_{c2} of 25.4 T and J_c of 2×10^5 A/cm^2 at 14 T [34]. A superconducting magnet wound with the V/Hf$_{0.4}$Zr$_{0.6}$ tape might be able to generate a magnetic field of 20 T at 1.7 K.

In the pseudobinary V$_2$Hf-V$_2$Nb system, the maximum values of T_c = 10.4 K and H_{c2}(4.2K) = 24.5 T are obtained at 27 at% Hf. The V$_2$(Hf,Nb) containing hafnium less than 17 at% can be deformed at room temperature. In the deformation process, arc-melted V$_2$(Hf,Nb) is rolled into a tape at room temperature with intermediate anneals at 1450°C, and then finally heat treated at 1000-1300°C. The addition of 4-8 at%Ti to V$_2$(Hf,Nb) improves both the workability and the superconducting properties. The titanium addition is also effective for decreasing the intermediate annealing temperature and the final heat treatment temperature. In the V-19at%Hf-8.3at%Nb-6at%Ti tape, T_c of 9.8 K, H_{c2} of over 22 T and an overall J_c of 1×10^4 A/cm^2 at 13 T have been obtained [35].

The Chevrel type ternary molybdenum chalcogenides, $M_xMo_6X_8$ (X = S, Se, Te), have a rhombohedral structure which is built up from cubic units of Mo_6X_8. The arrangement of Mo_6X_8 units leaves a space between them where the M atoms are situated. The ternary M atom is Pb, Sn, Ag, Cu, Zn, Mg or rare earth elements. The remarkable feature of this type of compounds is the extremely large dH_{c2}/dT and H_{c2}, as demonstrated in Fig.13 [36]. The H_{c2}(4.2K) of PbMo$_6$S$_8$ is about 50 T which is the highest value reported in the superconducting materials so far.

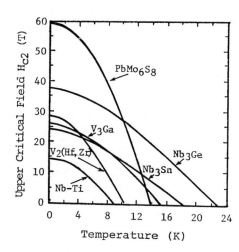

Fig.13 Upper critical field versus temperature for various high-field superconductors

Several attempts have been made to fabricate this extremely high H_{c2} compound into wires. The vapour reaction of molybdenum wire with first sulfur and then lead vapour to produce PbMo$_6$S$_8$ layer has been studied [37]. The drawing of powder filled tubes to fabricated PbMo$_{5.1}$S$_6$ wire has been studied, and J_c of 5×10^3 A/cm^2 has been obtained at 4 T [38]. Superconducting properties of the sputtered and co-evaporated CuMo$_6$S$_8$ films have been also studied [39, 40]. Improvements of relatively low J_c in this type of compounds are required for the practical high-field applications.

References

1. M.G. Benz and L.F. Coffin: Proc. 2nd Int. Cof. on Magnet Technology, 513 (1967)
2. K. Tachikawa: IEEE Publ. No. 72 CHO-682-5-TABSC, 371 (1972)

3. K. Tachikawa and Y. Tanaka: Japan. J. Appl. Phys. 6, 782 (1967)
4. S. Fukuda, K. Tachikawa and Y. Iwasa: Cryogenics 13, 153 (1973)
5. Y. Tanaka, K. Tachikawa and K. Sumiyama: J. Japan Inst. Metals 37, 835 (1970)
6. K. Tachikawa, Y. Tanaka, K. Inoue, K. Itoh and T. Asano: Cryogenic Engineering, (in Japanese), 11, 252 (1976)
7. K. Tachikawa: Proc. 3rd Int. Cryo. Engr. Conf.,(Illife Science and Technology Pub., Surrey, 1970) p. 339
8. K. Tachikawa, Y. Yoshida and L. Rinderer: J. Mat. Sci. 7, 1154 (1972)
9. K. Tachikawa, Y. Tanaka and Y. Iwasa: J. Appl. Phys. 44, 898 (1973)
10. M. Suenaga, O. Horigami and T.S. Luhman: Appl.Phys. Lett. 25, 624 (1974)
11. Y. Furuto, T. Suzuki, K. Tachikawa and Y. Iwasa: Appl. Phys. Lett. 24, 34 (1974)
12. K. Itoh and K. Tachikawa: Appl. Phys. Lett. 26, 67 (1975)
13. M. Ikeda, Y. Furuto, Y. Tanaka, I. Inoue and J. Tanii: Proc. Mag Tech. (MT-5), 715 (1975)
14. D.N. Cornish, H.L. Harrison, A.M. Jewell, R.L. Leber, A.R. Rosdahl, R.M. Scanlan and J.P. Zbasnik: presented at 8th Symposium on Engr. Problems of Fusion Research (San Francisco, 1979)
15. K. Tachikawa, Y. Tanaka, Y. Yoshida, T. Asano and Y. Iwasa: IEEE Trans. on Magnetics MAG-15, 391 (1979)
16. K. Togano, T. Asano and K. Tachikawa: J. Less-Common Metals 68,15 (1979)
17. H. Sekine, K. Tachikawa and Y. Iwasa: Appl. Phys. Lett. 35, 472(1979)
18. H. Sekine, T. Takeuchi and K. Tachikawa: to be published in IEEE Trans. on Magnetics (in 1981)
19. C.C. Tsuei: Science 180, 57 (1973)
20. S. Foner, E.J. McNiff, Jr., B.B. Schwartz, R. Roberge and J.L. Fihey: Appl. Phys. Lett. 31, 853 (1977)
21. J.D. Verhoeeven, D.K. Finnemore, E.D. Gibson, J.E. Ostenson and L.F. Goodrich: Appl.Phys. Lett. 33, 101 (1978)
22. K. Togano, H. Kumakura and K. Tachikawa: to be published in IEEE Trans. on Magnetics (in 1981)
23. S. Ceresara, M.V. Ricci, G. Pasotti, N. Sacchetti and M. Spadoni: IEEE Trans. on Magnetics MAG-15, 639 (1979)
24. S. Fukuda and K. Tachikawa: J. Japan Inst. Metals 39, 544 (1975)
25. M.R. Pickus, M.P. Dariel, J.T. Holthuis, J. Ling-Fai Wang and J. Granda: Appl. Phys. Lett. 29, 810 (1976)
26. J.R. Gavaler: Appl. Phys. Lett. 23, 480 (1973)
27. M. Kudo and Y. Tarutani: IEEE Trans. on Magnetics MAG-13, 331 (1977)
28. H. Kawamura and K. Tachikawa: Phys. Lett. A50, 29 (1974)
29. A.I. Braginski, J.R. Gavalar, G.W. Poland, M.R. Daniel, M.A. Janocko and A.T. Santhanam: IEEE Trans. on Magnetics MAG-13, 300 (1977)
30. A.I. Braginski, G.W. Roland and A.T. Santhanam: IEEE Trans. on Magnetics MAG-15, 505 (1979)
31. K. Inoue and K. Tachikawa: Appl. Phys. Lett. 18, 235 (1971)
32. K. Inoue and K. Tachikawa: J. Japan Inst. Metals 39, 1265 (1975)
33. K. Inoue and K. Tachikawa: IEEE Publ. No. 72 CH0682-5-TABSC, 415 (1972)
34. K. Inoue and K. Tachikawa: Appl. Phys. Lett. 29, 386 (1976)
35. K. Inoue, T. Kuroda and K. Tachikawa: IEEE Trans. on Magnetics MAG-15, 635 (1979)
36. ϕ. Fischer: Proc. Inter. Colloquium on "Physics in High Magnetic Fields" (CNRS, 1975) p. 79
37. M. Decroux, ϕ. Fischer and R. Chevrel: Cryogenics 17, 291 (1977)
38. T.S. Luhman and D. Dew-Hughes: J. Appl. Phys. 49, 936 (1978)
39. S. Alterovitz and J.A. Woollan: Solid State Commun. 25, 141 (1978)
40. J.A. Woollan, S.A. Alterovitz, K.C. Chi, R.O. Dillon and R.F. Bunshah: J. Appl. Phys. 49, 6027 (1978)

Research at the Nijmegen High Field Magnet Laboratory

H.W. Myron, C.J.M. Aarts, A.R. de Vroomen, and P. Wyder

High Field Magnet Laboratory and Physics Laboratory, Faculty of Science
University of Nijmegen
6525 ED Nijmegen, The Netherlands

The High Field Magnet Laboratory of the University of Nijmegen became operational in 1976 with the installation of two Bitter magnets. The present configuration of the laboratory includes a hybrid magnet which has been constructed and developed by the National Magnet Laboratory of the Massachusetts Institute of Technology; there the world's highest DC fields (30.1 tesla) has been reached with the Nijmegen hybrid. The purpose of the Nijmegen laboratory is to provide the highest DC fields available, and make the facility available to qualified users.

Interest has emerged in the study of the electronic structure of exotic materials by the dHvA effect and cyclotron resonance. A survey of experimental results obtained at the laboratory is made in this area. Some of the other research areas are mentioned, including industrial applications of magnetic fields.

The High Field Magnet Laboratory (HFML) of the Faculty of Science of the University of Nijmegen became operational in 1976. The purpose of the laboratory is to provide magnetic fields with a variety of specifications, and various spectrometers and other peripherals (like dilution refrigerators) at the disposal of qualified and interested scientists. The present magnet-configuration of the laboratory consists of two water cooled Bitter magnets with a maximum field of 15 tesla (T) and a hybrid magnet which can reach fields in excess of 25 T. The hybrid magnet has been constructed and developed by the Francis Bitter National Magnet Laboratory of the Massachusetts Institute of Technology where the highest continuous field of 30.1 T has been obtained with the "Nijmegen" hybrid [1]. Here we will not deal with the characteristics and physical plant of the laboratory since this has been described elsewhere [2]. The interested reader is referred to that publication of a complete description of the power supply, cooling and magnet systems as well as a general outline of the physical plant of the laboratory.

Plans are now in progress for an expansion of the laboratory which will include a variable gap radially cooled split coil magnet (fields of 12 T), a 6 MW 3 cm bore 19 T magnet and an additional hybrid system for DC fields of up to 35 T. However, these are not the only magnets available; through the various scientific departments of the Faculty of Science and the specialized research activities in which they are involved, a variety of auxilliary facilities have emerged. These in principle are available for general use, for example a wide range of superconducting coils including a 8 T split coil (with radial access), a 43 T pulsed field system used primarily for de Haas-van Alphen (dHvA) effect and a NMR facility with an ultra homogeneous field of 12 T [3].

In order to give a glimpse of the various scientific activities underway at the laboratory, we will outline some efforts in progress or which have been planned for the near future.

The electronic structure of exotic materials has been of interest to both the physicist and technologist because of their anomalous and wide ranging properties. Stimulated by this interest an effort has emerged in the study of the electronic structure by principally one of two techniques, the dHvA effect and cyclotron resonance. Recent studies of the electronic structure of PdSb have been made by both electronic structure calculations [4] and dHvA measurements [5] on the 15 T Oxford III system. PdSb has proven to be a prototype system for other isoelectronic NiAs systems. Typical results outlined in Table 1 show excellent agreement between theory and experiment. The designation of these orbits on a Fermi surface (FS) model is shown in Fig.1.

Table 1   dHvA frequencies along symmetry directions

Field direction	Orbit	Experiment [5]	Theory [4]
[0001]	α	3.15 ± 0.02	4.0
	μ	5.04 ± 0.05	4.9
	β	18.2 ± 0.1	16.9
	ν	22.7 ± 0.1	11.5
	σ	39.0 ± 0.4	35.7
	γ	71.7 ± 0.5	75.1
	λ	76.2 ± 0.5	75.1
[11$\bar{2}$0]	δ	2.32 ± 0.03	1.3
	ϵ	15.8 ± 0.2	18.9
	ϕ*	42.3 ± 1.0	38.3
	ψ	99.3 ± 1.0	98.6
[10$\bar{1}$0]	δ	1.98 ± 0.04	1.1
	ϵ_1	15.8 ± 0.3	17.6
	ϵ_2	18.9 ± 0.4	20.3

The study of the electronic structure of the transition metal dichalcogenides is extremely useful in helping to understand the origins of the charge density waves and associated phase transitions in these materials [6]. The FS of many of these materials has been measured and can be best described as cylindrical. Recently a new Group VI b compound 1T-$CrSe_2$ has been synthesized which has three sheets of FS, one of which is ellipsoidal and centered about the ΓA axis [8]. Unfortunately the synthesis of this material is indirect and leads to the fabrication of powdered samples; however, the observation of the dHvA effect in powdered specimens has become possible [9] when orbits have small radii compared to the dimension of the grains of the polycrystalline. For example, recent theoretical [10] and experimental [11] efforts on $TiBe_2$ show a spherical piece of FS; the observed dHvA orbits are close to the ones predicted. Typical results for $TiBe_2$ on the 43 T pulse field system are shown in Fig.2.

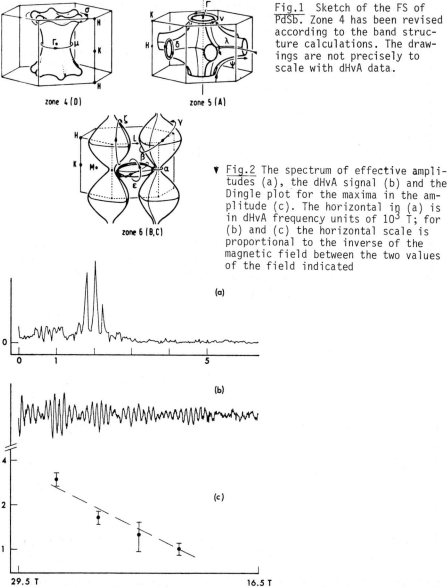

Fig.1 Sketch of the FS of PdSb. Zone 4 has been revised according to the band structure calculations. The drawings are not precisely to scale with dHvA data.

Fig.2 The spectrum of effective amplitudes (a), the dHvA signal (b) and the Dingle plot for the maxima in the amplitude (c). The horizontal in (a) is in dHvA frequency units of 10^3 T; for (b) and (c) the horizontal scale is proportional to the inverse of the magnetic field between the two values of the field indicated

Some members of the laboratory in collaboration with Dr. L. Esaki and Dr. L.L. Chang (IBM T.J. Watson Research Center) have studied the cyclotron resonance (CR) of superlattices [12]. Superlattices are structures having a periodic arrangement of layers made out of different materials. A typical system is made for example from a layer of InAs of thickness a on top of a layer of GaSb of thickness b and so forth. These layers are typically 30 to 250 Å thick thus the electrons in these systems show one-dimensional periodic modulation in addition to the normal three-dimensional potential. The

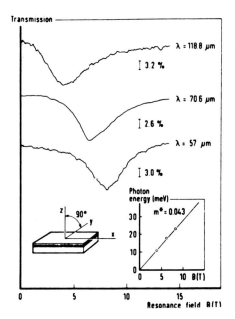

Fig.3 Transmission signal for InAs-GaSb superlattice as a function of magnetic field for different frequencies. The inset shows transition minima as a function of magnetic field and photon energy.

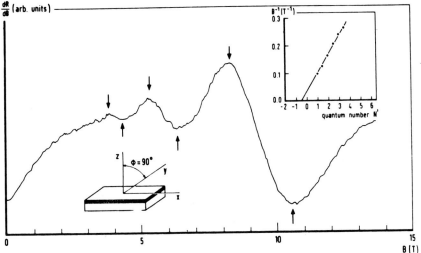

Fig.4 SdH oscillations in a superlattice, the insert shows the magnetic field at which extrema occur (see arrows) as a function of quantum number

periodicity of the superlattice gives rise to interesting effects which have been seen by cyclotron resonance through the effective masses [13]. Such results are shown in Fig.3. Fig.4 shows the oscillatory structure of the resistance characteristic of the Shubnikov-de Haas (SdH) effect, when plotted as a function of magnetic field. Extrema in the resistance occur when a Landau level crosses $E_F = (N + \frac{1}{2})\hbar\omega_c$. Assuming a two-dimensional density of states [14], the layer carrier density can be determined (6.7×10^{11} cm^{-2}).

Cyclotron resonance absorption of radiation by the free carriers of a semiconductor may have a pronounced effect on the DC conductivity; this is known as "cross modulation". Although cross modulation has been used to detect cyclotron resonance signals since very many years, the effect itself has not been analyzed and understood in detail. For the case of n-GaAs, it seems that in high magnetic fields cross modulation signals are caused by a change in free carrier density only [15]. This experiment was done by combining simple power transmission measurements, DC conductivity (a measurement of charge carriers and mobility) and the Hall effect (charge carriers alone) at the CR condition. The results for the induced Hall effect is shown in Fig.5.

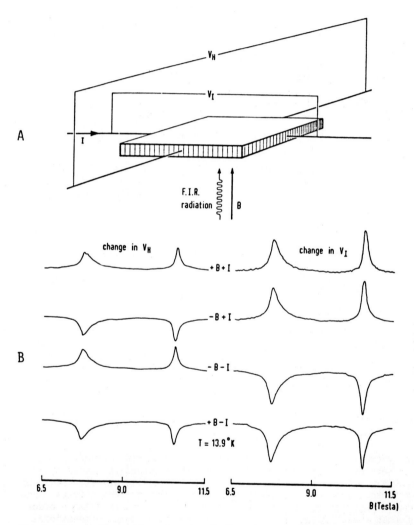

Fig.5 The configuration for the CR induced Hall effect data (A). B shows the measured change in V_I and V_H for λ = 70.6 μm as a function of magnetic field for different polarities of the current and magnetic field.

HGMS USING CRYOGENIC FLUIDS

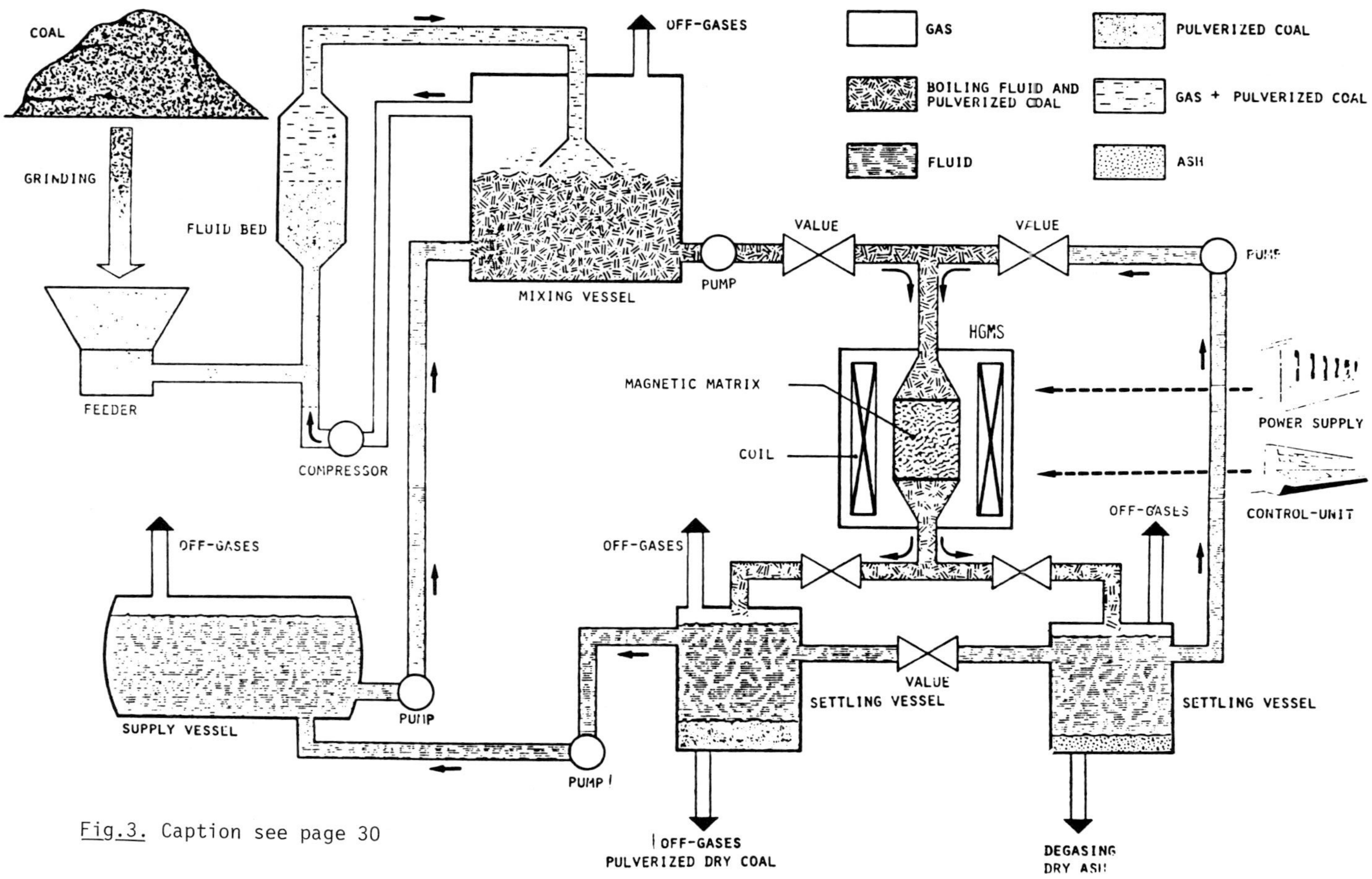

Fig.3. Caption see page 30

Recently, the Dutch government has been encouraging universities and engineering colleges to cooperate with industries in the uses of technology for industrial application. Such a program has recently been funded through our laboratory, in collaboration with Holec Industries, to study various aspects of high gradient magnetic separation (HGMS). Simply stated, HGMS is the separation of non magnetic particles from a mixture by using the gradient of the magnetic field created by a ferromagnetic filter. This process has commercial application in the purification of kaolin clay [16]. Preliminary results using liquid nitrogen as a carrier fluid in HGMS indicate a reduction of ash and sulfur content of pulverized coal as compared to earlier studies using water as a transport fluid. Experimental evidence [17] indicates that a substantial increase of transport velocity (using liquid nitrogen) does not drop performance. Extensive experimental work is planned in order to optimize and prove the feasibility of cryogenic HGMS. A possible arrangement of a pilot facility is shown in Fig.6.

There are additional projects at the laboratory in various stages of development. To name a few: study of pair breaking in strong coupling superconductors, transport properties of simple metals in high fields, study of magnetic breakdown, and point and double point contact spectroscopy in magnetic fields.

Acknowledgements

We would like to thank the following colleagues for communicating their results prior to publication: Drs. Bluyssen, Chang, van Deursen, Devillers, Dijkhuis, Esaki, de Groot, Kerkdijk, Koelling, Maan and Prof. Mueller. Part of this work has been supported by the "Stichting voor Fundamenteel Onderzoek der Materie" (FOM) with financial support of the "Nederlandse Organisatie voor Zuiver Wetenschappelijk Onderzoek" (ZWO).

References

1. Gloria B. Lubkin, Physics Today 30, # 12, 20 (1977).
2. K. van Hulst, C.J.M. Aarts, A.R. de Vroomen and P. Wyder, Journal of Magnetism and Magnetic Materials 11, 317 (1979).
3. These facilities are available through the research groups of Profs. P. Wyder, A.R. de Vroomen and C.W. Hilbers respectively.
4. H.W. Myron and F.M. Mueller, Phys. Rev. B 17, 1828 (1978).
5. M.A.C. Devillers and A.R. de Vroomen, Physica 95B, 183 (1978) and references therein.
6. See for example N.J. Doran, Physica 99B, 227 (1980).
7. C.F. van Bruggen, R.J. Haange, G.A. Wiegers and D.K.G. de Boer, Physica 99B, 166 (1980).
8. H.W. Myron (unpublished).
9. R.T.W. Meijer, L.W. Roeland, F.R. de Boer and J.C.P. Klaase, Solid State Commun. 12, 923 (1973); J.C.P. Klaase, R.T.W. Meijer and F.R. de Boer, Solid State Commun. 33, 1001 (1980).

Fig.6 (p.29) Process scheme for a cyclic cryogenic HGMS device. Pulverized coal is mixed in a vessel and subsequently transported through the HGMS equipment, while the magnetic field is switched on and collected in a settling tank. The washing off is done in the right hand part, while field is off

10. R.A. de Groot, D.D. Koelling and F.M. Mueller, J. Phys. C, Lett. (to be published).
11. A.P.J. van Deursen, A.R. de Vroomen and J.L. Smith, Solid State Commun. (to be published).
12. R. Dingle, Festkörperprobleme XV, 21 (1975); L. Esaki and L.L. Chang, Thin Solid Films 36, 285 (1976).
13. H. Bluyssen, J.C. Maan, P. Wyder, L.L. Chang and L. Esaki, Solid State Commun. 31, 35 (1979).
14. J.C. Maan, Ph.D. Thesis, University of Nijmegen (1979).
15. H.J.A. Bluyssen, J.C. Maan, T.B. Tan and P. Wyder, Phys. Rev. B (to be published).
16. For a recent overview of the problems in HGMS the reader is referred to the conference proceedings, Industrial Applications of Magnetic Separation, Y.A. Liu ed , IEEE, New York 1979.
17. J.I. Dijkhuis and C.B.W. Kerkdijk, IEEE - Magnetics (submitted).

Part II

Physics in Pulsed High Magnetic Fields

Strong and Ultrastrong Magnetic Fields: Development, State of the Art and Applications

F. Herlach

Katholieke Universiteit Leuven, Laboratorium voor Lage Temperaturen en Hoge-Veldenfysika, Celestijnenlaan 200 D
3030 Leuven, Belgium

1. Introduction

There are two major obstacles in the way of generating very strong magnetic fields : the ohmic heating of conductors and the magnetic stress [1]. The heating problem arises sharply above 15 T where superconducting coils become prohibitively expensive such that watercooled coils are more economical in spite of their enormous power consumption. To generate the presently feasible maximum of 20 T in a 5 cm bore, 10 MW are needed. Another 10 T can be added from a large superconducting coil surrounding the watercooled magnet in a hybrid system (Nijmegen, M.I.T., Grenoble) [2].

Pulsed magnets are not only less expensive by orders of magnitude, they are the only means for obtaining much higher fields. Because of some difficulties inherent in experimentation with pulsed fields, these are still not used as widely as it would make sense. Initially, many researchers find it hard to get used to single shot experiments where no adjustments can be made while the experiment is in progress, and faults can only be detected by careful data analysis. This increases the risk of misinterpretation of experimental results. Experience has shown that initial difficulties can be overcome with reasonable effort, and that most experiments can be accomodated in pulsed fields much better than one would have thought. Modern multi-channel data recording techniques permit efficient experimentation, and stability problems are greatly reduced as experimental parameters must only be stable for the duration of the field pulse. The restricted use of signal-enhancing integrating techniques is often balanced by the greater strength of the effects in higher fields, or it is circumvented by the use of other experimental techniques. A more fundamental problem is given by the eddy current heating of conductive samples. This can be reduced by using small samples or by increasing the pulse duration. There are now larger projects under consideration to provide fields of the order of 50 T for approximately one second [3]. This will extend the applicability of pulsed fields to experiments where longer relaxation times are involved, such as nuclear magnetic resonance in biological systems.

2. Nondestructive Pulsed Fields

2.1 Heating and Pulse Duration

The pulse duration Δt is determined by adiabatic heating of the coil to the maximum permissible temperature. This is calculated by integrating the power density of the Joule heating which leads to the equation

$$\int_0^{\Delta t} B^2 dt = 4\mu_o^2 a_1^2 (\alpha-1)^2 \gamma^2 \beta^2 f^2 D \int_{T_o}^{T} \frac{c_p}{\rho} dT = \frac{2\mu_o^2}{\pi} \frac{1}{a_1} \gamma^2 \beta \frac{\alpha-1}{\alpha+1} f\theta (T_o,T) W \quad (1)$$

$$\gamma = \frac{B}{\mu_o} \frac{a_1}{NI} = \frac{1}{2(\alpha-1)} \ln \frac{\alpha + \sqrt{\alpha^2 + \beta^2}}{1 + \sqrt{1 + \beta^2}} \sim \frac{1}{2\beta} \quad \text{for } \beta \gg 1$$

where B = magnetic induction at the center of the coil ; W = total energy delivered by the power supply ; I = current ; f = filling factor (conductor volume/winding volume) ; N = number of turns ; a_1 = inner radius of the coil ; a_2 = outer radius of the coil ($\alpha = a_2/a_1$) ; b = half the axial length of the coil ($\beta = b/a_1$) ; θ = the integral of the electrical conductivity $1/\rho$ weighted with c_p ; c_p = the specific heat ; D = the density ; Q = cross section of the conductor. Some values of the integrals are given in table 1.

Table 1 Characteristic functions related to the electrical heating of copper from liquid nitrogen temperature

Temperature	Enthalpy change	θ	$\int \frac{c_p}{\rho} dT$	Coil mass
K	J/g	$10^8 \Omega^{-1} m^{-1}$	$10^{13} J\, kg^{-1} \Omega^{-1} m^{-1}$	kg/MJ
100	5	3.5	0.18	200
150	20	2.2	0.4	50
200	37	1.6	0.6	22
300	74	1.2	0.9	14
400	114	0.9	1.1	9

The mass of the coil is determined by the total energy and the permissible temperature rise $T - T_o$. For a given energy supply, the factors α and β are therefore related through the condition for a constant winding volume

$$V = 2\pi a_1^3 (\alpha^2 - 1) \beta = \pi a_1 (\alpha + 1) NQ / f \quad (2)$$

Furthermore, the peak field is related to the inductively stored energy

$$W_m = \frac{1}{2} L I^2 = \frac{\pi}{16\mu_o} a_1^3 g \frac{\alpha+1}{\beta\gamma^2} B^2 \quad (3)$$

where g is the factor that accounts for the edge effects in the inductance [4]

$$L = \frac{\mu_o \pi}{8} a_1 \frac{(\alpha+1)^2}{\beta} N^2 g \quad (4)$$

$$g \sim 1 + 0.225 \frac{\alpha+1}{\beta} + 0.64 \frac{\alpha-1}{\alpha+1} + 0.42 \frac{\alpha-1}{\beta} \quad (5)$$

As a function of β, the peak field goes through a relatively flat maximum. The dimensioning of the coil is thus not critical and the choice of β may rather depend on the desired field homogeneity. In Fig.1, examples for the pulse duration as a function of the energy are given.

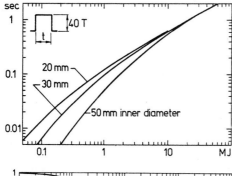

Fig.1 The maximum pulse duration for copper coils with different inner diameters as a function of the total energy. Coils are precooled with liquid nitrogen and allowed to heat up to 350 K. The filling factor is 0.8

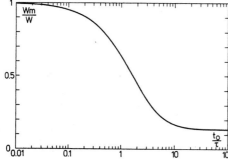

Fig.2 The fraction W_m of the total energy W delivered by the capacitor band is the inductively stored energy at peak field. This is plotted as a function of the half period of the discharge, divided by the time constant of the coil

In a capacitor discharge, W_m is given by the energy initially stored in the capacitor minus the ohmic losses incurred before peak field is reached. As shown in Fig.2, this depends on the ratio of the time constant of the coil

$$\tau = \frac{L}{R} = \frac{\mu_0}{4} \frac{a_1^2}{\rho} (\alpha^2 - 1) f g \tag{6}$$

to the pulse duration (half period t_o). For short pulses with $t_o/\tau < 0.1$, almost all of the energy is available to build up the magnetic field and is resistively dissipated only afterwards. In this regime, mechanically strong conductor materials with high resistivity can be used without precooling. For very long pulses with $t_o/\tau > 10$ the ohmic loss becomes dominant. In practice, if a long pulse duration is required, a compromise must be made regarding the resistive loss. A typical value is $W_m/W \sim 0.7$ at $t_o/\tau \simeq 1$. For $t_o/\tau \gg 1$, the condition for critical damping, $2\tau = \sqrt{LC}$ is approached where $W_m/W = e^{-2}$. The pulse duration can be stretched by using crowbar diodes ; these will lock the decay of the field pulse to the time constant $L/(R+R_c)$ where R_c is the resistance in the crowbar circuit. The longest possible decay is given by the time constant of the coil when $R_c = 0$. Maximum external energy dissipation in the crowbar resistance is obtained for $R_c = \sqrt{L/C}$. Both effects are fairly modest ; the temperature reduction is of the order of several ten degrees. For coils with larger damping, the crowbar circuit becomes quite inefficient as the switching occurs late in the pulse. A circuit with overcritical damping may have a much longer decay time than that given by the time constant of the coil. For a refined analysis of the coil performance, the skin effect and the magnetoresistance must be considered. Both effects result in a small reduction of the peak field but mainly in a radial temperature variation. Typically, at a field level of 40 - 50 T the innermost layer may become 50 - 100 degrees hotter than the outside. A few examples for the dimensioning of coils are given in table 2.

Table 2 Examples for the dimensioning of pulsed field coils

Energy	MJ	0.1	0.3	1	1	10	10	100	100
Field	T	40	50	50	40	50	40	50	70
Inner diameter	mm	20	20	20	30	20	50	50	50
Length	mm	90	100	140	150	340	300	800	800
Pulse duration[a]	ms	13	26	66	89	300	450	1600	700
Peak power	MW					50	40	100	200

a) for the first four examples, this is the rise time (quarter period) of the capacitor discharge ; for the others it is the time interval in which the field is kept constant where the total time interval is about twice as long. Coils are calculated for heating from liquid nitrogen to 300 K, with a filling factor of 0.8.

A pulse generator with a 100 kJ capacitor bank is an elegant, compact laboratory instrument for all experiments that can be completed in less than about 10 ms. One megajoule is a typical energy for a dedicated facility as they have been set up at Toulouse (1.25 MJ, 10 kV [5]) and Osaka (1.5 MJ, 30 kV [6]). For higher energies, capacitors are too bulky and are not really needed ; with the long pulse duration the energy can be supplied by a flywheel driven generator [7] or, under favorable circumstances, directly from the mains, with flexible pulse shaping by thyristors. The latter solution has been adopted for the 1.5 MJ, 6.6 MW magnet at Amsterdam which is a well designed prototype for all similar installations [8].

2.2 Mechanical Stress and Peak Field

The mechanical stress becomes a severe problem for the repeated operation of wire-wound coils above 35 T, and most coils fail suddenly between 40 and 50 T. If there is no substantial transmission of stress between the layers, the peak field as a function of the yield stress σ is given by [9]

$$B = \sqrt{2\mu_0\sigma} \; \sqrt{2} \; (1 - \frac{1}{\alpha}) \tag{7}$$

for a coil with constant current density. For hard copper, $\sqrt{2\mu_0\sigma}$ is close to 35 T and for the strongest alloys such as Be-Cu or maraging steel it is just a little over 50 T. The transmission of radial and axial stress within the coil is a complication which leads to surprising results. In particular, the circumferential stress is increased at the inner surface because of an additional "pull" from the outer layers which are expanding more as the radial expansion is proportional to the radius. Rigorous stress calculations are of little value for pulsed field coils because these are inhomogeneous composite structures where at least some of the components are likely to be stressed in excess of the plastic deformation limit, and the material properties under the combined attack of mechanical and heat shock are not well known. Thus, the fabrication of successful coils relies mostly on experience ; it is an art rather than a science.

It is possible to make coils with a current distribution that provides for an even distribution of stress throughout the coil [6]. For a coil without stress transmission between layers, the peak field is given by

$$B = \sqrt{2\mu_o \sigma} \sqrt{\ln \alpha} \qquad (8)$$

in the limit of a large number of layers. This equation is a good approximation for coils with as little as four layers [9]. In principle, this equation sets no limit to the peak field, but for practical values of α the gain over a constant current density coil is modest (table 3). These coils are machined from solid stock and therefore the number of turns is limited ; this results in a relatively short pulse duration in the range 100 µs - 1 ms. The large coils require much energy of which only a small part ends up in the experimental volume. In case of coil failure which becomes more likely as the megagauss limit is approached, the high energy may cause much havoc in the laboratory. Force-free coils, apart from the even much more complicated construction, appear to be impractical for the same reasons. In a force-free coil, the current at any point must flow in the direction of the field. This necessitates a distribution of high fields over extended volumes.

Table 3 Relative peak fields in coils with constant current density and in coils with evenly distributed stress, both without transmission of stress between layers.

α	2	3	5	10	20	50
$\sqrt{2}(1-\frac{1}{\alpha})$	0.71	0.94	1.13	1.27	1.34	1.39
$\sqrt{\ln \alpha}$	0.83	1.05	1.27	1.52	1.73	1.98

3. Ultrastrong Fields

Fields above the 100 T limit can be called "ultrastrong" with good reason. The violent destruction of the conductors is inevitable and proceeds at a rate which is related to the propagation of sound in the conductor material, pushing the time scale to the microsecond range. Generators for ultrastrong fields and their applications have been extensively discussed at earlier Würzburg conferences [10,11] and at the 2nd megagauss conference [12]. The following is only a brief summary and an update.

If the surface of a conductor is exposed to an ultrastrong magnetic field pulse, it will recede from the field at a speed given by the shock compression characteristics of the conductor material. This speed is approximately proportional to the 3/2 power of the magnetic field. Selected values are given in table 4. The energy that can be supplied by the pulsed power supply sets a limit to the field volume ; in combination with the speed of expansion this restricts the pulse duration to a few microseconds at the very best. A pulsed field penetrates into a conducting surface according to a diffusion equation. This process can be characterized by the flux diffusion speed $v_f = E/B$ and by the skin depth

$$a \sim \frac{v_f}{\nu} = \frac{\rho}{\mu_o v_f} = \sqrt{\frac{\rho}{\mu_o \nu}}, \quad \nu = \frac{1}{B}\frac{dB}{dt} . \qquad (9)$$

Due to the short pulse duration, the skin depth is usually small compared to the thickness d of the conductor. In this case, the Joule heating becomes independent of the resistivity. The energy transport related to the diffusion of magnetic flux is given by the product of the flux diffusion speed and twice the magnetic energy density ; as a consequence the energy density of the Joule heat is approximately equal to that of the magnetic field as long as the energy flow is unobstructed by discontinuities. The heating of different metals under this condition is given in table 4 ; the heating of a conductive sample with thickness d is reduced by a factor of the order 400 for d/a = 0.1 and by a factor 40000 for d/a = 0.02. The boiling point of most metals is reached at fields below 200 T. This does not set an absolute limit to the peak field ; the boiling occurs at the surface only and the phase boundary travels into the metal at a speed related to the energy input. BRYANT [13] has found steady-wave solutions of the flux diffusion equation where the speed of this interface is equal to the flux diffusion speed. These speeds are usually smaller than the particle speed due to the shock compression which is the dominant factor in limiting the peak field.

Table 4 Properties of metals subjected to pulsed megagauss fields

		Al[a]	steel[b]	Cu	Ta	W	Au	Pt
equipartition heating	°C/MG^2	1660	1020	1170	1730	1590	1590	1370
boiling point	T	130	160	150	180	200	130	170
particle speed at 500 T	km/s	3.5	1.8	1.7	1.2	1	1.1	0.9
particle speed at 1000 T	km/s	8.5	4.5	4.3	3.2	2.8	2.8	2.5

a) 2024 aluminium   b) AISI 304 stainless steel

For a megagauss generator, one can either use a stationary single turn coil and a power supply which delivers the required energy before the coil has essentially expanded, or one can compress magnetic flux by means of an imploding cylinder which must move at a speed that is higher than the added velocities of shock compression and flux diffusion. Both methods were introduced in the late fifties : the single turn coil by FURTH, LEVINE and WANIEK in 1957 [14] and explosively driven flux compression by FOWLER, GARN and CAIRD in 1960 [15]. It took a number of years until either method had been refined to the point where the megagauss fields could be used in experiments. The first meaningful solid state experiments were carried out at Los Alamos [16], mainly using a new type of simple flux compression device with peak fields between 100 and 250 T, according to the level of sophistication of the device. In these devices, a single turn coil is powered by an explosive-driven generator, consisting of two flat metal sheets which are slammed together like a bellows by strips of high explosive, thus squeezing the trapped flux into the coil [17]. The experiments were mainly optical ; Zeeman effect in emission and absorption, Faraday effect in semiconductors and insulators, including MnF_2 which has an antiferromagnetic transition at 101 T, and exciton spectroscopy. Many of these experiments were done at cryogenic temperatures. Magnetic fields up to 1000 T from cylindrical implosions were used by HAWKE et al. at Livermore to compress hydrogen isentropically into the metallic state [18].

Electrical measurements are much more difficult because of the strong electrical transients caused by the megagauss generator and by the rapidly rising field itself. Conducting and - to some degree - paramagnetic samples are heated and subjected to magnetically induced stress. The latter effect can be exploited in ultra high pressure experiments. In conducting samples, the eddy current heating follows the equipartition law when the skin depth is small compared to the sample dimensions. In some cases, extremely thin samples may be needed to reduce this heating to a tolerable level.

A breakthrough with the capacitor discharge method was made by FORSTER and MARTIN in 1967 [19] who demonstrated that the speed of the capacitor discharge is an essential factor and that lightweight single turn coils perform much better than the previously used thick-walled coils. Based on these findings, a practical 200 T generator was developed at the Illinois Institute of Technology [20]. This was used for a number of experiments as a megagauss target at the Stanford Linear Accelerator [21], and for infrared cyclotron resonance in semiconductors [22].

Although the coils explode violently, the unobstructed expansion of lightweight coils leaves the field volume undisturbed. This is the only method of megagauss generation that permits the repeated use of delicate samples.

Electromagnetically driven flux compression (EMC) was introduced by CNARE in 1966 [23]. It was applied to solid state experiments (cyclotron resonance, Faraday- and Zeeman-effect, exciton spectroscopy) by MIURA and his collaborators, with fields up to 280 T and at cryogenic temperatures [24]. The proper functioning of an EMC device, and in particular a small one, depends on a delicate balance of experimental parameters. The initial thickness of the imploding cylinder and the risetime of the driving field determine how much of the field penetrates into the cylinder as initial flux for the compression. On the other hand, the drive field must rise sufficiently fast to complete the energy transfer before the imploding cylinder decouples from the driving coil. In this regard, a z-pinch as used by ALIKHANOV et al. [25] is advantageous because the drive field follows the cylinder, and the initial flux is provided by an independent source. This becomes all less critical as the available energy increases, and then the implosion speed and thus the peak field can be much higher than with explosive propulsion which is limited by the speed of sound and the energy and momentum carried away by the detonation products. The final velocity of an electromagnetically driven conductor is only limited by the vaporization of the conductor caused by induced currents. This can be calculated by means of equation (1) ; the left hand side of this equation is proportional to the momentum per coil surface. Applied to a long cylindrical sheet, this yields for the speed

$$v = \tfrac{1}{2} \mu_0 d \int (c_p/\rho) \, dT \qquad (10)$$

where the integral includes the heat of melting and vaporization. At a given temperature, in the limit the boiling point, the speed is proportional to the sheet thickness d. For example, aluminium will theoretically burst at a speed of 25 km/s per mm thickness, and copper at 14 km/s. This has been confirmed in experiments [10,23] where observed implosion speeds came close to these values. To obtain magnetic fields of the order 1000 T or more, the basic requirement is a pulsed power supply capable of delivering several megajoules in fractions of a millisecond. Different types of suitable power supplies are now under construction for the next generation of controlled nuclear fusion experiments ; eventually some of these will become available for implosion work. Meanwhile, MIURA and his collaborators at the Institute for Solid State Physics (University of Tokyo) are already setting up a 5 MJ capacitor bank for this purpose. This will be another milestone in the development of megagauss physics.

4. Magnetoresistance in Small Gap Semiconductors

As an example for research in nondestructive magnetic fields, a selection of our recent results is given in Fig.3 [26]. The pulsed field apparatus is a 60 kJ, 3.5 kV capacitor bank with a mechanical switch. Wire-wound coils with internal steel reinforcement generate up to 45 T in 17 mm diameter with a half period of 10 ms. The coils undergo irreversible deformations above 42 T and are normally used below 40 T. Magnetoresistance in several small gap semiconductors was measured with a straightforward four-terminal d.c. method. The sample current was 1 mA in most experiments ; 10 mA and 0.1 mA were used occasionally to verify the absence of hot electron effects. The best records were obtained with 10 mA. The pickup voltages induced by the pulsed field in the sample leads were neutralized by means of a compensation circuit, consisting

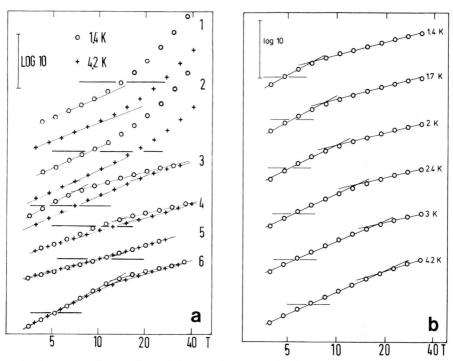

Fig.3 Transversal magnetoresistance, log {ρ(B) / ρ(0) - 1}, of several small gap semiconductors in strong pulsed magnetic fields

curve		1 & 2	3 & (b)	4	5	6
sample		$Hg_{1-x}Cd_xTe$		$Pb_{1-x}Sn_xTe$		PbTe
type; x		n; 0.135	n; 0.2	n; 0.2	p; 0.2	n
carriers at 77 K	$10^{16} cm^{-3}$	3.7	0.05	5	7	5
mobility at 77 K	$10^3 cm^2/Vs$	5.2	210	22	15.7	33.2
band gap at 4.2 K	meV	-50	64	81	81	189
baseline in plot	log	10	250	40	10	40

of an auxiliary pickup coil and an adjustable voltage divider. Initially, much trouble was caused by vibrations of the 50 μm platinum wires used to contact the samples ; this resulted in erratic, high frequency pickup signals. The problem has now been reduced to a tolerable level by redesigning the sample holders.

This research was inspired by the work of NIMTZ and his collaborators who had discovered the onset of a bend in the log-log plot of magnetoresistance vs. field in $Hg_{1-x}Cd_xTe$ at the upper end of the available field range. In our early experiments, it was confirmed that this is indeed indicative of a magnetic field induced phase transition : the continuation of the curve in the high field region was found to be another straight line with a different slope. This encouraged NIMTZ et al to study the phenomenon more extensively. They defined a critical field for the transition and found that this depends on the carrier density as it had been predicted for a magnetic field induced WIGNER condensation (localized ground state of the conduction electrons), while no other explanation has yet been found for this unusual transition. This is the most tangible evidence discovered so far for the elusive phenomenon of a WIGNER condensation. A review of these experiments and the underlying theory has been published recently [27].

In the present work, the transition has been followed as a function of temperature (Fig.3b) ; the observed behaviour is in qualitative agreement with theoretical expectations, i.e. the critical field decreases with decreasing temperature in this temperature range. The "half metallic" sample with x=0.135 behaves differently : it shows a dramatic increase of the magnetoresistance with both rising field and falling temperature. While this obeys a power law below 10 T, at higher fields it becomes nearly exponential. To be sure that this is not due to carrier freeze-out, we ought to measure the Hall effect. However, this is below the present threshold sensitivity of our apparatus (∼ 10 mV compared to a 10 V signal from the magnetoresistance). ·In any case, here is a group of interesting phenomena that will stimulate further experimental and theoretical research, and that falls neatly into the field range covered by nondestructive pulsed fields.

References

1 D.B. Montgomery, Solenoid Magnet Design (Wiley, New York 1969)
2 R.L. Aggarwal, A.J. Freeman and B.B. Schwartz, eds., Proc. Int. Conf. Solids and Plasmas in High Magnetic Fields, Cambridge, Mass. 18-20 Sept. 1978, J. of Magn. & Magn. Mater. Vol. 11, Nos 1-3 (North-Holland, Amsterdam 1979)
3 D.B. Montgomery in ref. [2], 293
4 V.G. Welsby, The Theory and Design of Inductance Coils (Mc Donald, London 1964)
5 S. Askenazy, Proc. Conf. The Application of High Magnetic Fields in Semiconductor Physics, Oxford, Sept. 11-15, 1978, p. 101
6 M. Motokawa, S. Kuroda and M. Date, J. Appl. Phys. 50, 7762 (1979)
   M. Date, J. of the Phys. Soc. of Japan 39, 892 (1975)
7 H.C. Praddaude and S. Foner, J. Appl. Phys. 50, 7771 (1979)
8 R. Gersdorf, F.A. Muller and L.W. Roeland, Rev. Sci. Instrum. 36, 1100 (1965)
9 J. Witters and F. Herlach, to be published
10 F. Herlach, 2nd Int. Conf. The Application of High Magnetic Fields in Semiconductor Physics, Würzburg, July 24 - Aug. 2, 1974, p. 84
11 N. Miura, G. Kido, K. Suzuki and S. Chikazumi, 3rd Int. Conf. The Application of High Magnetic Fields in Semiconductor Physics, Würzburg, Aug. 23 - 27, 1976, p. 441

12. Proc. 2nd Int. Conf. on Megagauss Magnetic Field Generation and Related Topics, Rosslyn, Va., May 29 - June 1, 1979
13. A.R. Bryant, Proc. Conf. on Megagauss Magnetic Field Generation by Explosives and Related Experiments, Frascati, Sept. 21 - 23, 1965, p. 183
14. H.P. Furth, M.A. Levine and R.W. Waniek, Rev. Sci. Instrum. $\underline{28}$, 949 (1957)
15. C.M. Fowler, W.B. Garn and R.S. Caird, J. Appl. Phys. $\underline{31}$, 588 (1960)
16. C.M. Fowler, Science $\underline{180}$, 261 (1973)
    C.M. Fowler, R.S. Caird, W.B. Garn and D.J. Erickson, IEEE Trans. Magn. $\underline{\text{Mag-12}}$, 1018 (1976)
17. F. Herlach and H. Knoepfel, Rev. Sci. Instrum. $\underline{36}$, 1088 (1965)
    See also papers by CAIRD et al and by BITSHENKOV et al in ref. [12]
18. R.S. Hawke, T.J. Burgess, D.E. Duerre, J.G. Huebel, R.N. Keeler, H. Klapper and W.C. Wallace, Phys. Rev. Lett. $\underline{41}$, 994 (1978)
19. D W Forster and C.J. Martin, Proc. Int. Conf. Les champs magnétiques intenses, Grenoble, Sept. 12 - 14, 1966 (CNRS Paris 1967) p. 361
20. F. Herlach and R. McBroom, J. Phys. E (Sci. Instrum.) $\underline{6}$, 652 (1973)
21. F. Herlach, R. McBroom, T. Erber, J. Murray and R. Gearhart, IEEE Trans. Nucl. Sci. $\underline{\text{NS18}}$, 809 (1971)
22. F. Herlach, J. Davis, R. Schmidt, H. Spector, Phys. Rev. B $\underline{10}$, 682 (1974)
23. E.C. Cnare, J. Appl. Phys. $\underline{37}$, 3812 (1966)
24. N. Miura, G. Kido, M. Akihiro, S. Chikazumi, J. Magn. & Magn. Mater. $\underline{11}$, 275 (1979)
25. S.G. Alikhanov, V.G. Belan, A.I. Ivanchenko, V.N. Karasjuk and G.N. Kichigin, J. Phys. E (Sci. Instrum.) $\underline{1}$, 543 (1968)
26. F. Herlach, G. De Vos, J. Kuppens and G. Nimtz, in preparation
27. G. Nimtz and B. Schlicht, Festkörperprobleme (Advances in Solid State Physics) $\underline{XX}$, 369 (1980)

Recent Topics on the Generation and Application of Pulsed High Magnetic Fields at Osaka

M. Date, M. Motokawa, K. Okuda, H. Hori, and T. Sakakibara

Department of Physics, Faculty of Science, Osaka University, Toyonaka Osaka, Japan

1. Introduction

There are a few ways to obtain very high magnetic fields of about 1 MOe mainly developed in the recent decades. The first success was obtained by FOWLER and others [1] by a flux compression due to chemical explosives and the second method developed by CNARE [2] is a similar compression due to electromagnetic forces. The flux compression methods can produce pulsed fields of 2 or 3 MOe [3] inside the compressing liner. However, specimens and a part of the apparatus are necessarily destroyed in each run. Moreover, the pulse width is of the order of 1 μs so that the study of metals is very difficult due to both heating and skin effect. It is also not convenient for precise measurements to see the increasing field only.
A new method of generating a high pulsed magnetic field up to about 1 MOe without destroying the coil has been developed at the Osaka University since 1970. The magnet consists of multi-layer coils which are designed so as to share the strong Maxwell stress within their tensile strength. A small 4-layer model magnet was tested and the maximum field of 1.07 MOe with the pulse width of 0.18 ms was obtained without destroying the coil. A three year project of constructing a high magnetic field laboratory based on the new idea was promoted by the Ministry of Education in Japan, and we now have a 1.5 MJ energy source with various kinds of pulsed magnets [4]. The laboratory is now partly open, not only for our university users but also for many domestic and foreign visiting researchers providing a maximum field of 0.7 MOe with a pulse width of about 1 ms and an accuracy of o.3%, temperature range($1 \sim 400 \pm 0.1$)K. The magnet system is still under construction and a 1 MOe magnet for low temperature use will be completed in a few years.
The purpose of the present paper is to show our essential points of the field generation techniques and to describe several typical experimental results recently obtained in our laboratory.

2. The Magnet System

Our magnet coils are made of maraging steel which consists of iron, nickel, cobalt, molybdenum and some other elements. The tensile strength is more than 200 kg/mm^2 which corresponds to the Maxwell stress of 0.7 MOe. One of the difficult problems of this material is the high electrical resistivity. It is

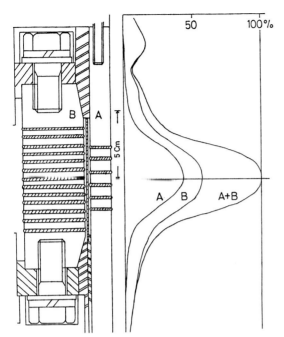

Fig.1 Cut view of the standard two-layer magnet. The coils A and B produce the fields A and B, respectively. The resultant field is shown by Curve A + B. An inhomogeneous field is seen near the coil connector (upper part)

about 50 μΩ-cm which is seven times larger than that of Be-Cu alloys. The coils are cut leaving strong ends and constructed as is partly shown in Fig.1 which also shows the field distributions produced by the inner and outer coils. Magnetic fields of about 0.7 MOe are produced with a field uniformity of more than 99% within a space of 5 mm. Natural mica sheets and polyimide laminates are used for insulating spacers. The inner diameter of the inner coil A is 20 mm and a liquid helium Dewar can be inserted with a liquid helium space of 5 mm in diameter. We are now testing the three layer magnet with the same inner diameter and the maximum field is expected to be 1 MOe. Five energy sources of 250 kJ capacitor bank are connected in parallel and the field is generated by closing the air-gap switch tubes of the five banks at the same time. The maximum voltage of the capacitors is 26 kV. The produced field is measured by the usual pick up coils and the absolute value of the field is calibrated by using the submillimeter electron spin resonance [5]. The accuracy is about 0.3%. Temperature increase due to the field generation is about 300°C after one shot so that one must wait about 20 minutes for the next run. The magnet is kept at room temperature and no cooling system is used because it is most simple and convenient for various kinds of experiments. All systems are operated by remote control.

3. Magnetization Measurements

Magnetization measurement is one of the most fruitful experiment for our magnet system. The measurement is done by using a pick up coil with flux compensation. Care should be paid to the eddy-current heating and the skin effect for metallic materials and to the magneto-caloric effect due to the adiabatic magnetization which is not negligible for paramagnetic materials. Several examples of the obtained results are shown in this section.

3.1 Ferrimagnetic Magnetization

There have been a lot of studies on the magnetization process in antiferromagnets. As for ferrimagnets, however, the complete magnetization curve has not been obtained because it needs a strong magnetic field. Recently we studied the magnetization of a Heisenberg ferrimagnet $Mn(CH_3COO)_2 4H_2O$ with MATSUURA et al. of the Faculty of Fundamental Engineering, Osaka Univ. and found a typical ferrimagnetic magnetization as is shown in Fig. 2 [6].

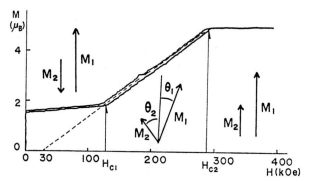

Fig.2 Ferrimagnetic magnetization of $Mn(CH_3COO)_2 4H_2O$

In usual ferrimagnets, the state is stable below H_{c1}. When an external field is between H_{c1} and H_{c2}, the spins show a canted state as is shown in Fig. 2 and it becomes ferromagnetic (paramagnetic) above H_{c2}. It is noted that the extrapolated magnetization curve of the spin canted state does not cross the origin and this is explained by introducing a biquadratic exchange interaction.

3.2 Field Dependent Slater-Pauling Curves

The Slater-Pauling curve expresses the concentration dependence of magnetic moments in 3d-alloys. Using a strong magnetic field, one can observe the increase in the saturation moment of these alloys induced by the high field magnetic susceptibility. Results of Ni-Mn and Ni-Cr alloys at 4.2 K are shown in Fig. 3 and the corresponding Slater-Pauling curves are given in Fig. 4. These data were obtained by OKUDA et al. in our group [7].

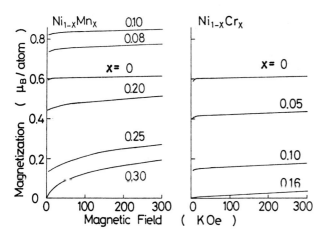

Fig.3 Magnetization curves of Ni-Mn and Ni-Cr alloys

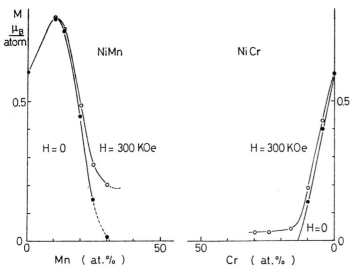

Fig.4 Field dependent Slater-Pauling curves of two alloys

3.3 Phase Diagram of FeF_2

The phase diagram of the fundamental antiferromagnet FeF_2 has not been determined because the anisotropy is so strong that the spin flop critical field is so far from the usual field region. JACCARINO of California Univ., Santa Barbara came to our laboratory and found the phase diagram given in Fig.5. In addition, a new phenomenon of single ion <u>exchange flopping</u> at multiples of the single bond exchange field was found in the system FeF_2-ZnF_2.

3.4 Metamagnetism in CoS$_2$ and CoSe$_2$ Mixtures

Metamagnetism in compounds such as FeCl$_2$ has been well known. However, a similar phenomenon recently found by ADACHI et al. of Nagoya Univ. in the spin fluctuating system of CoS$_2$ and CoSe$_2$ mixtures is not clear and interesting. Fig.6 shows the results. An important point is that the transition occurs in <u>paramagnetic</u> states [8].

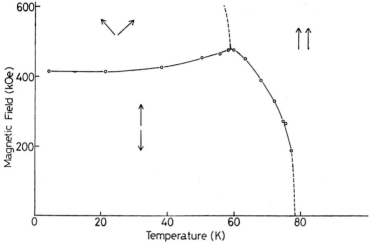

Fig.5 Magnetic phase diagram of FeF$_2$, H // c-axis

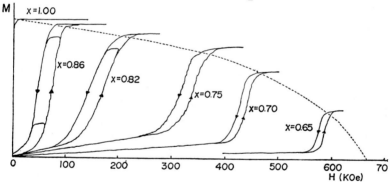

Fig.6 Metamagnetic transition in Co(S$_x$Se$_{1-x}$)$_2$ at 4.2 K

3.5 Low-Spin-High-Spin Transition in MnAs$_{1-x}$P$_x$

This compound is famous because it shows the low-spin-high-spin transition with increasing temperature. IDO of Tohoku Gakuin Univ. did the magnetization measurement and found that the transition also occurs under a magnetic field with a large temperature dependent hysteresis as is shown in Fig.7. The high spin and low spin states are observed in the paramagnetic and ordered states, respectively and a metamagnetic change corresponding to the low-spin-high-spin transition occurs in the intermediate region. Open and black circles show the transition for increasing and decreasing field, respectively. A large volume change for the transition is also observed[9]. The volume V is large in the high field region and $\Delta V/V$ is estimated to be about 10%.

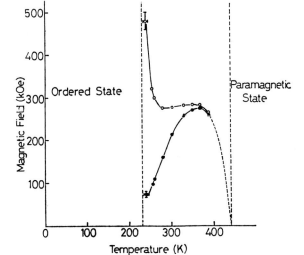

Fig.7 Low-spin-high-spin transition in $MnAs_{1-x}P_x$, x=0.1 Open and black circles are increasing and decreasing field data, respectively

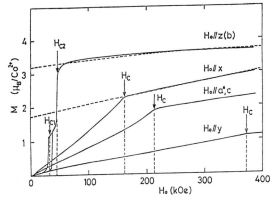

Fig.8 Magnetization curves of Ising like $CoCl_2 2H_2O$

3.6 Magnetization Studies of Low Dimensional Magnets

Magnetic properties of low dimensional magnets are one of the important problems in the present research of magnetism. We now show two examples of magnetizations in low dimensional systems. $CoCl_2 2H_2O$ is a typical Ising like antiferromagnet and the two step metamagnetism along the b-axis is well known. However, the transverse magnetizations have not been investigated completely because the Ising like magnet has a weak magnetization perpendicular to the Ising axis so that a strong magnetic field is needed to obtain complete magnetization curves. Fig.8 was obtained by MOLLYMOTO et al. at 4.2 K. H_c is the critical field above which the system is paramagnetic. A considerable increase in the magnetization above H_c is found and it is explained by the Van Vleck susceptibility and SH^3 term [10].

Large spin contractions expected in low dimensional magnets are found in one and two dimensional antiferromagnets by observing the whole magnetization curves. Three examples are shown in Fig.9. $Rb_2PbCu(NO_2)_6$ and copper benzoate are typical one dimensional magnets and CTS ($Cu(NH_3)_4SO_4H_2O$) is two dimensional. The nonlinear magnetization below H_c is due to the spin contraction and one can calculate the magnitude from the experimental results. It is about 58% for one dimensional magnets and 46% for CTS. These results are satisfactorily explained by the theory [11].

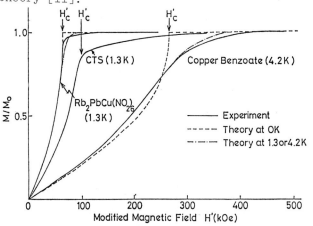

Fig.9 Magnetization of low dimensional magnets in the modified field scale with $H' = (g/2)H$

4. Optical Measurements

The optical Zeeman effect gives much information about the electronic structures. The applications to semiconductor physics will be shown by the NISHINA group in the present seminar so it is not discussed here. First, the Zeeman effect of the Na-D line is shown. The optical multichannel analyzer (OMA) is used and the method is shown in Fig.10. The light pulse and the OMA gate are open at the top of the magnetic field pulse so that the Zeeman effect is observed at maximum field. The Zeeman spectra of Na-D lines are also shown in Fig.10. The spectra at 500 kOe show a clear Paschen-Back effect. Similar experiments were done in a single crystal of ruby. As is well known, the R_1 and R_2 lines in ruby are sharp and are used for lasers. HORI et al. found the Zeeman spectra as is shown in Fig.11 [12]. There is a level mixing effect between R_1 and R_2 lines and the Paschen-Back effect is seen in the high magnetic field region. The level mixing occurs when the field is perpendicular to the c-axis.

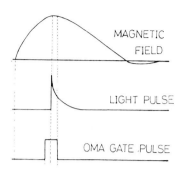
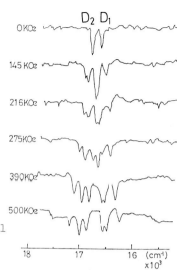

Fig.10 Time chart of the Zeeman spectrum measurements. The light pulse is applied to the Xe lamp at the field maximum. Experimental results for Na-D lines are shown on right

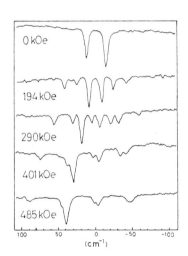
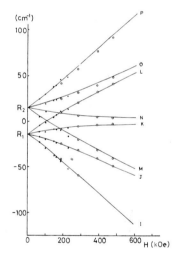

Fig.11 Zeeman spectra of R_1 and R_2 lines in ruby

5. Magneto-Resistance Measurements

The magneto-resistance of metals and alloys gives us a great deal of information about the electronic properties of conduction electrons. Usual metals show the positive magneto-resistance coming from the cyclotron motion of electrons. In some magnetic materials, however, a negative effect is expected because an external field quenches the spin fluctuation so that the resistivity is reduced. Results in some rare earth compounds will be presented by KASUYA of Tohoku Univ. at this seminar so it is

not reported here. We show the result for the helical spin magnet MnSi. This shows a large negative magneto-resistance due to the spin fluctuation (Fig. 12). About 70% of the resistivity is reduced with H_o = 400 kOe at 4.2 K [13].
Another Typical application is the determination of H_{c2} in high H_{c2} superconductors. The result for PbMo$_6$S$_8$ is given in Fig.13 which is the champion of all these materials [14].

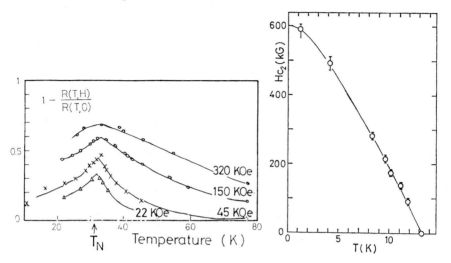

Fig.12 Magneto-resistance of MnSi   Fig.13 H_{c2} in superconducting PbMo$_6$S$_8$

6. Submillimeter Electron Spin Resonance

In the electron spin resonance (ESR) experiments at microwave frequencies the magnitude of the Zeeman energy is of the order of 1 cm^{-1}. Recent progress in laser techniques is now applicable to ESR in the submillimeter region up to about 100 cm^{-1}. For example, H$_2$O(λ=119 µm) and HCN(λ=337 µm) lasers are useful and the corresponding resonance field is 900 and 320 kOe, respectively for free electron spin with g=2.0. Constructions and applications of the spectrometer were done by MOTOKAWA of our Univ. and several interesting results are found [5]. Two important results are shown here. One of them is the discovery of the SH^3 term which is an additional new term of the effective spin hamiltonian in ionic crystals. This term is very small in the microwave frequency region. However, it becomes large under a strong magnetic field and it was observed in Co-Tutton salts by KURODA [15].
Another topic is the determination of exchange interaction in magnetic materials. Consider two spins S_1 and S_2 coupled by the exchange interaction J. If g-values of two spins are different to each other, they have different Zeeman energies in a magnetic field. When the difference in Zeeman energy is larger than the exchange energy, two well separated ESR lines are expected while an amalgamated narrowed line is observed under the

opposite condition. Fig.14 shows an example in CTS. When H∥b-axis(0°), two spins are equivalent and only one line is seen. However, the line becomes split by rotating the field. The exchange energy was found to be 0.24 K.

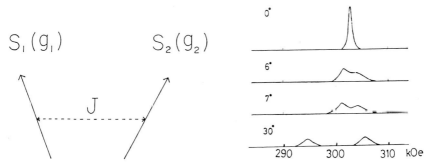

Fig.14 Exchange Splitting of ESR line in CTS

References

1. C.M.Fowler,W.B.Garn and R.S.Caird,J.Appl.Phys.31,588(1969)
2. E.C.Cnare,J.Appl.Phys.37,3812(1966)
3. S.Chikazumi,N.Miura,G.Kido and M.Akihiro,IEEE Trans.Mag.Mag-14,577(1978)
4. M.Date,IEEE Trans.Mag.Mag-12,1o24(1976)
5. M.Motokawa,S.Kuroda and Date,J.Appl.Phys.50,7762(1979)
6. M.Matsuura,Y.Okuda,M.Morotomi,H.Mollymoto and M.Date,J.Phys. Soc.Japan 46,1031(1979)
7. K.Okuda,H.Mollymoto and Date,J.Phys.Soc.Japan 47,1015(1979)
8. K.Adachi,M.Matsui,Y.Omata,H.Mollymoto and M.Date,J.Phys.Soc. Japan 47,675(1979)
9. H.Ido,T.Sakakibara and M.Date, to be published
10. H.Mollymoto,M.Motokawa and M.Date,J.Phys.Soc.Japan49,108(1980)
11. H.Mollymoto,M.Motokawa,M.Date,J.Phys.Soc.Japan 48,1771(1980)
12. H.Hori,H.Mollymoto and M.Date,J.Phys.Soc.Japan 46,908(1979)
13. T.Sakakibara and M.Date, to be published
14. K.Okuda,M.Kitagawa,T.Sakakibara and M.Date,J.Phys.Soc.Japan 48,2157(1980)
15. S.Kuroda,M.Motokawa and M.Date,J.Phys.Soc.Japan 44,1797(1978)

Explosive Generation of High Magnetic Fields in Large Volumes and Solid State Applications

C.M. Fowler, R.S. Caird, D.J. Erickson, B.L. Freeman, and W.B. Garn

Los Alamos Scientific Laboraory, University of California
Los Alamos, NM 87545, USA

Abstract

Various methods of producing ultra-high magnetic fields by explosive flux compression are described. A survey is made of the kinds of high magnetic field solid state data obtained in such fields by various groups. Preliminary results are given for the magnetic phase boundary that separates the spin-flop and paramagnetic regions of MnF_2.

1. Introduction

This year marks the twentieth anniversary of the first description of devices that produced ultra-high magnetic fields (1000-1500 T) by explosive flux compression techniques [1]. These devices rely upon the explosive cylindrical implosion of a thin walled conducting cylinder, usually called a liner, that contains an initial magnetic field. To the extent that flux is conserved within the liner, the initial magnetic field is amplified inversely as the square of the liner radius as it implodes. Since these early experiments similar results have been achieved at several other laboratories. However, as will be noted later, these systems have been exploited very little as research tools. On the other hand a certain amount of scientific information has been obtained in the range of 100-200 T with other types of explosive flux compression systems. There are some distinct advantages to these systems that partially offset the difficulties in using explosives. The initial magnetic fields required can be supplied by relatively slow capacitor banks; they can generate large fields in volumes substantially larger than those obtained by other methods; they can develop much larger fields than those obtained by other methods, at least to date.

A brief description of various high field systems with representative field volumes and time histories is given in section 2. Included here are speculations on the magnitudes of ultra-high fields that might be produced within the next few years. In section 3 we present the results of recent measurements made to obtain the temperature dependence of the magnetic phase boundary that separates the spin-flop and paramagnetic regions in MnF_2. Very preliminary results are described that were obtained from a new rotating mirror spectrograph with optics designed for the near ultraviolet.

2. High Field Systems

Fields to 250 T

Figure 1 shows schematically the two-stage system most commonly used at Los Alamos. The first stage consists of two strip generators connected in series

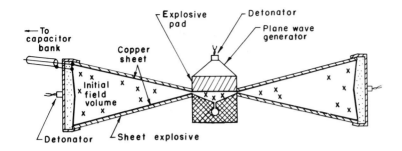

Fig.1 Schematic drawing of a two stage high field generator.

with the second stage but fired in parallel. We describe first the strip generators since they are also used individually to generate fields up to 150 T. In this case, the load coils are normally made from 51 mm square brass bar stock, 76 mm long. A hole of the appropriate diameter is drilled through the block. The output ends of the generator plates are connected to the load by soldering them to the faces formed by machining a slot from one side of the block to the hole. The explosive layers consist of two sheets each of Dupont Detasheet C-8. The sheets are about 340 mm long and 95 mm wide, into the plane of the figure, with a combined mass of approximately 1.6 kg. The angled copper plates are 1.6 mm thick, 340 mm long and are separated by 100 mm at the current input end. To enhance the structural strength of the plates they are bent upwards about 20 mm on the long sides to form troughs that, in turn, hold the explosive strips. The inside dimensions of the troughs are about 100 mm.

Figure 2 shows characteristic field vs time curves obtained for several load coil diameters. The peak fields obtained for a given system are remarkably consistent, usually varying no more than one or two percent for the larger load coil diameters. Not plotted on Fig.2 are results for 9 mm D coils. Here, peak fields are somewhat more variable, ranging from 140 to 150 T. An interesting variant of this class of generators has been developed by HERLACH et al. [2]. Here, the explosive is sandwiched between two sheets of copper which are, in turn, driven outward to contact stationary outer conductors. The outer conductors can therefore be made quite massive, even surrounded by concrete, to make them quite resistant to distortion from magnetic forces. BICHENKOV [3] has achieved conversion efficiencies of explosive to magnetic energy approaching 15% with similar systems. We have tried several shots with devices of this type. Their performance is at least equal and in some particulars superior to those shown on Fig.1. However, they are considerably more expensive to fabricate.

We resume our description of the two-stage system used for higher fields. The second stage and load coil are machined from a single piece of brass bar stock, usually 76 mm deep, into the plane of the paper, 180 mm wide and 76 mm high. The triangular cavity, or second stage, is normally about 150 mm wide and 32 mm high at the triangle apex. Initial magnetic fields are supplied by a capacitor bank and fill both first and second stage cavities as well as the cylindrical load coil. The second stage is initiated at such a time as to continue flux compression into the load coil at the end of the first stage compression. Our second stage explosive system consists of a plane wave

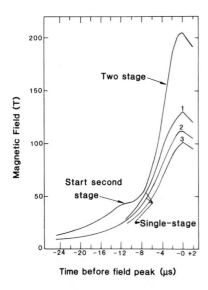
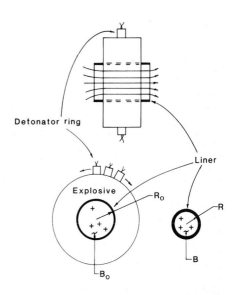

Fig.2 Field vs. time plots for single- and two-stage systems. Load coil diameters for curves 1-3 are 11.1, 15.9 and 19.1 mm.

Fig.3 Schematic drawing of a cylindrical implosion system.

initiator and a high explosive pad, normally 76 mm high, 76 mm deep and 155 mm long. The explosive used most often is a plastic bonded type, 95% HMX and 5% binder. The P081 plane wave initiator is available from Los Alamos. HERLACH and his collaborators [2] have used a related two-stage system. They achieved simultaneous plane initiation by using a metal flyer plate to impact the explosive.

A typical field vs time plot for this system is given in Fig.2 for a field coil 15.9 mm in diameter and 76 mm long. This record may be compared with that obtained from a single-stage device with the same sized load coil that is also shown. The single-stage devices are quite forgiving in that about the same peak fields are obtained from similar systems, even with variations of several percent on metal thickness, explosive thickness and width, the angle between the copper plates and the initial energy supplied to the system. We have found that the two-stage devices are more demanding. In our systems, for example, performance seems to degrade if the magnetic fields at the start of the second stage depart appreciably from the range of 35-45 T. Generally, peak fields obtained with a given system vary from shot to shot by 2-3%, and up to 5% for the smaller diameter, higher field shots. Typical peak fields for other load coils are 170 T and 240 T for diameters of 19.1 and 9.5 mm respectively.

Cylindrical Implosion Systems: fields to 1500 T

Since the first paper appeared on the generation of ultra high magnetic fields by flux compression many papers have been written on the subject. Some of these papers have described experimental attempts to develop similar systems.

Most have been devoted to theoretical studies of factors that affect the system performance. While it is not within the scope of this report to survey this work, the interested reader can refer to KNOEPFEL [4] to get a feeling for the state of the subject up to 1970. Owing in part to the difficulties inherent in making workable systems, some attempts to generate ultra large fields were not completely successful. This naturally, and correctly, led to questions about the field measurement techniques employed. The controversy now appears to be resolved since several other groups have succeeded in generating similar results. However, one result of the controversy was that almost all the early experiments done with these fields had as a primary objective that of furnishing an alternative field measurement technique to check the results obtained from the magnetic induction probes normally used. Even before the controversy arose, we also felt the need to verify probe readings. Consequently measurements were made of the Faraday rotation in quartz, using the Hg green line as the light source [5]. Owing to light intensity problems, the signals from the photomultiplier tube that was used to monitor the transmitted light was too weak to allow good resolution beyond 220 T. To this value, however, the fields calculated from the rotation agreed with those obtained from the probe. Incidentally, those results may be the first reported "solid state" data obtained in megagauss fields. Somewhat later CAIRD and coworkers [6] noted small non-linearities in the rotation when the Hg blue line was used. This may have resulted from impurities in the quartz. High field Faraday rotation data were subsequently reported by HERLACH et al. [7] for crown glass using Hg blue light. Again agreement with field measurements obtained by induction probes was obtained. For some years the highest field data reported was obtained at 510 T by GARN et al. [8] in which the Zeeman splitting of the sodium D lines was observed photographically and measured to be 164 Å. Again, the objective of verifying the probe response was met.

Factors affecting the performance of cylindrical implosion systems can be seen from Fig.3. Two views are shown of the initial configuration consisting of a conducting liner surrounded by a ring of explosive with initiation system and an initial magnetic field, B_o, within the liner. A view of the liner at some time after explosive initiation is shown at the lower right. Virtually every parameter indicated has some influence on the magnetic field compression dynamics: type and dimensions of the explosive charge and liner and value of the initial field. To obtain ultra-high fields consistently, great care must be exercised in almost every phase of the experiment. Uniformity of both liner and explosive are required and close machining tolerances must be maintained. Proper initiation of the explosive is of utmost importance. Care must also be taken when introducing the initial magnetic field, otherwise undesirable motion of metal components in the system may arise from the accompanying magnetic forces. Discussions on this latter point may be found in [4], [5], and [9]. A useful discussion of various initiation systems has been given by NAKAGAWA et al. [10]. The first hydrodynamic calculations on such systems treated the liners as perfect conductors [1]. Among the predictions was the existence of a "turn-around" radius. This radius is that which the imploded liner achieves at which the magnetic fields get so large that their pressures are sufficient to stop and then reverse the direction of liner motion. As might be expected, smaller initial fields lead to larger final fields, but at smaller radii. These idealized calculations agreed reasonably well with experiment in a number of respects. Generally speaking, when the initial fields were small, the peak fields occurred at radii smaller than the probes and the probes were therefore destroyed while the fields were still rapidly rising. For large initial fields predicted peak fields agreed fairly well with measured fields. Somewhat later KIDDER [11] used a much more sophisticated magneto-hydrodynamic code to calculate the compression dynamics. The results were that about the same maximum fields should occur but at a somewhat smaller radius, since an appreciable part of the initial flux penetrated into the liner.

Perhaps the most thorough experimental analysis of these systems has been given by BESANCON et al. [12]. Starting with initial magnetic fields of about 5 T, they consistently produced fields in the range 1000-1200 T even though the fields were rapidly increasing at the time of probe destruction. Field compression ratios of about 200 were obtained with copper liners and compare favorably with those obtained by CAIRD et al. [6], who used stainless steel liners but employed substantially smaller probes. As noted by BESANCON et al. [12], flux diffusion into stainless steel is substantially greater than in copper. HAWKE and his coworkers [13], with similar systems, used the magnetic fields as pressure sources to squeeze materials in a nearly isentropic manner. Although the nature of the experiment did not permit direct measurement of the peak fields, the generation of pressures in the multimegabar range implied fields greater than 1000 T. The discussion of high magnetic fields as pressure sources does not properly belong to this conference. However, for those who might be interested, a short discussion of high field compression of both solids and plasmas may be found in [14].

Most recently, PAVLOVSKII et al. [15] quote production of fields of 500 T in coils of diameter 15 mm and 1500 T in coils of 7 mm diameter. This same team [16] has also just recently reported Faraday rotation data obtained with one of these systems at fields of about 950 T. To our knowledge, this is the highest magnetic field solid state data yet obtained.

We conclude this discussion with one other explosively driven cylindrical device, described by DI GREGORIO et al. [17], that has been used for high magnetic field solid state research. It employs a thin-walled cylindrical liner overlaid with explosive inside of which a magnetic field has been induced. Here, however, the explosive is initiated at one end of the cylinder, around its circumference. The implosion is thus quasi-cylindrical and produces a time-varying magnetic field that moves along the axis of the liner with explosive detonation velocity. Fields in the neighborhood of 400 T have been produced with such devices. GUILLOT and LE GALL [18] have observed interesting high field magneto-optic properties of yttrium-iron garnet, presumably using such systems.

Fields Greater Than 1500 T

In a semipopular article, BITTER [19] discussed the possibility of generating fields of 5000 T and higher using chemical explosives. He also outlined a number of interesting experiments that might be done in such fields. He did not consider the dynamics of flux compression, but instead concentrated upon means of generating the initial flux required, and to some extent upon the amount of explosive needed based upon rather optimistic energy conversion efficiencies. It is now well known that the liner dynamics and flux diffusion play dominant roles in the flux compression process [4]. Figure 4 gives calculated peak fields obtained for various initial system parameters assuming ideal flux compression but including liner dynamics. The coordinates are peak magnetic field and radial compression ratio. Parameters include various values of initial magnetic field and different liner materials and thicknesses. One of the principal dynamical limitations at present is that of

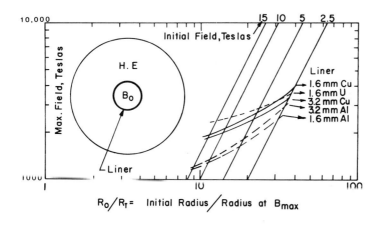

Fig.4 Idealized calculations giving maximum field and compression ratio for different liners and initial magnetic fields, B_o. The dimensions of the HMX explosive charge were fixed at 305 mm OD and 107 mm ID.

maintaining cylindrically symmetric implosions. At present, we think that our liner implosions remain reasonably symmetric up to radial liner compressions somewhat greater than 20 to 1. This ratio is based upon observed magnetic field amplifications somewhat greater than 200 to 1 with stainless steel liners, and an estimate that at least half of the initial flux had diffused into the liner at the time of probe destruction. According to Fig.4 if the symmetry can be extended to radial compressions of 35 to 1, then fields of order 2500 T should be produced with copper liners, provided that no more than two thirds of the initial flux is lost by diffusion into the liner. The result of increasing the outer diameter of the explosive charge is to raise somewhat the curves for the specific liners studied. This has the effect of reducing the symmetry required to achieve a given final field, but also requires a corresponding increase in the initial field.

BROWNELL [20] has calculated several configurations using a one-dimensional magneto-hydrodynamic code somewhat improved over that first employed by KIDDER [11]. As it turns out, peak magnetic fields occur somewhat in advance of minimum liner radius, because of the rapidity of flux diffusion at this time. Some of these calculations predicted peak fields near 4000 T in diameters of 4-5 mm with duration times for fields greater than 90% of peak of 0.2-0.25 µs. However, as BROWNELL points out, the results must be treated with great caution in view of the uncertainties in characterizing the physical properties of the materials at the extreme conditions required in the calculations. A final remark may be made about scaling the dimensions of the system as, for example, by doubling all dimensions. The dynamical time scale would be approximately doubled, while flux diffusion into the liner would vary as a lower power over most of the implosion. In this sense, higher peak fields would be expected, although the initial energy and the amount of explosive required per unit length would be quadrupled, and the time for the development of perturbations would be doubled.

We finally present some idealized calculations for systems analogous to the electromagnetic implosion devices employed by MIURA and his associates [21].

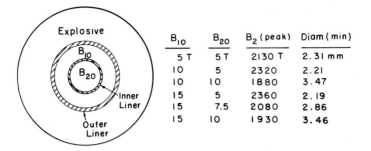

B_{10}	B_{20}	B_2 (peak)	Diam (min)
5 T	5 T	2130 T	2.31 mm
10	5	2320	2.21
10	10	1880	3.47
15	5	2360	2.19
15	7.5	2080	2.86
15	10	1930	3.46

<u>Fig.5</u> Idealized calculations for combined explosive-electromagnetic flux compression system. The dimensions of the HMX explosive charge were fixed at 235 mm OD and 108 mm ID. The wall thickness of both aluminum liners is 1.6 mm, and the OD of the inner liner is 51 mm.

As is seen in Fig.5, an inner liner is added to the system shown in Fig.4. The experimental region is centered inside the inner liner, which has an initial magnetic field B_{20}. This liner is then driven magnetically by the field between the two liners, with initial value B_{10}. The explosive system and initial liner dimensions are not varied. Peak fields and turnaround diameters are given for various initial field values, B_{10} and B_{20}. The calculations are completely idealized in that flux inside the liners is conserved. Only a few calculations have been made. It is likely that results similar to those shown in Fig.4 can be predicted with other materials and configurations. A few calculations have been made where several concentric liners have been placed in one flux compression system, each liner in turn, being driven by its adjacent field. However, no striking improvements in high field generation have as yet been obtained.

3. Solid State Data: ≤200 T

Examples of data obtained with cylindrical implosion systems were mentioned in the preceding section. We mention here a few examples obtained from strip systems or two stage systems, generally in fields in the 100-200 T range. Since this report is limited to explosively produced fields we do not include, for example, the beautiful cyclotron resonance experiments in megagauss fields of HERLACH et al. [22] and of MIURA and his collaborators [21]. These experiments usually involved the analysis of laser light reflected from the surface of various materials placed in high fields obtained by electromagnetic liner implosion. They could also be carried out in explosively produced fields, perhaps with some advantages, but have not been done to our knowledge. Almost all of the information in this category has been obtained either by DRUZHININ and his associates [23] or by the Los Alamos group. Generally speaking, the first group employed the Faraday effect using a single wavelength of light obtained from a laser. Most of the work reported by the Los Alamos group also involved optical effects both in the Faraday mode and in direct transmission. Through use of a rotating mirror spectrographic camera this work is now carried out over the entire visible spectrum in each shot. Additionally, much of the work is done at cryogenic temperatures. Examples of Faraday rotation data may be found in [24]. The advantages of broad band coverage are apparent here in that the Verdet coefficients are obtained as a

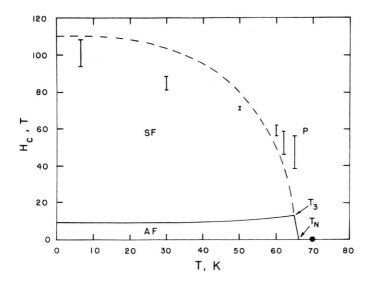

Fig.6 Possible boundary separating the spin-flop and paramagnetic phases of MnF_2. The dash curve is based upon a highly simplifed molecular field model.

continuous function of wavelength in each shot. Non-linearity in the rotation, as evidenced at higher fields for some materials, can also be detected more easily, in our opinion. An example of a direct transmission experiment can be found in [25]. Here, the exciton spectrum of GaSe has been obtained to about 200 T, and a calculation is presented that is in good agreement with the observed data.

Some years ago, CAIRD et al. [26] studied the optical absorption in MnF_2 in high fields at 7 K. They observed that much of the absorption structure present at lower fields disappeared at a field of about 100 T. Previous work had strongly indicated that most of the absorption lines arose from magnon-assisted transitions. Thus, the low temperature disappearance of these lines at high field was consistent with the interpretation that the material had become completely paramagnetic. We have recently carried out similar experiments at higher temperatures with the intention of mapping out the magnetic phase boundary that separates the spin-flop and paramagnetic regions. The resulting data are plotted in Fig.6. The uncertainties shown reflect the spread in values obtained by different observers in their inspection of the original records. As expected, the fading out of the lines became more gradual as the temperature increased, thus making it more difficult to estimate the fields associated with their disappearance. Not shown on the plot are uncertainties in the temperatures. The values listed are possibly in error by as much as two or three degrees particularly at the higher temperatures. The point at about 7 K represents the spread obtained from several shots. For comparison purposes the dashed curve of Fig.8 shows a simple molecular field model calculation of H_c with temperature, bounded by the triple point and approximately twice the exchange field (110 T) at 0 K.

The difference signal from two matched $\dot{B}$ probes was obtained for one of the 7 K experiments. One of the probes looped the MnF_2 crystal, while the other encircled a magnetically inert sample. The spin-flop transition signal at about 9.5 T was quite pronounced, but stray noise, probably resulting from imperfect probe matching, prevented meaningful interpretation of the records at fields much higher than 20 T. Thus our efforts to detect the second order phase change to the paramagnetic state were unsuccessful. Although we recognize that the experiments may be difficult, we plan to make more attempts of this nature at higher sample temperatures with the hope of narrowing the uncertainties in the fields presently assigned. Based upon earlier developmental experimental results, we are optimistic that better matching of the probes can be obtained. In these experiments, matched probes encircling identical magnetically inert materials were tested in high field producing devices that had very large time rates of change of field. In some cases the difference signals of the matched probes remained less than one percent of the individual probe signals, at fields in excess of 100 T, where the fields were changing at rates of order 10^9 T/s.

As will be seen from a study of the references cited in this paper, the diagnostics employed in pulsed megagauss field studies has been rather limited. However technological advances continue to increase diagnostic capabilities. Examples include a wide range of lasers as light sources, and continued improvement in electronic detectors, such as image intensifiers. We conclude this paper with a preliminary report concerning the performance of a new sweeping image spectrograph. It is patterned after the spectrograph we have used in the past for broad band coverage in the visible spectrum, as described by BRIXNER [27]. However, in this new instrument, Brixner has modified the optics for the near ultraviolet. The reflection grating is blazed for 200 nm and is presently mounted to allow coverage on the film plane from 200 to 400 nm, thus overlapping the coverage of the older spectrograph (380-660 nm). To date we have been able to record Hg arc lines shorter than 240 nm, although the shorter wavelength lines are of necessity, considerably attenuated by the optical system. A more serious problem may exist with our present light source, a tube of argon gas shocked with an explosive detonator. This source should be rich in the near ultraviolet, but our photographic records show a large absorption band around 280 nm, and very little light is recorded for wavelengths shorter than this band. We are continuing investigations on this effect, as well as others designed to study the overall camera attenuation as a function of wavelength. One room-temperature Faraday rotation experiment has been done up to a field of about 100 T, mainly to test the camera performance. The sample was ZnO, and both light and magnetic field were parallel to the c-axis of the crystal. The test was not a very stringent one since this material has an absorption cutoff almost in the visible spectral region. The ultraviolet polarizers used in the experiment were obtained from the 3M Company, Industrial Optics Division, in St. Paul, MN.

*Work supported by the U.S. Department of Energy.

<u>References</u>

1. C. M. Fowler, W. B. Garn, R. S. Caird: J. Appl. Phys. 31, 588 (1960).
2. F. Herlach, H.Knoepfel, R. Luppi: In <u>Proceedings of the Conference on Megagauss Magnetic Field Generation by Explosives and Related Experiments</u> ed. by H. Knoepfel, F. Herlach (Euratom, Brussels, 1966) p. 287.

3. E. I. Bichenkov: Sov. Phys. Doklady 12, 567 (1967).
4. H. Knoepfel: Pulsed High Magnetic Fields (North-Holland Publishing Co., Amsterdam-London, 1970).
5. C. M. Fowler, R. S. Caird, W. B. Garn, D. B. Thomson: In High Magnetic Fields, ed. by H. H. Kolm, B. Lax, F. Bitter, R. G. Mills (M.I.T. Press and John Wiley and Sons, New York, London, 1962) p. 269.
6. R. S. Caird, W. B. Garn, D. B. Thomson, C. M. Fowler: J. Appl. Phys. 35, 781 (1964), and Ref. 2, p. 101.
7. F. Herlach, H. Knoepfel, R. Luppi, J. E. van Montfoort: Ref. 2, p. 471.
8. W. B. Garn, R. S. Caird, D. B. Thomson, C. M. Fowler: Rev. Sci. Instr. 37, 762 (1966).
9. C. M. Fowler, R. S. Caird, W. B. Garn: LA-5890-MS (1975) document available from the Los Alamos Scientific Laboratory, Los Alamos, NM, USA 87545.
10. Y. Nakagawa, Y. Syono, T. Goto, J. Nakai: Sci. Rep. Ritu A Vol. 25, No. 1 (1974), Research Institute for Iron, Steel and other Metals, Tohoku University, Sendai, Japan.
11. R. E. Kidder: Ref. 2, p. 37.
12. J. E. Besancon, J. Morin, J. M. Vedel: C. R. Acad. Sc. Paris, t 271, Ser. B, 397 (1970).
13. R. S. Hawke, D. E. Duerre, J. G. Huebel, R. N. Keeler, W. C. Wallace: J. Appl. Phys. 49, 3298 (1979).
14. C. M. Fowler: Science 180, 261 (1973).
15. A. I. Pavlovskii, N. P. Kolokolchikov, M. I. Dolotenko, A. I. Bykov: Prib. Tekhn. Eksper., No. 5, 195 (1979).
16. A. I. Pavlovskii, N. P. Kolokolchikov, O. M. Tatsenko, A. I. Bykov, M. I. Dolotenko, A. A. Karpikov: Paper presented at Second International Conference on Megagauss Field Generation and Related Topics, May 29-June 1, 1979, Washington, DC. Proceedings in press.
17. C. di Gregorio, F. Herlach, H. Knoepfel: Ref. 2, p. 421.
18. M. Guillot, H. le Gall: In Magnetism and Magnetic Materials-1976. AIP Conf. Proc. No. 34, ed. by J. J. Becker, G. H. Lander, p. 391.
19. F. Bitter: Sci. Amer. 213, 65 (1965).
20. J. H. Brownell, Los Alamos Scientific Laboratory: Private communication.
21. N. MIURA, G. Kido, M. Akihiro, S. Chikazumi: J. Magn. and Magn. Mtls 11, 275 (1979). (This article is a recent rather brief survey of the high field work in progress by this group. Among others, it contains numerous references to other work accomplished by the group including both production and application of megagauss fields.)
22. F. Herlach, J. Davis, R. Schmidt, H. Spector: Phys. Rev. B10, 682 (1974).
23. A. I. Pavlovskii, V. V. Druzhinin, O. M. Tatsenko, R. V. Pisarev: JETP Lett. 20, 256 (1974); V. V. Druzhinin, A. I. Pavlovskii, A. A. Samokhvalov, O. M. Tatsenko: JEPT Lett. 23, 234 (1976); V. V. Druzhinin, G. S. Krinchik, A. I. Pavlovskii, O. M. Tatsenko: JETP Lett. 22, 130 (1975).
24. W. B. Garn, R. S. Caird, C. M. Fowler, D. B. Thomson: Rev. Sci. Instr. 39, 1313 (1968); C. M. Fowler, R. S. Caird, W. B. Garn, D. J. Erickson, B. L. Freeman: J. Less-Common Metals 62, 397 (1978).
25. C. H. Aldrich, C. M. Fowler, R. S. Caird, W. B. Garn, W. G. Witteman: In press, Phys. Rev. B.
26. R. S. Caird, W. B. Garn, C. M. Fowler, D. B. Thomson: J. Appl. Phys. 42, 1651 (1971).
27. B. Brixner: Rev. Sci. Instr. 38, 287 (1967).

Generation of Megagauss Fields by Electromagnetic Flux Compression

N. Miura, G. Kido, H. Miyajima, K. Nakao, and S. Chikazumi

Institute for Solid State Physics, University of Tokyo, Roppongi, Minato-ku
Tokyo, Japan

>Pulsed high magnetic fields in the megagauss range have
>been produced by the electromagnetic flux compression using
>a condenser bank of 285 kJ. A larger condenser bank with
>a total energy of 6.5 MJ (5 MJ for primary current and 1.5 MJ
>for the seed field) is being installed to generate even
>higher fields. Non-destructive fields in the submegagauss
>range have also been produced. Generated high fields were
>successfully used for various solid state experiments.

1. Introduction

Since FOWLER et al. first reported the generation of magnetic fields higher than 1000 T (10 MG) by the explosive method in 1960 [1], several techniques have been developed for generating very high magnetic fields in the megagauss range ("megagauss fields"). In 1966 CNARE reported the generation of fields up to 2 MG by the magnetic flux compression by an electromagnetically driven metal ring (liner) [2]. The electromagnetic flux compression has several advantages over the explosive flux compression. i) A relatively little part of the apparatus for generating the fields is destroyed by each pulse. ii) successive repetition of the generation is easier. iii) The coil system and the triggering system of the pulse can be readily combined with other experimental apparatus for solid state physics. These advantages are particularly important for applying the generated megagauss fields to solid state research.

In the Institute for Solid State Physics, the University of Tokyo, we have developed techniques for generating the megagauss fields [3-5]. By using a condenser bank of 285 kJ, high fields up to 280 T have been generated. The megagauss fields are very useful for solid state physics, because the influence of the fields on electrons in solids is extremely large in the megagauss fields. The megagauss fields have been conveniently employed for various solid state experiments. On the basis of the accumulation of experience in the present system, we have started a new project for generating even higher fields. Larger condenser banks with a total energy of 6.5 MJ are being installed. Non-destructive fields up to about 45 T have also been produced by using a condenser bank of 32 kJ [6,7]. For obtaining higher and longer fields in a non-destructive way, a condenser bank of 200 kJ is being built.

In this paper, we will describe techniques for generating the pulsed high fields by our present systems, as well as the outline of our new systems.

2. Electromagnetic Flux Compression

The coil system for the electromagnetic flux compression is schematically shown in Fig.1. By a large pulsed primary current (about 1-2 MA) supplied from the condenser bank, the secondary current of almost the same magnitude is induced in the liner (a ring of Cu or Al) in the opposite direction. The electromagnetic force between the two opposite current is so strong that the liner is squeezed rapidly towards the center. During the process of the squeezing, the magnetic flux consisting of the seed field produced by a pair of injection coils is compressed by the motion of the liner. As the total magnetic flux inside the liner is almost conserved, the field at the center becomes very high, as the diameter of the liner becomes sufficiently small.

We use a main condenser bank of total energy of 285 kJ (30 kV) to supply the primary current and a sub-bank of 47 kJ (3.3 kV) to inject the seed field. Ten ignitrons (Mitsubishi MI-3300F) are used in parallel connection to switch the primary current. They replaced 12 older ignitrons (Mitsubishi MI-3300E) which were connected in six parallel channels consisting of two ignitrons of series connection [3]. By changing the ignitron bank to the new one, the total inductance of the primary circuit was reduced from 326 nH to 171 nH. The maximum current of 1.6 MA can be obtained from the main condenser bank. The primary coil is made from a steel plate of the thickness of 20-26 mm. The wave-shaped slots with insulating plate in them are necessary to improve the uniformity of the field in the primary coil, reducing the effect of the feed gap. This type of primary coil can be repeatedly used, but it is broken after 8-10 shots of experiments by the strong electromagnetic force. It is essential for the application of the megagauss fields to solid state experiments that the primary coil is not destroyed over several shots. In the present system we are using liners of copper 66 mm in outer diameter, 1.0-1.6 mm in thickness and 20 mm in length. We usually inject a seed field of about 3 T in the liner. Fig.2 shows an example of the experimental traces of the observed high field pulse.

The process of the electromagnetic flux compression can be well simulated by a computer calculation [8]. Fig.3 shows an example of the results of the

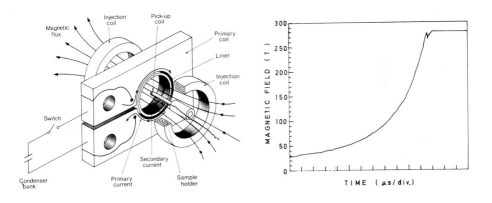

Fig.1 Coil system for the electromagnetic flux compression.

Fig.2 A recorder trace of the observed high field pulse.

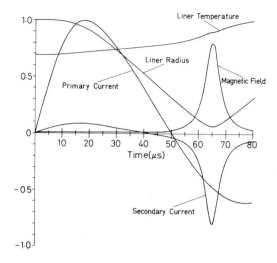

Fig.3 An example of the computer simulation of the electromagnetic flux compression using the main condenser bank of 1 MJ. The maximum primary current: 2.7 MA, the maximum secondary current: 18.7 MA, the maximum magnetic field: 754 T and the maximum liner temperature: 435 T.

computer simulation, taking into account the magnetic field distribution in the primary coil and liner, the change of the resistivity of the liner due to the temperature rise and so on. The parameters were chosen in accord with the experiment using a part of the new condenser bank system: namely, we assumed that the energy of the main condenser bank is 1 MJ (40 kV); the seed field is 3.64 T; the diameter, the thickness and the length of the liner are assumed to be 100 mm, 5 mm and 30 mm, respectively; the inner diameter and the thickness of the primary coil are 108 mm and 40 mm. The calculated maximum of each quantity is: primary current 2.7 MA, the secondary current in the liner 18.7 MA, the magnetic field 754 T, and the liner temperature 435 K. The maximum velocity of the liner is 1.05 km/s. The calculation predicts that the maximum field is obtained in a volume with the diameter of 6.5 mm. There is a tendency that as the higher seed field is injected, the maximum field is decreased but the available volume of the high field is increased [8,9]. Such computer simulation is very useful for designing a coil system.

3. Outline of the New Condenser Bank

Since 1979, construction of a new condenser bank system of 6.5 MJ has been started by Nichicon Capacitor Ltd. The outline of the new condenser bank is schematically shown in Fig.4. The main bank is divided into two, the one with the energy 1 MJ and the other with 4 MJ. Each of the divided banks can be operated independently using two separate control systems to energize a smaller coil system 1 and a larger coil system 2, respectively. When we require 5 MJ for the main bank, the 1 MJ bank is also connected to the coil system 2 by switching SW 2. In this way, we can easily select and operate one of the two coil systems with different size depending on the experiment. The whole main bank of 5 MJ consists of ten unit banks of 0.5 MJ. We can store the energy in any combination of these units by controlling SW 1. When SW 1 is turned off for a certain unit, the triggering of the gap switch for the same unit is also stopped, and we can make the unit completely dead. Thus we can vary the capacitance of the main bank, which is useful choosing a proper experimental parameter depending on the requirement of the field

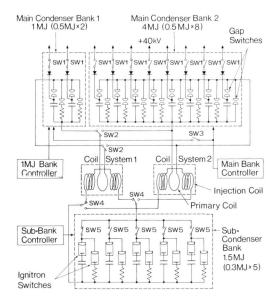

Fig.4 Schematic diagram of the new condenser bank system.

intensity. The capability of the partial use of the condenser bank is also useful for searching a wrong part when some trouble occurs. Twelve spark gap switches for start are used for each unit bank, so that 120 gaps are used in the entire main bank of 5 MJ.

The primary current is clamped after reaching the maximum by the diversion (crowbar) circuit, in order to absorb the unnecessary energy of the electric current by the diversion resistance inserted in series with the gap switch for diversion. This is important to lower the load for the primary coil prolonging its life. The same number of the gap switches for diversion are used as for start. The sub-condenser bank of 1.5 MJ is used for injecting a seed field. The sub-bank can be connected either to the coil system 1 or 2 by controlling a large switch SW 4. The sub-bank consists of 5 unit banks of 0.3 MJ. It is again possible to select the combination of used units by SW 5. For the sub-condenser bank, 15 and 20 ignitrons (MI-3300 E) are used for the start and for the crowbar circuit, respectively.

The triggering system for the start and diversion switches in the main condenser bank is shown in Fig.5. Every 12 gap switches in the main circuit are triggered by a submaster gap switch, and submaster gap switches are triggered by a master gap. Two master gap switches are installed for each of the start and diversion, one for the 1 MJ bank and one for the 4 (or 5) MJ bank. For triggering the gaps in the main circuit, a pulse of 60 kV peak voltage is superposed to the gaps by firing the submaster gap. At the same time, the discharge between the center electrode and one of the electrodes of the gap is started to reduce the jitter time among the gaps. Ferrite disks are used as saturable reactance to apply the superposed voltage effectively across the gaps. The gap switches are operated under a pressure of 1-5 atm. in sealed cases depending on the charging voltage in the bank. The maximum current of 6 MA can be obtained from the main condenser bank of 5 MJ. The project requires four years in total to complete the installation of the system, but

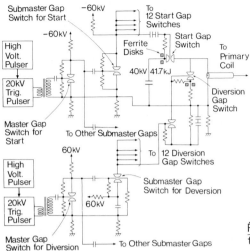

Fig.5 Triggering system for the gap switches.

a preliminary experiment has been started using a condenser bank of 1 MJ and the coil system 1, which have been almost completed.

By using the completed new condenser bank system, it is expected to produce very high field (hopefully higher than 700 T). However, as the stored energy of the condenser bank is enlarged and the primary current is increased, the electromagnetic force becomes extremely large. As a result, there are many difficulties in building a primary coil and a collector plate for the primary current to avoid their destruction. In fact, the sizes of the coil system and the collector plates in the new system are much larger than in the present system. On the other hand, as the number of condensers and switches connected in parallel is increased for a larger bank, the impedance of the condenser bank itself including that of switches, cables and collector plates is decreased in comparison with the impedance of the coil. For instance, the inductance of the 5 MJ bank using 120 gap switches will be about 30 nH, whereas in the present system of 285 kJ using 10 ignitron switches, the inductance of the bank is 90 nH. On the contrary, the inductance of the primary coil for the new system will be much larger than that for the present system. Design of the coil system as well as the preliminary experiments are in progress. The new condenser bank and the coil systems will be installed in a building with the total area of 30×48 m, which is being built for the new project.

4. Generation of Non-destructive Sub-megagauss Fields

For conducting solid state research in the megagauss fields, it is also important to perform preliminary experiments in a non-destructive field before the sample is put into the destructive megagauss fields. Since the KAPITZA's pioneering work [10], various types of magnets have been developed for pulsed high magnetic fields. FONER and his collaborators designed solid helix magnets made of beryllium-copper alloys [11]. This type of magnet has a mechanically strong structure and suitable for generating non-destructive sub-megagauss fields [6]. On the other hand, the duration of the field is usually about a few hundred microseconds which is not long enough for some experiments. In addition, a great effort and skillful technique are required to machine the coil. Multi-turn wire-wound-coils can generally produce longer pulsed fields (longer than a millisecond) [12].

We generate pulsed high fields up to 45 T using a condenser bank of 32 kJ (3.3 kV) and a magnet the cross-section of which is shown in Fig.6.

The coil is operated at liquid nitrogen temperature. The coil is wound from rectangular copper wire 1.8×3.5 mm^2 which has three-fold insulation layers of polyvinyl formyl, polyimide tape and glass fiber tape. The wire is wound on a bobbin made of glass epoxy (FRP). After the winding of the copper wire, tape of glass fiber is tightly wound on it, and the coil is impregnated with epoxy resin which is solidified under a high pressure. The high pressure is obtained in a high pressure apparatus with oil as shown in Fig.7. The epoxy resin is sealed in the coil with a rubber tube tightly fixed around the outer surface of the coil. A pressure of 40-50 atm. is obtained in the oil by pushing a piston from the top with force of 6-8 tons. The oil is heated up to about 140°C by a heater from the outside. This method of solidification has brought a good result concerning the strength of the magnet.

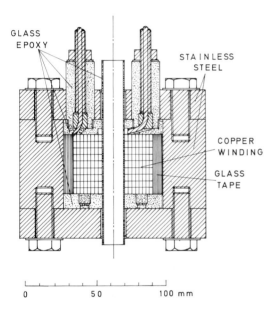

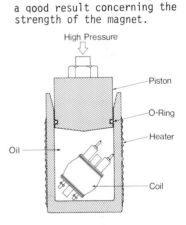

Fig.6 Cross-section of non-destructive pulse magnet for long duration.

Fig.7 Apparatus for solidifying epoxy resin impregnated in a magnet under high pressure.

After the epoxi resin is solidified, the outer surface is machined with a lathe and the coil is pushed into a cylinder of stainless steel. The outer diameter of the inner coil is machined slightly larger than the inner diameter of the stainless steel cylinder. Therefore, the inner coil is cooled down to liquid nitrogen temperature, while the stainless steel cylinder is heated up to about 300°C, before the coil is pushed into the cylinder. Owing to the thermal expansion and shrinkage, the coil is tightly fixed in the cylinder. By using this type of magnet combined with a 32 kJ condenser bank, we can produce a pulsed field up to about 45 T with a rise time of 1-2 ms. The produced fields have been conveniently used for transport measurements [7,13].

In our new project, a larger condenser bank with a total energy of 200 kJ is being constructed by Shizuki Electric Company and installed for non-destructive high fields. The circuit diagram of the 200 kJ bank is shown in Fig.8. The bank consists of four unit banks of 50 kJ. A unique characteristic point of this bank is that the maximum charged voltage and the total capacitance of the bank can be selected by controlling the switch SW 1. Namely we can utilize it either as a faster bank of 10 kV, 4 mF with series connection or as a slower bank of 5 kV, 16 mF with parallel connection. The capability of the selection of the capacitance is very useful to use the bank for various types of magnets. For example, the faster bank is suitable for a Foner-type magnet to produce higher non-destructive field, where the effect of resistance is important. On the other hand, the slower bank is suitable for a wire-wound magnet to produce long pulsed field. The discharge current is clamped by crowbar diodes to obtain long decay time of the pulse. Switches SW 2 should also be switched simultaneously whenever the SW 1 is controlled. Generally, it takes a long time to recool the magnet which has been heated up at each pulse. Therefore, we can select from four magnets by switching SW 3. Therefore, we can easily operate four magnets in turn as they are connected to the terminal. The switches SW 1, SW 2, SW 3 are controlled by a pneumatic control.

This new condenser bank together with magnets with larger size will extend the field range as well as the pulse duration. Consequently, we hope that it will provide a powerful means for studying solid state physics in high magnetic fields.

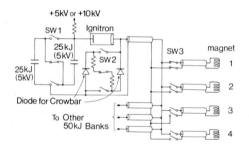

Fig.8 Circuit diagram of the 200 kJ condenser bank for the generation of non-destructive sub-megagauss fields.

References

1. C. M. Fowler, W. B. Garn and R. S. Caird: J. Appl. Phys. <u>31</u> (1960) 588
2. E. C. Cnare: J. Appl. Phys. <u>37</u> (1966) 3812
3. N. Miura, G. Kido, I. Oguro and S. Chikazumi: *Physique sous Champs Magnétiques Intenses* (Proc. Int. Colloquium on Physics in High Magnetic Fields) (CNRS, 1975) P.345
4. S. Chikazumi, N. Miura, G. Kido and M. Akihiro: IEEE Trans. Magn. <u>MAG-14</u> (1978) 577
5. N. Miura, G. Kido, M. Akihiro and S. Chikazumi: J. Magn. Magn. Mater. <u>11</u> (1979) 275
6. K. Suzuki and N. Miura: J. Phys. Soc. Japan <u>39</u> (1975) 148

7. K. Hiruma, G. Kido and N. Miura: Solid State Commun. $\underline{31}$ (1979) 1019
8. N. Miura and S. Chikazumi: Japan J. Appl. Phys. $\underline{18}$ (1979) 553
9. H.G. Latal: Ann. Phys. (USA) $\underline{42}$ (1967) 352
10. P. Kapitza: Proc. Roy. Soc. (London) $\underline{A105}$ (1924) 691, $\underline{A115}$ (1927) 658
11. S. Foner and H. H. Kolm: Rev. Sci. Inst. $\underline{28}$ (1957) 799, S. Foner and W. G. Fisher: ibid. $\underline{38}$ (1967) 440
12. H. A. Jordaan, R. Wolf and D. de Klerk: Phys. Letters $\underline{44A}$ (1973) 381
13. K. Hiruma, G. Kido, K. Kawauchi and N. Miura: Solid State Commun. $\underline{33}$ (1980) 257

Application of Megagauss Fields to Studies of Semiconductors and Magnetic Materials

G. Kido, N. Miura, M. Akihiro, H. Katayama, and S. Chikazumi

Institute for Solid State Physics, University of Tokyo, Roppongi, Minato-ku Tokyo, Japan

Various measurements have been carried out for semiconductors and magnetic materials in ultra-high magnetic fields generated by an electromagnetic flux compression method. Cyclotron resonance in n-GaP was observed at the wavelength of 119 μm and the longitudinal mass was determined as $m_\ell^* = 4.8\ m_0$. Magneto-optical exciton spectra were measured for GaSe, PbI_2 and CdS. An anomalous diamagnetic shift of the ground state was observed for each material. Transitions of the spin configuration were observed in TbIG and GdIG by both Faraday rotation and reversible magnetic susceptibility measurements, and the spin phase diagrams were investigated.

1. Introduction

Recently, several kinds of techniques have been developed to apply megagauss fields for the investigation of solids [1~7]. The difficulties in experiments with megagauss fields are caused by both the short pulse duration time of several microseconds and the destruction of the sample in every single shot. Generally, optical measurements are the most suitable techniques for the observation in a rapidly rising pulsed magnetic field, because of the good time resolution. In fact, most of the reported experimental results in megagauss fields were obtained by measuring the changes of the transmitted light intensity through the sample. In the visible region, FOWLER et al. built a system for the measurement of magneto-optical streak spectra by using a fast rotating mirror [1]. In the infrared region, monochromatic laser beams are available for the short time experiments with sufficient output intensity. A cyclotron resonance experiment was first attempted by HERLACH using an CO_2 infrared laser and a single turn coil up to 160 T [2]. We have developed techniques for generating megagauss fields by electromagnetic flux compression and for applying them to solid state experiments [3,4]. Techniques have also been developed for calibrating the field intensity with high accuracy [5]. As a result, reliable data can be obtained in various kinds of measurements for semiconductors and magnetic materials in megagauss fields: 1) Time resolved magneto-optical spectra using an image converter camera and an image intensifier [6]. 2) Infrared cyclotron resonance using a variety of laser beams of different wavelengths. 3) Reversible magnetic susceptibility by modulating the field in r-f region. Our generation system of the megagauss field is presented in the preceeding paper. In this paper, we will present the latest experimental results with megagauss fields since the Oxford conference in 1978 [7].

2. Cyclotron Resonance in GaP

GaP has a conduction band structure similar to that of Si, but the band minimum of GaP is "camel's back" shaped, which is caused by the absence of inversion symmetry [8]. The minima of the conduction bands are in the <100> direction close to the X point which is a saddle point. The energy separation between the conduction band minima and the X point is only a few meV [9,10]. The constant energy surface of the conduction band has a very large anisotropy for GaP, and there has been a controversy regarding the anisotropy factor $K = m^*_\ell/m^*_t$ [11,12]. The far-infrared cyclotron resonance experiments were carried out in magnetic fields up to 120 T.

2.1 Measurement of the Far-Infrared Cyclotron Resonance

The schematic diagram of the system for measuring the far-infrared cyclotron resonance in fields up to a few megagauss is shown in Fig.1. The megagauss fields are generated by a flux compression method by using a pair of injection coils, a primary coil and a liner. The details of the techniques are described in the preceeding paper. The intensity of the field is measured by integrating the voltage induced in a pick up coil wound on the sample holder. The sample was cooled by cold helium gas and the temperature of the sample was monitored by a copper-constantan thermocouple. Far-infrared radiation was generated by a pulsed water vapor laser with wavelengths of 119 and 220 μm. The discharge length and the inner diameter of the laser tube were 1800 mm and 60 mm, respectively. The mixture of He and H_2O gas was energized by discharging the energy stored in a capacitor of 0.25 μF through a resistor (1-5 kΩ). Stabilization of the pulsed laser operation was achieved by superposing a dc current of several tens of milliamperes to the pulsed current. The fluctuation of the laser output was suppressed to be less than 2 % for each pulse by the superposition. A thin polyethylene sheet was inserted in the resonator at the Brewster angle to obtain linearly polarized radiation. Otherwise, an H_2O laser emits radiation which has an arbitrarily circularly polarized component for each discharge. The circularly polarized radiation of definite direction was obtained by inserting a quarter wave plate made of quartz crystal in the optical path. The laser beam was

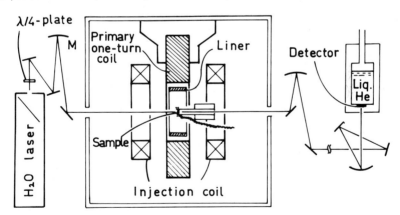

Fig.1 Block diagram of the system for measurement of infrared cyclotron resonances in the ultra-high magnetic fields

focused on the sample with a large confocal mirror M. The transmitted radiation was led to an infrared detector by the mirror system and was detected by the intrinsic photoconduction of Ge(Ga) cooled down to 6 K. Signals of the detector and the magnetic field intensity are memorized in transient recorders (Biomation; 8100).

Measurements in the fields below 40 T were performed by the system shown in Fig.2. Multi-turn coils were utilized for the non-destructive pulsed fields, which were energized by a capacitor bank of 32 kJ. The rise time of the field ranged from 0.6 to 2.5 ms according to the design of the coils. An HCN and a DCN laser were operated in cw mode. The temperature of the sample was precisely controlled by a heater device in the temperature range of 77-300 K. The radiation from the laser was transmitted by means of a light pipe system. Calibration of the field was carried out by measuring the ESR signal of pink ruby (Fig.4). The established g-value for the center of three peaks is 1.98 which corresponds to the cyclotron mass parameter of $m^* = 1.01\ m_0$. The pink ruby used for the measurements was kindly supplied by Mr. Hirose (Shinkosha Co). The error of the field measurement was less than 1 %.

<u>Samples</u> Sliced bulk crystals of n-type GaP grown at Sumitomo Metal Mining Co. were used for the cyclotron resonance measurements at 119 μm in the megagauss range. The carrier concentration of the sample ranged from 0.8 to 1.2×10^{17} cm^{-3}. Experiments at $\lambda = 337$ μm have been performed using another <100> epitaxial sample grown at Mitsubishi Monsanto Chemical Co. The thickness of the epitaxial layer is about 0.1 mm.

2.2 Results and Discussion

The transmission signals of GaP are shown in Fig.3 as a function of the magnetic field at the wavelength of 119 μm. Two absorption peaks were observed with the magnetic field applied to the <100> axis of GaP. The first peak

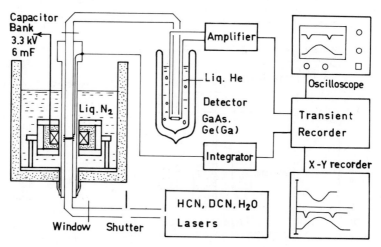

<u>Fig.2</u> Block diagram for the measuring system of infrared cyclotron resonance in the non-destructive pulsed magnetic fields

which positions at the lower field is ascribed to the transverse mass m^*_t, and the second peak to $\sqrt{m^*_\ell m^*_t}$, where m^*_ℓ is the longitudinal mass. This is the first observation of the second absorption peak. The resonant fields are 23 T and 100 T. The anisotropy factor $K = m^*_\ell/m^*_t$ is determined to be 19 using these values. However, $K = 19$ is considerably larger than $7.9^{+3.2}_{-2.0}$ estimated from the angular dependence at λ = 119 μm [11]. The absorption intensity of the second peak is about 2 times larger than the first one with the linearly polarized radiation. On the contrary, when the left circularly polarized light was used, the second peak exhibited a smaller absorption than the first one. For the right circularly polarized light, only the second absorption peak appeared. Although it may seem strange, it reflects the large anisotropy and also the transition between Landau-levels in the "camel's back" structure of the conduction band.

Cyclotron resonance signals in GaP at λ = 337 μm are shown in Fig.4. The <100> axis of the sample was inclined by θ to the magnetic field in the (100) plane. Two absorption peaks were found in the range of 15° < θ < 30°. For θ = 0 that is when B// <100>, only a single peak was observed and the transverse mass m^*_t was determined to be 0.252 + 0.003. The cyclotron mass m^* for the ellipsoidal constant energy surface is described by θ in the following formula;

$$(m^*_t/m^*)^2 = 1 - (1 - 1/K)\sin^2\theta. \quad (1)$$

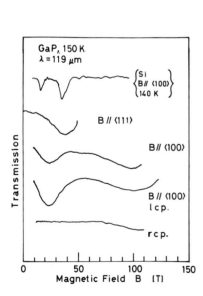

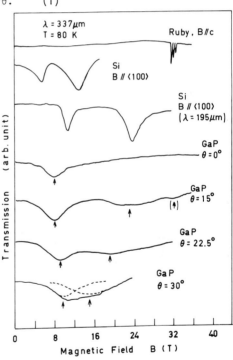

Fig. 3 Cyclotron resonance at the wavelength 119 μm for n-GaP and n-type Si.

Fig.4 Cyclotron resonance absorption for epitaxial n-GaP and n-Si. θ is the angle between the <100> direction and the magnetic field.

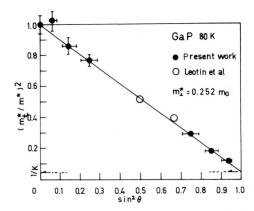

Fig.5 The obtained $(m^*_t/m^*)^2$ are plotted against $\sin \theta$. The straight line corresponds to the anisotropy factor K = 20. λ = 337 μm.

For the second peak, θ should be replaced by $(90°-\theta)$ in Eq. (1). The observed (m^*_t/m^*) are plotted against $\sin^2 \theta$ in Fig.5. The anisotropy factor K = 20 ± 10 was derived from the crosspoint of the straight line. It is shown in Fig.4 that the intensity of the second peak reduces with decreasing θ and seems to disappear at $\theta = 0$. This is in good contrast to the observation for λ = 119 μm. In the case of Si, the relative absorption intensity ratio of the second peak to the first did not change by the photon energy of radiation appreciably. This difference between Si and GaP is caused by the "camel's back" structure of the conduction band for GaP.

3. Exciton Spectra in Megagauss Fields

The exciton spectra of GaSe, PbI$_2$ and CdS were observed in ultra-high magnetic fields. A very fast time resolving optical spectrometer constructed by using an image converter camera (IMACON 700 with special large cathord; John Hadland) was used to measure the magneto-optical spectra [6]. The fundamental problem of the hydrogen atom in a high magnetic field and the diamagnetic shift of exciton lines were investigated.

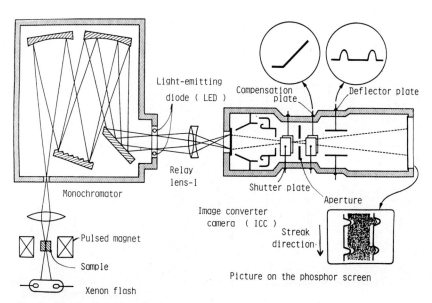

Fig.6 Block diagram of the time resolved spectrometer assembled with an image converter camera.

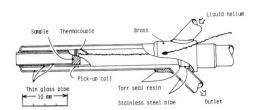

Fig.7 Sample holder for cooling the sample down to 4 K. The ends of the double glass tube are fused together after pick-up coil and thermocouple have been inserted. The holder is placed in the vacuum space for thermal isolation.

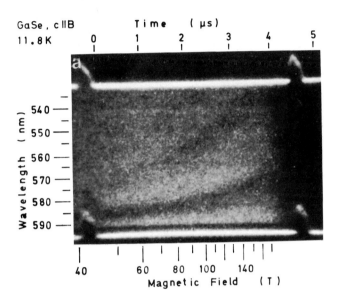

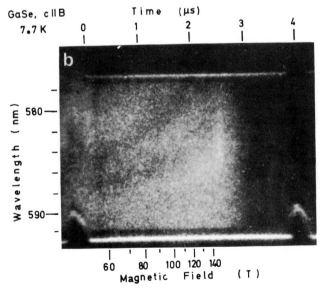

Fig.8 The example of a streak photograph showing the magneto-optical spectra of exciton for GaSe.
(a) Spectra obtained in a wide wavelength region.
(b) Spectra with wavelength scale enlarged to obtain the ground state lines with high resolution. The markers at both ends of the photograph are employed to correlate the magnetic field intensity and the time axis

3.1 Experimental

The time resolved spectrometer used for the experiment is schematically illustrated in Fig.6. The image converter camera was used for the streak camera. The light transmitted through the sample in the pulsed magnetic field is focused at the entrance slit of the monochromator. The entrance light is dispersed by the grating and the spectra are formed at the horizontal exit slit where a photographic plate is usually placed. The dispersion was altered by using different gratings. The horizontal slit has two light emitting diodes on the both sides of the slit, which were used for the calibration of the wavelengths. The images of the spectra and the two diodes are formed on the cathode of the image converter camera by the relay lens I. This horizontal image is streaked vertically on the phosphor screen by applying an increasing voltage between the compensation plates. The image intensifier was used together with the image converter camera when the light intensity was insufficient. In order to correlate the magnetic field intensity to the position of streak direction on the picture, the vertical streak is modulated horizontally by the double pulsed voltage. The magnetic field and the modulating voltage were memorized in the transient recorders. The sample was cooled by liquid helium with using a micro cryostat, as shown in Fig.7. The temperature is measured by a gold(iron)-copper thermocouple. The achieved low temperature ranged from 4 to 20 K.

<u>Sample</u> The experiment was performed on the leaves of GaSe cleaved from a single crystal grown in a Bridgman furnace. The GaSe crystal was kindly prepared by Dr. Sasaki at Tohoku University. The samples of PbI_2 for the measurements were 140 nm in thickness prepared by vacuum deposition on a quartz plate at the Central Research Laboratory of Hitachi Co. Flakes of the CdS crystal were grown by a vapor growth technique.

3.2 Results and Discussion

1. GaSe Figure 8 shows an example of the streak photograph of GaSe. Three absorption lines are found in Fig.8(a). The strong absorption line around the wavelength of 585 nm corresponds to the ground state line of N=1. The lines for N=2 and N=3 and additional weak absorption lines between N=1 and N=2 are also seen in the picture. The ground state lines show a diamagnetic shift up to the field as high as 100 T, and show a broadening in the higher magnetic field region. In order to obtain a better resolution for the N=1 line, the wavelength scale was enlarged as shown in Fig.8(b) by using a grating with finer grooves. The energy splitting ΔE between two lines increases linearly with field. The corresponding g-value was estimated to be 2.9, which is in good accordance with $g = 2.7 \pm 0.2$, reported by MOOSER and SCHLÜTER [13]. On the Contrary to the ground state, the observed

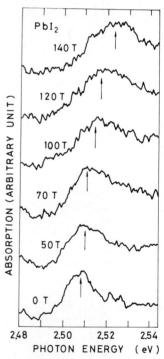

Fig.9 Absorption curves for PbI_2 at different magnetic fields obtained by microphotodensitometer.

excited states N=2 and N=3 do not exist on the extrapolation of the diamagnetic shift for the hydrogenic 2s and 3s states. They seem to be associated with the interband transition between the Landau-levels. Although the lowest Landau transition is above the 2s and 3s states the observed lines N=2 and N=3 can be extrapolated to them by comparing the present result with the previous data obtained by AOYAGI et al. [14]. It was found experimentally that in the megagauss field region the Landau-like states are dominant in GaSe.

b) PbI_2 The absorption spectra at each magnetic field shown in Fig.9 are converted from the streak photograph by using a microphotodensitometer. The shift is almost proportional to the square of the magnetic field up to 100 T. The reduced mass of the exciton is determined as $0.21\ m_0$ from the shift value of 5.4 meV at 100 T. In the field range above 100 T, a larger increase of the shift was observed, which cannot be explained by the existing theoretical field dependence for the 1s state.

c) CdS   The absorption spectra without the field were observed at 4.2 K for several flakes of CdS. An exciton peak was found at 285.4 nm (2.5543 eV) for E C. The shift of the A-exciton is observed from 486 nm to 484 nm corresponding to the increase of the field from 50 T to 120 T. This value is considerably smaller than the theoretically expected value from the hydrogen model. This shift can be interpreted by taking into account the spin Zeeman energy [15] and the quenching due to the electron phonon interaction [16].

4. Transition of the Spin Configuration in TbIG and GdIG

It is well known that in ferrimagnetic garnets, one can bring about a transition from collinear to the canted magnetic structure by the application of a high magnetic field and changing temperature. The transition of the spin configurations in TbIG and GdIG was studied by using ultra-high magnetic fields. The choice of the garnet was based on the fact that TbIG is very strongly anisotropic and presents large magnetic and magnetooptical effects, whereas GdIG has isotropic character. The transition was observed by the measurements of both Faraday rotation and reversible magnetic susceptibility [17,3].

The unit cells of garnets contain eight molecules of $R_3Fe_5O_{12}$ (R; rare-earth ion) with 24 Fe^{3+} ions in tetrahedral d-sites, 16 Fe^{3+} in the octahedral a-sites and 24 Tb^{3+} in the dodecahedral c-sites. The magnetic moments of the ferric ions at the d-sites are ordered antiferromagnetically to those

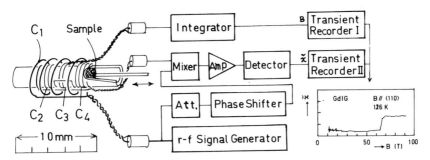

Fig.10   Block diagram of the system for measuring high frequency reversible susceptibility in megagauss field.

sites. The exchange interaction between rare-earth ions and the ferric ions is much weaker than the principal magnetic interaction of superexchange coupling between Fe^{3+} ions in the a- and d- sites. Therefore, the collinearity between the rare-earth and ferric sublattice is expected to be disturbed first by externally applied magnetic fields.

4.1 Experimental

A YAG(Nd^{3+}) laser (λ=1.06 μm) was used as a light source for the observation of Faraday rotation signals. Two Faraday rotation signals with 90° phase difference were observed simultaneously by using Walaston prism as an analyser. Si-photodiodes were used for the detection of the light beams. Single crystals of TbIG and GdIG trown by the flux methods were made into disks of 2 mm in diameter and 0.4-0.6 mm in thickness. The samples were kindly supplied by Dr. Dillon, Jr. and Dr. van Uitert of Bell Laboratories.

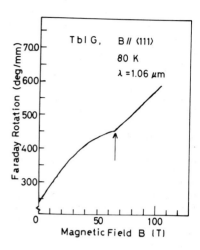

Fig.11 Faraday rotation angle for TbIG. The arrow indicates the transition point of the spin configuration.

The experiments were performed on samples with (111), (110) and (100) surface of bulk crystals, whose surfaces were optically polished.

The reversible magnetic susceptibility was measured by modulating the magnetization with a high frequency magnetic field. Figure 10 shows a

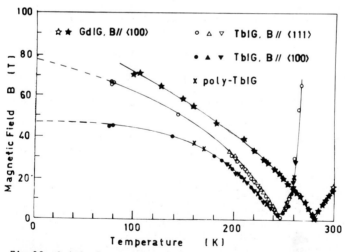

Fig.12 Critical magnetic field of the spin transition from the collinear phase to the canted phase for TbIG and GdIG as a function of temperature. ○,● and ☆ are determined by Faraday rotation; △ and ▲ by time derivative of magnetization; ▽,▼,★ and X by reversible magnetic susceptibility.

block diagram of this system. The modulation field of about 60 MH$_z$ is generated by a high power oscillator, which is operated in pulse mode in order to avoid heating the sample. Coils C_2 and C_4 are wound oppositely for cancelling the large induction voltage due to the pulsed magnetic field. The sample is inserted in one of the coils C_2 and C_4, after the induced r-f signal in the coils is balanced at the mixer part by adjusting the attenuator and the phase shifter carefully. The deviation of the r-f voltage due to the sample is amplified and detected, which corresponds to the reversible susceptibility. Typical experimental result is displayed in Fig. 10.

4.2 Results and Discussion

An example for the Faraday rotation in TbIG is shown in Fig. 11. A break appears on the signal at 66 T where the spin configuration changes from the collinear phase to the canted phase. The critical magnetic field is also determined by measuring the time derivative of the magnetization dM/dt. Figure 12 shows the temperature dependent critical magnetic fields of the spin transition in TbIG and GdIG determined by Faraday rotation, dM/dt and magnetic susceptibility. The magnetic fields were applied both in the easy axis <111> and the hard axis <100> directions. In the region up to 40 T, non-destructive magnetic fields are used in the experiments. The critical field for the <111> direction is much larger than that for <100> direction, indicating a large anisotropy in the transition field in TbIG. The spin phase diagrams determined by these methods were in good agreement with each other for every direction. The transition point for the polycrystal of TbIG, which is not transparent, was also determined by the reversible susceptibility. On the contrary, reflecting a very small anisotropy of GdIG the critical field seems to be independent of the direction of the external magnetic field applied to the crystal.

References

1. C.M. Fowler: Science 180 (1973) 261.
2. F. Herlach, J. Davis, R. Schmidt and H. Spector: Phys. Rev. B10 (1974) 682.
3. S. Chikazumi, N. Miura, G. Kido and M. Akihiro: IEEE trans. Magnetics 14 (1978) 577.
4. N. Miura, G. Kido, M. Akihiro and S. Chikazumi: J. Mag. & Mag. Materials 11 (1979) 275.
5. G. Kido, N. Miura, K. Kawauchi, I Oguro and S. Chikazumi: J. of Phys. E (Sci. Inst.) 9 (1976) 587.
6. G. Kido, N. Miura and S. Chikazumi: Proc. 13th Int. Conf. on High Speed Photogragh and Photonics, Tokyo (1978) pp. 552.
7. N. Miura, G. Kido, H. Katayama and S. Chikazumi: Int. Conf. of the Application of the High Magnetic Fields in Semiconductor Physics, Oxford (1978) pp. 223.
8. F.H. Pollak, C.W. Higginbotham and M. Cardona: Proc. 8th Int. Conf. Phys. Semiconductors, Kyoto (1966) pp. 20.
9. P. Lawaetz: Solid State Commu. 16 (1975) 66.
10. P.J. Deam and D.G. Thomas: Phys. Rev. 150 (1966) 690.
11. K. Suzuki and N. Miura: Solid State Commu. 18 (1976) 233.
12. J. Leotin, J.C. Ousset, R. Barbaste, S. Askenazy, M.S. Skolnick, R.A. Stradling and G. Poiblaud: Solid State Commu. 16 (1975) 363.
13. E. Mooser and M. Schlüter: Il Nuovo Chimento 18B (1973) 164.
14. K. Aoyagi, A. Misu and S. Sugano: J. Phys. Soc. Japan 18 (1963) 1448.
15. H. Venghaus. S. Suga and K. Cho: Phys. Rev. B16 (1977) 4419.
16. G. Behnke, H. Büttner and J. Pollmann: Solid State Commu. 20 (1976) 873.
17. F.M. Yang, N. Miura, G. Kido and S. Chikazumi: J. Phys. Soc. Japan 48 (1980) 71.

Part III

Cyclotron Resonance and Laser Spectroscopy of Semiconductors

Far-Infrared Spectroscopy of Semiconductors in High Magnetic Fields

A.M. Davidson, P. Knowles, P. Makado, and R.A. Stradling
Physics Department, University of St. Andrews, KY16 9SS, U.K.

S. Porowski and Z. Wasilewski
Institute of High Pressure Physics, 01-142 Warsaw, Poland

1. Introduction

The parameter (γ) which determines the effect of an applied magnetic field (B) on shallow impurity states is the ratio of the zero point cyclotron energy $(1/2\hbar eB)/m^*$ to the binding energy of the impurity (13.6 $m^*/m\varepsilon^2$) where m^* is the effective mass and ε the dielectric constant. The shallow donors in InSb have been used as a model system for investigating high field effects because of the ability to prepare high-purity samples relatively easily, the low effective mass, and a spherical conduction band which avoids the theoretical complexity of anisotropic bands. The parameter γ is equal to unity at a field of 0.2T and at about this field a metal-insulator phase transition occurs at liquid helium temperatures in high-purity material. At higher fields, sharp transitions between impurity levels can be observed spectroscopically in the far-infrared region of the spectrum and values of γ of about 100 can be achieved in steady fields available in the laboratory [1]. The only situation where such high effective fields are found with completely free electrons is close to the surface of collapsed stars where γ values of 1000 ($\equiv 10^8$T) can be found near to pulsars [2].

This paper reports the first study of the cyclotron resonance and impurity spectra in InSb as a function of hydrostatic pressure, investigates the variation with magnetic field of the highly excited states of the donors in GaAs in the intermediate field regime ($\gamma \sim 1$) where the advantages of employing inter-excited-state transitions are discussed, and finally the central cell structure of the shallow donors in InSb, InP and GaAs are compared.

2. The Effect of Hydrostatic Pressure on the Donor Spectra in InSb

With the donors in InSb the $1s \to 2p_{-1}$ (000→010) transition becomes so sharp at fields of the order of 10T that central cell structure arising from different chemical donor contaminants can be observed [3]. Very recently an extensive study of this central cell structure has been made. In nominally undoped material up to four residual donors are found and these may be seen in the experimental spectrum shown in Fig.1. These residual donors are referred to as A to D with donor A lying deepest. It has been suggested that A and B may arise from the presence of oxygen and silicon contamination [4,5]. The central cell shifts for tin, selenium and tellurium have also been determined. The relative order of the chemical shifts for the individual group four and group six substitutional donors is of importance in assisting the interpretation of similar studies of the shallow donors in other III-V compounds such as GaAs and InP (Sect. 4). The pressure experiments reported here extend these earlier experiments and assist the identification of the central cell components.

The pressure was generated at room temperature in a Be-Cu cell containing a sapphire window to admit the infrared radiation. The cell contained a light

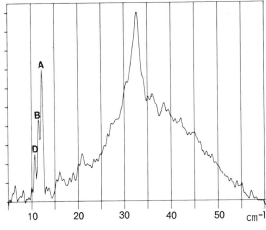

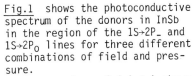

Fig.1 shows the photoconductive spectrum of the donors in InSb in the region of the $1S \rightarrow 2P_-$ and $1S \rightarrow 2P_0$ lines for three different combinations of field and pressure.

A comparison of (a)-(c) shows a pronounced narrowing of the $1S-2P_-$ lines on increasing pressure and field. In (c) the line widths are instrumentally limited on the $1S-2P_-$ lines for donors B-D.

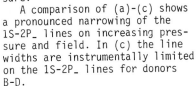

(a) is taken at zero pressure and 10T and shows a spectrum similar to those in reference [3] and the separation of donors A and B is 0.8 cm^{-1}

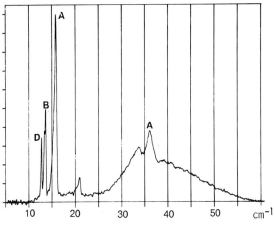

(b) is taken at 4.9 kbar and 9T. The integrated intensity of donor A is about four times that of B and their separation is 2.5 cm^{-1}. As donor A splits off from the remaining donors, it becomes resolved on the broader $1S-2P_0$ line at 35 cm^{-1}

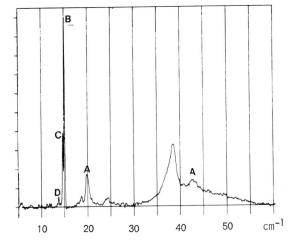

(c) is taken at 4.9 kbar and 12.5T. Donor A now has less than half of the integrated intensity of B and their separation is 5 cm^{-1}

hydrocarbon as the pressure transmitting medium. The cell was then sealed and placed inside the cryostat of a 13T superconducting magnet and cooled to liquid helium temperatures. The photoconductive signal from the sample itself was used to detect the infrared radiation from a Michelson interferometer with the "cross-modulation" effect producing the cyclotron resonance and the sharp donor transitions being generated by the "photothermal effect".

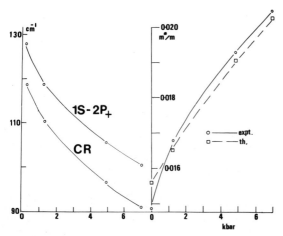

Fig.2 The left-hand side shows the variation with pressure of the position of 1S→2P+ line impurity shifted cyclotron resonance and the cyclotron resonance line in InSb at a field of 2T. The right-hand side shows the cyclotron effective mass at 2T deduced from experiment together with the variation expected for a pressure coefficient of 0.15eV/kbar and a band edge mass of $0.0139m_o$.

As can be seen from Fig.2 the cyclotron resonance line was followed to pressures of 7 kbar. The band-edge mass was deduced following the method employed by PALIK et al. [6]. The experimental variation of band edge mass with pressure was within 20% of that predicted from the previously determined pressure coefficient of the band gap of 0.015eV/kbar. Both the cyclotron resonance line and the impurity lines narrowed on increasing the pressure. At 2T and 7 kbar the cyclotron resonance line width corresponded to an $\omega\tau$ value in excess of 200. The remarkable sharpness of the lines shows that the pressure remains extremely hydrostatic when the pressure-transmitting liquid freezes, as, even if all the observed width of the cyclotron resonance line were attributed to a variation in stress over the sample, the stress inhomogeneity would be less than 2%.

As the effective mass increased with increasing pressure both the 1s-2p- and 1s-2p₀ lines moved as expected to higher energy and the 1s-2p+ moved to lower energy. However, the A component of the 1s-2p- moved at a much higher rate than the other three components as can be seen from Fig.3. At a pressure of 5 kbar the shifts are five times as great as the other components. Furthermore the intensity of the A component behaved highly anomalously with respect to the other three, decreasing rapidly with increasing magnetic field at pressures of the order of 5 kbar. This behaviour is the opposite of the thermal depopulation effect normally observed in this field range where the intensity of the deeper lying components (A) grow progressively at the expense of the shallower (B,C and D).

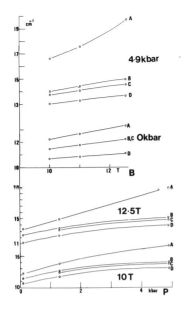

Fig.3 shows the variation with field and pressure of the four central cell components of the 1S→2P_ line for the donors in InSb. The A component varies rapidly with respect to the other components.

In addition, at pressures of the order of 5 kbar, an A component split off from the 1s-2p$_0$ line and had a similar dependence of intensity and position on field. These results suggest that the ground state of the A is deepening anomalously with increasing field while, at the same time, the oscillator strength of the transition is decreasing. An effect of pressure is to bring several resonance states within the conduction band closer to the band edge [7,8] and an effect of the magnetic filed is also to reduce the separation of the conduction band edge and these resonance states by the zero point cyclotron energy. In this respect, magnetic field and pressure are acting in a complementary manner. The lowest of these resonance states can be populated at a pressure of 7 kbar at the highest fields available [8]. Indeed, changes in the infrared spectrum of the donors are observed when this state is populated and these will be the subject of a future publication. It has been suggested that the lowest of the resonance states is associated with oxygen [5]. It is now proposed that component A of the central cell spectrum is due to the same impurity and that the anomalous behaviour in position and intensity of this component is due to the repulsion of the donor states associated with the k=0 minimum due to coupling with the resonant states as they approach the band edge. The increasing admixture of states away from the band edge into the A levels would then account for the decrease in intensity of the observed lines. It is planned to extend the range of pressures to confirm the preliminary interpretation outlined above. These experiments should add fresh insight into the nature of resonance states and the association of levels with particular band extrema.

3. Zeeman Splitting of Impurity States in the Intermediate-Field Regime ($\gamma \sim 1$)

When the energies of the impurity states are required for comparison with effective mass theory, there are substantial advantages to be gained from measurements of transitions solely between excited states as compared to more conventional experiments involving transitions from the 1S ground state to the excited states. In excited-state studies it becomes possible to observe in first order the excited states of the same parity as the ground state (i.e., the nS and nD levels) for which an electric dipole transition from the ground state is parity forbidden. The perturbations of the S-like excited states from their effective mass positions by central cell effects are considerably smaller than for the 1S ground state.

Difficulties in interpretation associated with complex band structures can be avoided by studying the donors in a material with the conduction band edge located at the centre of the zone although a narrow-gap material should not be chosen as non-parabolic effects would introduce additional complexities. The most suitable material for studying Zeeman effects in the intermediate-field regime is GaAs which has a relatively wide band gap and can be prepared with fewer electrically active impurities than any other semiconductor apart from the indirect-gap elemental semiconductors Si and Ge. In an earlier series of experiments with this material [9,10], nearly thirty different transitions between excited states were observed and twelve transitions were positively identified. The experimentally determined variation with field agreed with that predicted from variational calculations to 1% or better implying an accuracy on the calculated energies of the states themselves approaching 0.1%. These experiments have now been extended and a further six transitions have been observed. The interpretation has been assisted by the appearance of further calculations for some of the more highly excited states [11] and four more transitions have now been positively identified. As can be seen from Table 1 all the allowed $n=2\rightarrow2$ and $2\rightarrow3$ transitions have in fact been detected

Table 1   Identified $n=2\rightarrow2$ and $2\rightarrow3$ inter-excited-state transitions

	$2P_-$	$2P_0$	$2P_+$
$2S$	B_2	M	B_3
$3S$	*	E	*
$3D_{-2}$	I_1	X	X
$3D_{-1}$	B_1	**	X
$3D_0$	D	G_2	*
$3D_{+1}$	X	A_2	B_1
$3D_{+2}$	X	X	A_n

* outside energy range studied

** too flat in energy to be detected in magneto-optical study at constant frequency

X   forbidden transition

$J = 3P_- \rightarrow 3S$ $I_2 = 3D_{-2} \rightarrow 4F_{-2}$
$K = 3D_{-1} \rightarrow 3P_{-1}$ $N = 3D_{-2} \rightarrow 4F_{-3}$ $R = 4F_{-2} \rightarrow 4D_{-2}$

The line designation is that employed by Skolnick et al. [10].

except those which are outside the energy range studied or which have insufficient slope with field to be detected by the experimental technique of working with a fixed-frequency source and sweeping the field.

A number of the newly identified transitions are shown in Fig.4 which illustrates an advantage of excited-state transitions in addition to those listed in the introductory paragraph of this section. As has been reported elsewhere [12], it is possible to produce considerable changes in population of the excited states by illumination with near-band-gap radiation. These changes are reflected in changes in intensity of the observed lines. The population of the excited states is normally small at temperatures close to 4 K although the matrix elements are generally larger than for transitions from the ground state. Consequently the intensities of the excited-state transitions are very sensitive to small changes in population induced by changes in the intensity of background radiation falling on the sample. As the electronic system is no longer in equilibrium with the lattice the capture cross-sections of the states concerned can be estimated. In Fig.4 the gains for the two recordings have been set so that the apparent strength of line 'P' is the same in both cases. The upper recording is taken with about twice the power of the band-gap radiation falling on the sample compared with the lower while the lattice remains at 4.2 K in both cases. It can be seen that the cyclotron resonance is about an order of magnitude stronger in the upper recording as would be expected because of the increased number of free carriers. The other lines increase in strength by about a factor of two or three except for line 'P' (unidentified), $2P_0 \rightarrow 2S$ and $2P_- \rightarrow 2P_0$. This result demonstrates that the $2P_0$ level has a long lifetime and suggests that the populations of the $2p_-$ and $2p_0$ are approaching equality. The $2P_0 \rightarrow 2S$ line is sufficiently sharp that central cell structure is becoming apparent. Even sharper in energy, although not so obvious in the recording as the field, is the experimental variable, the $3D_{-2} \rightarrow 4\Gamma_{-3}$ transition (line N). This line has a slope with field of 0.42 cm^{-1} T^{-1} giving a halfwidth of 0.03 cm^{-1} which is of the same order as line widths involving the n=2 states. Also of comparable width are the $3D_{-2} \rightarrow 4F_{-2}(I_2)$ and $4F_{-2} \rightarrow 4D_{-2}$ (R) lines. Although the principle quantum number n is large for the $3D_{-2}$, $4F_{-3}$, $4F_{-2}$ and $4D_{-2}$ states, they remain sharp because they are pinned below the conduction band edge. Thus, transitions between these states figure prominently in the inter-excited-state spectrum.

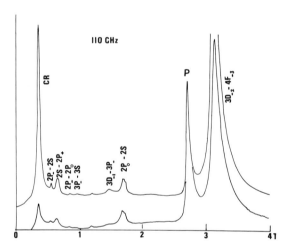

Fig.4 Photoconductive response obtained by sweeping the field of a high-purity GaAs sample at 4.2 K to radiation at 110 GHz. The cyclotron resonance and inter-excited-state transitions are labelled. In both recordings band-gap radiation is falling on the sample but in the upper case the power of this background radiation is about double that in the lower recording

Although the main features of the millimetre and submillimetre spectrum of the shallow donors in GaAs are well understood, one unanswered question remains. When any uncompensated semiconductor containing a dilute assembly of shallow donors is photoexcited some of the electrons initially excited into the conduction band may, at low temperatures, recombine with neutral donors or acceptors to produce D$^-$ or A$^+$ centres. These states are electrical analogues of the negatively charged hydrogen ion H$^-$, i.e., a proton with two electrons attached, and have been used as the basis of a fast and sensitive far-infrared detector. Such centres may also be formed on thermal excitation of the band minima. Recently evidence has been put forward for the existence of D⁻ centres in CdS, a single valley material [13]. The experiment was similar to those reported in this section for GaAs except that transmission was employed rather than photoconductivity at several fixed values of sub-millimetre wavelengths. No deliberate optical pumping was employed. Weak structure was observed at fields below the cyclotron resonance signal which was believed to be associated with the transfer of a single electron from the singlet ground state to the N=1 Landau level leaving a neutral donor in its ground state. The identification was based primarily upon agreement between predicted and observed values of photodetachment threshold frequency against magnetic field and account had to be taken of broadening effects on the states concerned.

However, a group of the strongest of the inter-excited state transitions lies in the same region of wavelength and field as the structure attributed to the D⁻ state in CdS and agreement is as good if the observed line was in fact A_4. The theoretical zero-field intercept for example for the set of n→3 to 4 transitions is 0.049R* whereas that for the D⁻ to conduction band threshold is 0.055R* which is well within experimental error. Furthermore, the experimental conditions required for the formation of D⁻ states of either thermal or optical excitation to locate the second electron on a single donor site are identical to the conditions for populating the excited states of the neutral donor as required for the observation of the inter-excited-state transitions. However, it should be noted that the excited states responsible for the lines identified to date do lie slightly deeper than the D⁻ band (in zero field the n=3 states lie at 0.11R* and the n=4 states at 0.063R* as compared to the D⁻ binding energy of 0.055R*). The lifetime of the D⁻ state for the absorption of an acoustic phonon and transfer to the conduction band may therefore be rather less than for the excited states detected so far, requiring that the temperature be rather lower for the build up of significant occupancy of the D⁻ states. The observed widths of the lines indicate that the natural lifetimes of the lower lying n=3 and 4 states are quite long ($\geq 10^{-9}$s) and the optical pumping experiments show that their own occupancy can be changed quite easily even at 4.2 K.

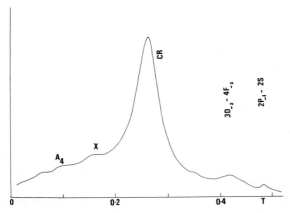

Fig.5 Photoconductive response obtained by sweeping the field of a high-purity GaAs sample at 1.5 K to radiation at 110 GHz. The cyclotron resonance (CR) and inter-excited-state transitions are labelled. X remains unidentified. Its zero field intercept on a plot of photon energy against field is close to an n = 3 → 4 transition (0.049R*) and to D⁻ (singlet) → conduction band (0.055R*).

A further puzzling feature of the initial report of D⁻ transitions in CdS is that only the D⁻ to N=1 Landau level threshold was reported. However, magneto-optical measurements with Ge [14] demonstrate that the photodetachment transitions to the N=0, 2 and 3 Landau levels are as strong as that to the N=1 Landau level. None of these transitions were reported in the CdS measurements. In general, multivalley conduction bands would be expected to help stabilize the D⁻ states. In order to substantiate the initial report of D⁻ states in a material with a k=0 conduction band, a careful search was made for D⁻-associated structure amongst the inter-excited-state spectrum. The observed intensities of the transitions are quite sensitive to temperature as would be expected. At the lowest temperature studied to date (1.5 K) there is a line (marked X in Fig.5) below the cyclotron resonance line which is in approximately the correct position for the D⁻ to N=1 Landau level transition. However, none of the other lines detected had the correct dependence upon field for a transition from the D⁻ singlet state to N=0 Landau level. It is therefore likely that X is an n=3→4 transition and the report of the observation of D⁻ states in CdS therefore remains unsubstantiated with GaAs.

4. Central Cell Structure of Shallow Impurities in III-V Compounds

Central cell effects or chemical shifts are the small corrections to the effective mass model of the common donors arising from the non-vanishing amplitude of the ground-state wavefunction at the origin. In the III-V compounds with direct energy gaps central cell corrections amount to $\lesssim 100\mu V$ (1 cm⁻¹). Only far-infrared spectroscopy has yet had the resolution to resolve completely the individual components present in the spectrum of material prepared without deliberate doping. The observation of central cell structure therefore permits the identification of the chemical contaminants introduced by various growth processes. The combination of far-infrared spectroscopy and photoconductive techniques permits extremely high sensitivity and in experiments with high-purity germanium the detection limit has been established as fewer than 10^7 impurities cm⁻³ [15,16].

It has recently been shown that the relative magnitude of the central cell corrections determined to date can be satisfactorily explained on the assumption that the dominant property of the impurity concerned is its electronegativity [17]. Thus the relative order of the different contaminants found for one material can therefore be used with some confidence to predict the positions for other materials. With this in mind experiments are in progress with back doped samples of InP, GaAs and InSb to effect a comparison of the relative order of the chemical shifts of different impurities and also to determine the residual contaminants in high-purity material grown by different techniques. The most extensive series of results exists for the donors in GaAs [18]. In these earlier experiments it was found that vpe material from three sources all contained the same three residual donors, one of which appears to be Si and the other two have been provisionally identified as C and Ga vacancy. LPE material also contained three donors which were found to be Si, Sn and Pb and alkyl material contained Pb and C. Experiments have now been extended to vpe material grown by Bell Telephone Laboratories and by L.E.P. Samples from these two sources contained the same three impurities found with other vpe material.

With InSb the central cell shifts have been measured for Se, Te and Sn impurities [3]. Sn is found to be a residual contaminant in zone-refined and pulled material but not in LPE growth. Se and Te appear not to be residual contaminants. The two deepest donors are provisionally attributed to oxygen and silicon. As discussed in section 2, experiments are now in progress with hydrostatic pressure above 5 kbars in a attempt to increase the resolution of the central cell structure further.

The most extensive spectrum of central cell effects for any material has been observed for nominally undoped n-InP at 4T. At 4.2 K, eight components appear on the 1s-2p$_+$ transitions [19]. On cooling to 1.3 K and illuminating the sample with band-gap radiation, a narrowing of the component lines by up to an order of magnitude was achieved and with some samples the 1s-2p$_+$ line was resolved into eleven well-defined peaks which differ in energy by 0.1 - 1.8 cm^{-1} [19] at 119μm laser wavelength.

Samples back doped with S, Si and Sn have been studied. The Sn-doped sample was prepared from undoped material by irradiation with thermal neutrons to convert In by a nuclear reaction [20]. The presence of such a multitude of components to the central cell spectrum makes experiments with back-doped samples very difficult as the lines inevitably broaden and shift with doping. However, as a result of these measurements, it appears that Si produces an increase in the strength of line D originally reported by Stradling et al. [21], Sn increases the strength of line F and S produces a growth of signal close to the strong line F [22].

Figure 6 shows the relative position of the three impurities so far identified with InP in relation to two recently investigated high-purity vpe samples. With no less than twelve lines detected so far in the central cell spectrum of InP it is clear that there are more components than can be accounted for by the likely single substitutional impurities. The central cell structure of the donors in CdTe showed ten components [23]. Strong evidence exists that complexes can generate additional electronic states in the elemental semiconductors, Si and Ge [15,24]. It would appear that a significant number of the lines observed with the compound semiconductors are produced by complexes. However, the relative order of the simple substitutional donors so far identified with the III-V materials is the same.

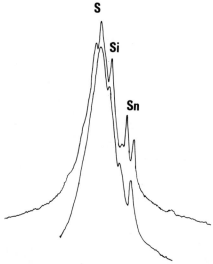

Fig.6 The central cell spectrum of the shallow donors in two high-purity InP samples taken with 118μm radiation. The abscissa is magnetic field, the line centre is about 3.7T and the components are spread over a field range of about 0.1T. Experiments with back-doped samples show that peaks due to S, Si and Sn are in about the positions indicated but twelve components have been detected to date in other samples [19] and no precise identification is possible at this time

Acknowledgements We should like to thank Dr. R.J. Nicholas and Mr. S. Sessions for assistance with the laser experiments shown in Fig.6, Dr. F. Kuchar for providing the transmuted sample of InP and Dr. D. Robertson for the gift of the vpe InP samples.

References

1. R. Kaplan: Phys. Rev. 181, 1154 (1969)
2. M. Ruderman: J. Magn. Magn. Mater. 11, 269 (1979)
3. R. Kaplan, R.A. Cooke, R.A. Stradling: Solid State Commun. 26, 741 (1978). F. Kuchar, R. Kaplan, R.J. Wagner, R.A. Cooke, R.A. Stradling, P. Vogl: J. Phys. D (to be published)
4. L. Konczewicz, E. Litwin-Staszewska, S. Porowski: Proc. Int. Conf. on Narrow Gap Semiconductors, Warsaw (1977), p. 211
5. I.M. Langer: In *New Developments in Semiconductor Physics*, Proceedings, Szeged 1979, ed. by F. Beleznay, J. Giber, G. Ferenczi, Lecture Notes in Physics, Vol. 122 (Springer, Berlin, Heidelberg, New York 1980), p. 123
6. E.D. Palik, G.S. Picus, S. Teitler, R.F. Wallis: Phys. Rev. 122, 475 (1961)
7. E. Litwin-Staszewska, S. Porowski, A.S. Filipchenko: Phys. Status Solidi B48, 525 (1971)
8. L. Kmowski, M. Baj, M. Kubalski, R. Piotrzkowski, S. Porowski: In Proc. Int. Conf. on Phys. of Semiconductors, Edinburgh 1979, p. 417
9. P.E. Simmonds, J.M. Chamberlain, R.A. Hoult, R.A. Stradling, C.C. Bradley: J. Phys. C7, 4165 (1974)
10. M.S. Skolnick, A.C. Carter, Y. Couder, R.A. Stradling: J. Opt. Soc. Am. 67, 947 (1977)
11. C. Aldrich, R.O. Greene: Phys. Status Solidi B93, 343 (1979)
12. E.M. Gershenzon, G.N. Goltzman, N.G. Ptitsyna: JETP. Lett. 25, 539 (1977)
13. D.M. Larsen: Phys. Rev. Lett. 42, 742 (1979)
14. M. Taniguchi, S. Narita: J. Phys. Soc. Jpn. 47, 1503 (1979)
15. E.E. Haller, B. Joos, L.M. Falicov: Phys. Rev. B21, 4729 (1980)
16. M.S. Skolnick, R.A. Stradling, J.C. Portal, S. Askenazy, R. Barbaste, K. Hansen: Solid State Commun. 15, 1281 (1974)
17. H.P. Hjalmarson, P. Vogl, D.J. Wolford, J.D. Dow: Phys. Rev. Lett. 44, 810 (1980)
18. R.A. Cooke, R.A. Hoult, R.F. Kirkman, R.A. Stradling: J. Phys. D11, 945 (1978)
19. R.A. Stradling: Proc. NATO Advanced Study Institute on New Developments in Magneto-Optics, Antwerpen, 1979
20. F. Kuchar, E.J. Fantner, G. Bauer: Phys. Status Solidi A24, 513 (1974)
21. R.A. Stradling, L. Eaves, R.A. Hoult, N. Miura, P.E. Simmonds: Proc. Conf. on GaAs and Related Compounds, Boulder 1972, p. 65
22. R.F. Kirkman: Ph.D. Thesis, Oxford University (1976)
23. P.E. Simmonds, R.A. Stradling, J.R. Birch, C.C. Bradley: Phys. Status Solidi B64, 195 (1974)
24. M.C. Ohmer, J.E. Lang: Appl. Phys. Lett. 34, 750 (1979)

Submillimeter-Magneto Spectroscopy of Semiconductors

M. von Ortenberg and U. Steigenberger

Max-Planck-Institut für Festkörperforschung
Hochfeldmagnetlabor Grenoble 166 X
38042 Grenoble-Cedex, France

1. Introduction

Submillimeter-magneto spectroscopy has become an extremely powerful tool in the investigation of electronic energy levels in semiconductors. The outstanding success of this method started already twenty-seven years ago, however, not by application of submillimeter radiation, but of microwaves [1]. There are many reasons for the now still increasing application of submillimeter-magneto spectroscopy in semiconductor physics. The two most important ones are the tremendous progress in high magnetic field and laser technology during the recent years and the wide versatility of this method in combination with other experimental techniques. Especially the second point has raised the interest of many physicists and the original simple transmission experiment of a bulk sample slab has been replaced by more sophisticated setups combining different experimental methods. In the following we report was an example the results of two different types of investigation: firstly of the combination of high pressure technology at low temperature with cyclotron resonance measurements, and secondly of a useful modification of the strip-line technique making this kind of investigation suitable for a larger variety of semiconductor materials.

2. The Effect of External Uniaxial Stress on the Cyclotron- and Impurity Resonances in Tellurium

In a series of fundamental papers HENSEL et al. have demonstrated that an external uniaxial stress as an additional parameter in cyclotron-resonance experiments provides considerably more information on the energy-band structure of the investigated material than the usual unmodified resonance experiment[2,3]. The stress induced shift of the resonance line and eventually an additional splitting are directly correlated with the deformation potentials and the symmetry of the crystal.

For tellurium, however, most of the stress dependent experiments have been performed under hydrostatic pressure, because hydrostatic pressure is more easily realised in an experiment than uniaxial stress. So far under hydrostatic pressure transport measurements, including the Shubnikov-de Haas effect, and interband absorption experiments have been performed[4]. The essential results consist in both a variety of the energy-band structure and changes in the effective carrier concentration[13,23].

Because of the special form of the compliance tensor for tellurium, hydrostatic and uniaxial stress are complementary[6]. Whereas for hydrostatic pressure the helical chains are pressed together and slightly expanded along the symmetry axis, for uniaxial stress along this axis holds exactly the opposite. It is evident that from a combination of measurements of both under hydrostatic and uniaxial stress, the deformation dependence of any stress induced effect can be derived[7,8].

The essential effect of stress on the energy-band structure of tellurium is a modification of the famous "camel-back" of the upper valence band for the motion parallel to the c-axis. For tellurium uniaxial stress P_C can only be applied parallel to this symmetry axis of the crystal because of the low activation energy of dislocations in this material. In earlier experiments of the cyclotron-resonance in tellurium under uniaxial stress for the magnetic field parallel to the c-axis (B||c), the change of the "camel-back" became not effective, since the quantized motion of the electrons perpendicular to the field and hence to the c-axis is not affected hereby[9]. That is why we used in our present investigation the configuration of the external magnetic field perpendicular to the c-axis (B⊥c). The combination of the two conditions (P||c) and (B⊥c) requires that the stress application is perpendicular to the bore of the magnet. Therefore the construction of a sophisticated mechanical stress apparatus was necessary. Using different wavelengths of conventionally and optically pumped molecular-gas lasers (λ=337,394,496,570 µm) we measured the magneto transmission of pure tellurium single crystals at helium temperatures up to B=20 Tesla. As parameter we varied the uniaxial stress P_C up to the destruction of the sample, usually at about 2.5 k-bar. Special caution was paid to the preparation and shaping of the samples to avoid interference effects.

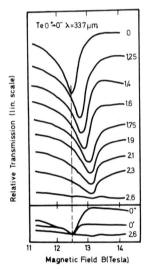
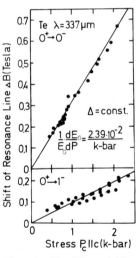

Fig. 1 The stress dependence of the 0^+-0^- transition in tellurium

Fig. 2 The line shift of the cyclotron resonance lines as a function of the stress P_C

In Fig.1 we have summarized the data of the $0^+ - 0^-$ transition using 337 µm-wavelength radiation. For clarity each transmission curve is shifted by an arbitrary offset on the vertical scale. The parameter indicates the stress P_C in k-bar. There are two dominant features characterizing the experimental results: firstly, there is a considerable shift of the resonance line towards higher magnetic fields with increasing stress application indicating an increase of the cyclotron mass; secondly, the even more striking effect is a decrease of the line intensity and distortion of the line shape under stress up to the complete vanishing of the resonance line. The stress application is reversible in the sense that immediately after stress release even at low temperatures the transmission minimum shifts back to its original position (curve 0' in the lower part of Fig.1). The line intensity, however, remains considerably reduced. After one hour annealing at He-temperature the curve 0" is recorded. After warming up to room temperature the original transmission curve 0 is reproduced. Stress application up to about $P_C=1.8$ k-bar is completely reversible, so that immediately after stress release the original $P_C=0$-curve is reproduced.

It is convenient to decouple the different features of the experimental results for the interpretation. Whereas the line position is uniquely determined by the energy-band structure, both line intensity and shape are rather complex quantities. To investigate theoretically the influence of external stress on the energy-band structure of tellurium we have to establish an appropriate hamiltonian including pressure dependent terms. The most convenient procedure for this purpose is the "method of invariants" introduced by Bir and Pikus[10]. For tellurium we have constructed a 6x6 hamiltonian considering the coupling between the two H_6-conduction bands, the H_4 and H_5 valence bands, and the two H_6 lower valence bands. In terms of group theoretical coupling we obtain the following hamiltonian:

$$\mathcal{H} = \begin{vmatrix} \mathcal{H}^{\Gamma_4} & \mathcal{H}^{\Gamma_4 \Gamma_6^*} & \mathcal{H}^{\Gamma_4 \Gamma_5^*} & \mathcal{H}^{\Gamma_4 \Gamma_4^+} \\ \mathcal{H}^{\Gamma_4 \Gamma_6} & \mathcal{H}^{\Gamma_6} & \mathcal{H}^{\Gamma_6 \Gamma_5} & \mathcal{H}^{\Gamma_6 \Gamma_4} \\ \mathcal{H}^{\Gamma_4 \Gamma_5} & \mathcal{H}^{\Gamma_6 \Gamma_5} & \mathcal{H}^{\Gamma_4} & \mathcal{H}^{\Gamma_5 \Gamma_4^+} \\ \mathcal{H}^{\Gamma_4 \Gamma_4} & \mathcal{H}^{\Gamma_6 \Gamma_4} & \mathcal{H}^{\Gamma_5 \Gamma_4} & \mathcal{H}^{\Gamma_4} \end{vmatrix} \quad (1)$$

By comparison with the results of the stress dependent $\vec{k}\cdot\vec{p}$-perturbation theory it is possible to reduce the number of the still unknown stress dependent parameters to two, namely to the stress dependence of the energy gap E_G between the conduction and upper valence band, and of the splitting 2Δ between the two upper valence bands. This does not mean, that all other parameters in the hamiltonian are stress independent, but that they have a simple stress dependence of the form $(1-\varepsilon_{jj})$, where ε_{jj} is the dilatation. A detailed discussion of the derivation will be given elsewhere[11]. For the numerical determination of the two stress dependent parameters E_G and Δ we used in a consistent way four different cyclotron-resonance lines. Extensive numerical computations inclusive the diagonalization of $\mathcal{H}$ for $B_\perp c$ were necessary. In Fig.2 we represent the experimental data and the theoretical fits for the $0^+ - 0^-$ and $0^+ - 1^-$ transition using 337μm-wavelength radiation. Considering the fact that four transitions have been fitted like this by only two parameters the agreement is excellent. The results of this fit procedure are a rather strong linear increase of the energy gap E_G with increasing uniaxial stress P_C and a completely stress independent spin-orbit splitting parameter Δ, namely

$$\frac{1}{E_G} \cdot \frac{\partial E_G}{\partial P_C} = \frac{2.39 \cdot 10^{-2}}{\text{k-bar}} \quad , \quad \frac{1}{\Delta} \cdot \frac{\partial \Delta}{\partial P_C} = 0 \quad . \quad (2)$$

The increase of the gap E_G with uniaxial stress P_C is not surprising, since NEURINGER found a corresponding decrease under hydrostatic pressure, which is complementary to the uniaxial stress application[12]. The second result of a stress independent spin-orbit splitting parameter Δ is rather striking at first sight, since so far the pressure dependence of this parameter has been considered as the essential origin of the stress dependent change of the "camelback" under hydrostatic[13] and uniaxial[15] pressure. Of course, it might be that the present result under uniaxial stress is caused accidentally by the superposition of the two contributions of the deformations parallel and perpendicular to the c-axis. We believe, however, that the result reflects much more the fundamental fact, that spin-orbit coupling is a relativistic effect and becomes dominant only in those parts of an attractive potential, where the kinetic energy is very high. This situation is encountered only in the direct vicinity of the core of the lattice constituents. The core potential, however, should not be sensitive to small distortions of the lattice. We would like to emphasize, that this aspect is a completely novel feature of the stress dependence of the "camel-back" and is in strict contrast to earlier estimations of the spin-orbit energy under hydrostatic and uniaxial stress [13,5]. In the historical development in the determination of $(1/\Delta)(\partial \Delta/\partial P_c)$ we find unambiguously the tendency to smaller values of this quantity, which supports our consideration [1].

Usually the 6x6-hamiltonian can hardly be handled and the subspace of the two upper valence bands can be separated resulting in the "COUDER"-model [14]:

$$\mathcal{H} = \begin{vmatrix} \Delta - \alpha_z k_z^2 - \alpha_\perp k_\perp^2 & \Pi k_z \\ \Pi k_z & -\Delta - \alpha_z k_z^2 - \alpha_\perp k_\perp^2 \end{vmatrix} . \qquad (3)$$

The stress dependence of the four parameters is then given by:

$$\frac{1}{\alpha_\perp} \cdot \frac{\partial \alpha_\perp}{\partial P_c} = -4.36 \times 10^{-2} \text{ k-bar}^{-1} \qquad (4)$$

$$\frac{1}{\alpha_z} \cdot \frac{\partial \alpha_z}{\partial P_c} = 1.10 \times 10^{-2} \text{ k-bar}^{-1} \qquad (5)$$

$$\frac{1}{\Pi} \cdot \frac{\partial \Pi}{\partial P_c} = 0.79 \times 10^{-2} \text{ k-bar}^{-1} \qquad (6)$$

$$\frac{1}{\Delta} \cdot \frac{\partial \Delta}{\partial P_c} = 0 \quad . \qquad (7)$$

In Fig.3 we have plotted by the full dots and solid curve the intensity of the $0^+- 0^-$ transition as a function of the uniaxial stress. Here the intensity is defined by the area under the resonance line in the absorption constant. For comparison we have also included in the same figure the stress dependence of the resonance position and the line width. Whereas the resonance position

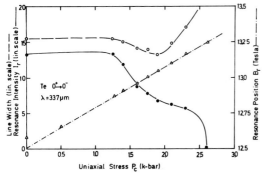

Fig. 3 The line parameters of the 0^+-0^- transition as a function of the uniaxial stress P_c

is an absolutely linear function of the uniaxial stress the intensity decrease exhibits two different thresholds at about 1.2 and 2.2 k-bar. The decrease in intensity goes down to the complete vanishing of the line. This is achieved at about $P_C=2.5$ k-bar just before the mostly obtainable maximum stress value. The onset of the intensity decrease is accompanied by a small narrowing of the resonance line with a succeeding significant broadening.

The intensity of the resonance is theoretically determined by the square of the interaction matrix element and the population of the initial and final states[15]. We have corroborated that a pressure induced change of the matrix element is not responsible for the observed strong decrease. Therefore, pressure induced population changes are to be discussed.

In principle there are three different mechanisms for pressure induced population changes in tellurium. Firstly, there is the possibility that an additional valley of the upper valence band is shifted, so that it becomes populated by holes. From energy-band calculations one expects this valley to be located at the Γ-point of the Brillouin zone[16]. No reliable theoretical or experimental values of the energy difference to the band edge at the H-point are, however, available so far. The zero stress energy separation of the two valence band extrema should be larger than about 10 meV, since up to this energy the valence band has been investigated by Shubnikov-de Haas experiments and no anomaly due to a second valley has been reported[17]. The required stress shift is then of the same order of magnitude and would result in a stress dependence of the gap at the Γ-point of

$$\frac{1}{E_G^\Gamma} \frac{\partial E_G^\Gamma}{\partial P_G} \geq 10^{-2} \text{ k-bar}^{-1}$$

which is comparable to our value for the gap at the H-point under uniaxial stress and also comparable to the value given by NEURINGER for hydrostatic stress[12]. The significant problem remains, however, that a population of the Γ-point valley should be detectable by means of new cyclotron resonance lines, which increase in intensity in the corresponding way as the intensity of the lines at the H-point decreases. So far no new lines have been observed, which could account for such an explanation. The proposed explanation strikes, however, because of its simplicity.

The second possible interpretation uses a "pressure freeze-out" effect of impurities. By a stress induced increase of the impurity binding energy the

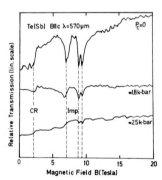

Fig. 4 The impurity levels are also depopulated with increasing uniaxial stress P_C

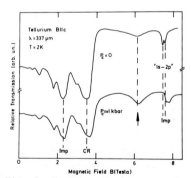

Fig. 5 So far only one transition has been observed which increases with stress

free-carrier states should become depopulated. To investigate whether such
a mechanism could account for our observations in tellurium we concentrated
on the impurity resonances for $B||c$, because in this configuration the impuri-
ty resonances are much more significant than for $B\perp c$. In Fig.4 we have plotted
the transmission curves of Sb-doped Te-samples in Voigt configuration using
570 μm-wavelength radiation. The parameters indicate again the stress value.
Beyond the cyclotron resonance at about 2 Tesla additional resonance lines
are observed in agreement with photoconductivity measurements of VON KLITZING
[18]. These lines are interpreted as internal impurity transitions. The asto-
nishing result is now, that these lines also vanish with increasing uniaxial
stress. It seems that all states associated with the valley at the H-point of
the Brillouin zone become depopulated. In addition to the intensity decrease
with increasing uniaxial stress a shift of the resonance lines towards higher
magnetic fields is observed. This fact is quite surprising, since from the
usual hydrogen-like impurity model one expects that the impurity energy is
increased for increased values of the mass parameter as found experimentally
for both configurations $B||c$ and $B\perp c$. It seems worth being mentioned that the
value of the shift of the impurity energy is rather small for uniaxial stress
in contrast to the theoretical results of THANH and SUFFCZYNSKI[19] as well
as the experimental data of TANI and TANAKA[20]. To resolve this discrepancy
we calculated the binding energy of the bonding and antibonding state of a
shallow impurity in tellurium. Extending the method of THANH[21] we included
in the hamiltonian both the stress and magnetic field dependence of the band
structure. Since in the presence of strong magnetic fields the variational
function of THANH is not any more realistic, we made the following ansatz for
$B||c$:

$$\mathcal{H}\psi_\pm = \begin{vmatrix} \Delta - \alpha_z k_z^2 - \frac{\hbar\omega_c}{2} + V(x^2+y^2,z^2) & \Pi k_z \\ \Pi k_z & -\Delta - \alpha_z k_z^2 - \frac{\hbar\omega_c}{2} + V(x^2+y^2,z^2) \end{vmatrix} \psi_\pm \cdot \exp\left\{\frac{-(x^2+y^2)}{2l^2} - b^2 z^2\right\} \quad (8)$$

and ψ_+, ψ_- are given for the bonding and antibonding state respectively by

$$\psi_+ = \begin{pmatrix} c_+ + \alpha \cdot \cos k_m z \\ i\beta \cdot \sin k_m z \end{pmatrix} \quad \psi_- = \begin{pmatrix} i\alpha \cdot \sin k_m z \\ c_- + \beta \cdot \cos k_m z \end{pmatrix} \quad (9)$$

Here the k_z-dependent coupling parameters α and β are given by

$$\alpha = \frac{\gamma}{\sqrt{1+\gamma^2}}, \quad \beta = \frac{1}{\sqrt{1+\gamma^2}}, \quad \gamma = \frac{\Pi k_m}{\sqrt{\Delta^2 + \Pi^2 k_m^2} - \Delta} \quad (10)$$

k_m represents the k-vector of the maximum in the "camel-back". As variational
parameters we used $c_\pm$ and b. We avoided a third variation, namely of l^2 by
considering only the high magnetic field limit by setting

$$l^2 = \frac{\hbar}{eB} \quad . \quad (11)$$

The results of the computations are plotted in Fig.6. There is an increasing
splitting of the bonding and antibonding state with increasing magnetic field.
This splitting, however, is slightly decreased after application of uniaxial
stress. The stress, however, decreases the binding energy of both states, but
affects the bonding state more than the antibonding. This is due to the stress
dependence of the parameter Π, which makes the two maxima of the "camel-back"
to reject. In comparison with the results of THANH and SUFFCZYNSKI for hydrostatic pressure, the pressure induced changes for the binding energies are
almost one order of magnitude smaller for uniaxial stress[19].

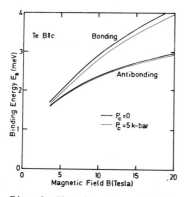

Fig. 6 The bonding state is more influenced by stress than the antibonding

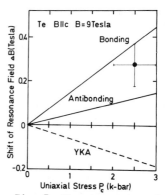

Fig. 7 The stress dependence of the observed impurity line favors the bonding state

From the extensive numerical results of our computations we were able to derive directly the magnetic field shift of the impurity resonance as function of uniaxial stress. The results are plotted in Fig.7. Again there should be in the experiment a linear dependence of the magnetic field shift on stress. It demonstrates again that the bonding state is about three times more sensitive to stress than the antibonding. We have included in the same plot one point of the experimental results after preliminary evaluation. This data seems to favor the bonding state. It should be noticed that the simple YAFET-KEYES-ADAMS theory without considering the detailed "camel-back" structure of the upper valence band predicts for tellurium a shift of the impurity resonance to lower magnetic fields with increasing uniaxial stress[22].

Concerning the shallow impurities in tellurium we can so far draw the following conclusions: The uniaxial stress distorts the upper valence band in such a way that the shift of all associated energy levels, quasi-free or bound, is consistently understood and reproduced by theory. Since, however, the binding energy of shallow impurities decreases with uniaxial stress, there is no "stress freeze-out" effect and shallow impurities cannot be accounted for the observed strong depopulation effects.

In some of our Sb-doped samples we observed for $B||c$ an additional absorption line as shown in Fig.5. This line cannot be explained by quasi-free carriers or shallow impurities including hydrogen-like and helium-like models[18]. This additional resonance is located at about 6 Tesla using 337-μm-wavelength radiation. The surprising property of this resonance, however, is that the intensity of the line increases with increasing uniaxial stress. Since this line is not present in all of our samples, it cannot be attributed to a band-structure effect. It might be possible that it is somehow associated with lattice defects, namely dislocations. These kind of lattice defects seem to play an important role in the presence of hydrostatic pressure, however, with the apparent opposite tendency; EREMETS et al. explain the intensity increase of their intervalence-band absorption by the hole-attractive potential of the dislocation[23]. Dislocation effects, however, should be similar for hydrostatic and uniaxial stress. Possibly charged dislocations lined up in a Hertz-grid arrangement could account for the decrease in the background transmission in our experiment. In strongly defected tellurium a decrease of transmission by a factor of 250 has been observed, however, no population change of the resonance levels[24].

Summarising our results we can make the following conclusions: because of the sensitivity to small changes in the energy of electronic states submillimeter-magneto spectroscopy is an excellent tool to investigate the influence of external pressure on the energy-band structure of semiconductors. For tellurium we were able to present exact values for the stress dependence of the $\vec{k}\cdot\vec{p}$-parameters. Using these values not only the stress dependence of all cyclotron resonance lines is reproduced, but also of the shallow impurity lines. It seems that in the past the distortion of the "camel-back" has been overestimated. At least under uniaxial stress the changes near the H-point of the Brillouin zone are not so significant. The strong depopulation effect of the states near the H-point has raised the question, whether the Γ-valley is slightly shifted by uniaxial stress so that it becomes populated. Concerning this question no decision can be made so far.

For a complete understanding of the pressure effects in tellurium, especially to solve the problem whether under hydrostatic stress the distortion of the energy-band structure is really so much larger than under uniaxial stress, the present experiments are to be repeated under hydrostatic pressure. Since hydrostatic and uniaxial stress are complementary there should also be no depopulation observable, if this effect is due to the energy-band structure. The preparations for this investigation in cooperation with the Academy of Science of the USSR are in progress.

3. Polarization Effects in the Strip-Line Transmission

The strip-line is a special optical arrangement for the investigation of highly conductive materials by submillimeter-magneto spectroscopy using lasers as radiation source[25]. It consists of a microscopic waveguide, one wall of which is made of the sample material. The attenuation of this waveguide, the strip-line attenuation, is therefore directly determined by the dielectric properties of the wall material. Since the strip-line attenuation depends strongly on the geometrical ratio λ/D, where λ is the radiation wavelength and D the strip-line width, it is recommendable to use monochromatic radiation.

In the presence of an external magnetic field there are three favorite strip-line configurations: the parallel, perpendicular, and surface configuration. The one most often used is the parallel configuration, because the strip-line propagation vector is parallel to the magnetic field and therefore very well applicable with an ordinary, non-split coil magnet. In the parallel configuration, however, the attenuation is dominated by the ordinary Voigt mode[26]. For semiconductors with spherical Fermi surface, like InSb, but also with non-spherical but compensating multi-valley Fermi surface, like PbTe, the ordinary Voigt mode is well determined by the component ε_{zz} of the dielectric tensor. This means that only tilted orbit resonances are detectable. For an ellipsoidal energy surface of a single valley as for Cd_3As_2 also the other components of the dielectric tensor contribute to the ordinary Voigt mode and screen the resonance. This implies that without the exact knowledge of the plasma frequency the resonance position cannot be determined. Mostly, however, the plasma frequency is not known at all.

The basic consideration is now to find such an experimental arrangement in the strip-line, that modifies the radiation modes in the sample material in such a way, that the screening is suppressed. This can easily be achieved by a linear polarizer in form of a Hertz-grid at the semiconductor surface. In this case in the vacuum part of the strip-line only a pure TM-mode can propagate. In the semiconductor material directly below the surface the two normal modes are superimposed in such a way that the resulting electric field vector is parallel to the z-direction, as indicated in Fig.8. From the theoretical point of view this means that we have modified the boundary conditions at the vacuum-semiconductor interface :

$$E_x = E_x' = 0 \qquad E_z = E_z' \qquad H_x = H_x' \ . \tag{12}$$

Straightforward calculation yields the following set of equations for the strip-line propagation constant h:

$$ik_{y2}(k_{yo}C_0 + hB_0 - k_{y1}B_0 - hB_1)\sin k_{y2}D + k_0^2(C_1 - C_0)\cos k_{y2}D = 0 \tag{13}$$

$$C_{0/1} = \{[k_0^2 \varepsilon_{yy} - h^2][k_0^4 \varepsilon_{yx}\varepsilon_{xz} - (k_{y0/1}^2 + h^2 - k_0^2 \varepsilon_{xx})(hk_{y0/1} - k_0^2 \varepsilon_{yz})]\}/ \cdot \tag{14}$$
$$\{[hk_{y0/1} - k_0^2 \varepsilon_{yz}][(h^2 - k_0^2 \varepsilon_{yy})\varepsilon_{xz}k_0^2 - k_0^2 \varepsilon_{xy}(hk_{y0/1} - k_0^2 \varepsilon_{yz})\}$$
$$+ k_0^2 \varepsilon_{yz}/(hk_{y0/2} - k_0^2 \varepsilon_{yz})$$

$$B_{0/1} = [k_0^4 \varepsilon_{yx}\varepsilon_{xz} - (h^2 + k_{y0/1}^2 - k_0^2 \varepsilon_{xx})(hk_{y0/1} - k_0^2 \varepsilon_{yz})]/ \tag{15}$$
$$[(h^2 - k_0^2 \varepsilon_{yy})\varepsilon_{xz}k_0^2 - k_0^2 \varepsilon_{xy}(hk_{y0/1} - k_0^2 \varepsilon_{yz})]$$

$$h^2 + k_{y2}^2 = k_0^2 \tag{16}$$

$$k_0^4 \varepsilon_{xz}\varepsilon_{yx}(hk_{y0/1} - k_0^2 \varepsilon_{zy}) + k_0^4 \varepsilon_{zx}\varepsilon_{xy}(hk_{y0/1} - k_0^2 \varepsilon_{yz}) - k_0^2 \varepsilon_{zx}\varepsilon_{xz}(h^2 - k_0^2 \varepsilon_{yy}) \tag{17}$$
$$-(hk_{y0/1} - k_0^2 \varepsilon_{zy})(hk_{y0/1} - k_0^2 \varepsilon_{yz})(h^2 + k_{y0/1}^2 - k_0^2 \varepsilon_{xx}) - k_0^4 \varepsilon_{xy}\varepsilon_{yx}(k_{y0/1}^2 - k_0^2 \varepsilon_{zz}) = 0 \ .$$

With the usual approximations we obtain for the intensity attenuation constant:

$$K = \text{Im} \left\{ \frac{k_0(C_0 - C_1)}{D(k_{yo}C_0 + hB_0 - k_{y1}C_1 - hB_1)} \right\} \ . \tag{18}$$

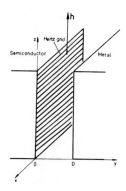

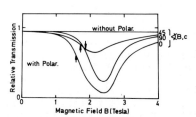

Fig. 8 Schematic of the modified strip-line with polarizer at the semiconductor surface

Fig. 9 The simulated strip-line transmission for Cd_3As_2 demonstrates the importance of the polarizer

For the limit $C_0 \gg C_1, B_0, B_1$ we obtain the well-known attenuation formula for the parallel configuration [25,26]

$$K \approx \text{Re} \left\{ \frac{1}{D\sqrt{\varepsilon_{zz}}} \right\} \tag{19}$$

In Fig. 9 we have plotted the simulations of the strip-line transmission according to (13)-(17) for the material parameters of Cd_3As_2. Without polarizer there is no resonance structure observable because of the large screening. With polarizer, however, the simulated spectra exhibit a significant resonance structure. The parameters of the curves indicate the angle between the c-axis and the magnetic field. The resonance position as found in the dielectric tensor is indicated by arrows. It agrees again very well with the position of the turning point in the slope before the transmission minimum. It should be noticed that even for the high-symmetry configuration B//c the resonance modulation is the strongest despite the absence of the tilted orbit resonance. This means that the parallel strip-line configuration has now become applicable to a much larger variety of semiconductor materials and the resonance in the dielectric tensor can directly be detected without any perturbing screening effects.

References

1) G.Dresselhaus, A.F.Kip and C.Kittel, Phys.Rev. 92, 827 (1953)
2) K.Suzuki and J.C.Hensel, Phys.Rev. B9, 4184 (1974)
3) J.C.Hensel and K.Suzuki, Phys.Rev. B9, 4219 (1974)
4) see, for references, Yu.V. Kosichkin:" Optical Properties of Tellurium Under High Pressure", in The Physics of Selenium and Tellurium, Proc. Int. Conf., Königstein, F.R. Germany, May 28-31,1979, ed. by E. Gerlach, P. Grosse, Springer Series in Solid-State Sciences, Vol. 13 (Springer, Berlin, Heidelberg, New York 1979) p.96
5) K.H. Hermann, J.Phys. Chem. 230, 123 (1965)
6) P. Grosse: Die Festkörpereigenschaften von Tellur, Springer Tracts in Modern Physics, Vol. 48 (Springer, Berlin, Heidelberg, New York 1969)
7) U. Steigenberger, M.v.Ortenberg, E. Bangert:" Submillimeter Cyclotron Resonance in Uniaxially Stressed Tellurium", in The Physics of Selenium and Tellurium, Proc. Int. Conf., Königstein, F.R. Germany, May 28-31, 1979, ed. by E. Gerlach, P. Grosse, Springer Series in Solid-State Sciences, Vol. 13 (Springer, Berlin, Heidelberg, New York 1979) p.119
8) U.Steigenberger and M.v.Ortenberg, phys.stat.sol.(b) 93, K139 (1979)
9) M.v.Ortenberg and R.Ranvaud, Solid State Commun. 13, 333 (1973)
10) G.L.Bir and G.E.Pikus: 'Symmetry and Deformation Effects in Semiconductors', Moscow (1972)
11) U.Steigenberger, E.Bangert and M.v.Ortenberg, Proceedings of the 5th Int.Conf. on Infrared and Millimeter Waves', Würzburg 1980
12) L.J.Neuringer, Phys.Rev. 113, 1495 (1959)
13) V.B.Anzin, M.S.Bresler, I.I.Farbstein, E.S.Itskevich, Yu.V.Kosichkin, V.A.Sukhoparow, V.W. Telepnev and V.G.Veselago, phys.stat.sol.48,531(1971)
14) Y.Couder, M.Hulin and H.Thomé, Phys.Rev. B7, 4373 (1973)
15) M.v.Ortenberg and K.J.Button, Phys.Rev. B16, 2618 (1977)
16) J.Treusch and R.Sandrock, phys.stat.sol. 16, 487 (1966)
17) E.Braun, L.J.Neuringer and G.Landwehr, phys.stat.sol.(b) 53, 635 (1972)
18) K.von Klitzing, Phys.Rev. B21, 3349 (1980)
19) D.Thanh and M.Suffczynski, Solid State Comm. 11, 189 (1972)

20) T. Tani, S. Tanaka:"Pressure Effect on the Impurity State and Impurity Conduction in Tellurium", in The Physics of Selenium and Tellurium, Proc. Int. Conf., Königstein, F.R. Germany, May 28-31, 1979, ed. by E. Gerlach, P. Grosse, Springer Series in Solid-State Sciences, Vol. 13 (Springer, Berlin, Heidelberg, New York 1979) p.142
21) D.Thanh, Solid State Commun. 9, 631 (1971)
22) Y.Yafet, R.W.Keyes, and E.N.Adams, J.Phys. Chem.Sol. 1, 137 (1956)
23) M.Eremets, Yu.V.Kosichkin, A.Shirokov:"Pressure Influence on the Intervalence Band Absorption in Te", in The Physics of Selenium and Tellurium, Proc. Int. Conf., Königstein, F.R. Germany, May 28-31,1979, ed. by E.Gerlach, P. Grosse, Springer Series in Solid-State Sciences, Vol. 13 (Springer, Berlin, Heidelberg, New York 1979) p.113
24) M.v.Ortenberg, IEEE - MTT 22, 1081 (1974)
25) M.v.Ortenberg in 'Infrared and Millimeter Waves',ed.K.J.Button, Vol. 3 p. 275 (1980)
26) M.v.Ortenberg, Infrared Phys. 18, 735 (1978)

Four-Wave Spectroscopy of Shallow Donors in Germanium

R.L. Aggarwal

Francis Bitter National Magnet Laboratory†, and
Department of Physics, Massachusetts Institute of Technology
Cambidge, MA 02139, USA

1. Introduction

In recent years, four-wave mixing spectroscopy (FWMS) with tunable lasers has emerged as a powerful technique for the study of optical excitations in gases, liquids and solids. In a simple version of FWMS, two laser beams, say, at frequencies ω_1 and ω_2 interact in the medium of interest to generate radiation at the frequency $\omega_4 = 2\omega_1 - \omega_2$. All the processes leading to output at ω_4 can be derived by a third-order nonlinear susceptibility tensor $\chi^{(3)}_{ijkl}(-\omega_4, \omega_1, \omega_1, -\omega_2)$ which is a polar tensor of 4th order. It should be pointed out $\chi^{(3)}$ does not vanish identically for any symmetry group. In contrast, the second-order nonlinear susceptibility tensor $\chi^{(2)}$ is identically zero for all media except non-centrosymmetric crystals. Thus, phenomena involving $\chi^{(3)}$ can be studied in all media: gases, liquids and solids.

The output power of radiation generated at ω_4, $P\omega_4$, increases quadratically with $\chi^{(3)}$ and is given in CGS units under phase-matched conditions by [1]:

$$P\omega_4 = \frac{1024\,\pi^4\,\omega_4^2}{3n_1^2 n_2 n_4 c^4} |\chi^{(3)}_{eff}|^2 P\omega_1^2 P\omega_2 \cdot \frac{\ell_{eff}^2}{A^2} \cdot T_1^2 T_2 T_4 \qquad (1)$$

where n_1, n_2, and n_4 are the respective indices of the medium at frequencies ω_1, ω_2 and ω_4, respectively. $\chi^{(3)}_{eff}$ is the effective third-order nonlinear susceptibility coefficient. A is the area of the beams where they interact. $P\omega_1$ and $P\omega_2$ are the incident laser powers at frequencies ω_1 and ω_2. The effective interaction length ℓ_{eff} in (1) is always less than the distance over which the beams coincide when there is a finite wave vector mismatch of the interacting beams $\Delta k = |2\vec{k}_1 - \vec{k}_2 - \vec{k}_4|$. T_1, T_2 and T_4 are the single-surface transmission coefficients at ω_1, ω_2 and ω_4, respectively. $\chi^{(3)}$ exhibits resonance behavior when the difference frequency, $\Delta\omega = \omega_1 - \omega_2$, is tuned through a Raman-active transition of the medium. Therefore, a measurement of ω_4 as a function of $\Delta\omega$ yields the Raman spectrum of the medium.

---

*Work supported by the Air Force Office of Scientific Research.
†Supported by the National Science Foundation.

The above method of FWMS, one of the techniques of coherent Raman spectroscopy, has several advantages over incoherent Raman scattering [2]. The most important advantage is that the value of P_{ω_4} in FWMS can be many orders of magnitude larger than in the spontaneously scattered radiation. Other advantages of the ω_4 signal are its spatial coherence and high resolution. Thus, samples in which the spontaneous Raman scattering is intrinsically weak or masked by fluorescence and black-body radiation can be studied in FWMS.

We have applied the high resolution technique of FWMS to the study of the valley-orbit split 1s ground state of shallow donors in germanium, using Q-switched CO_2 lasers [1,3]. Valley-orbit splitting of the 1s ground state of group V donors (P, As, Sb, Bi) in germanium has been previously studied by a number of investigators. By observing transitions allowed in infrared absorption from the 1s levels to a common p level, Reuszer and Fisher [4] made an accurate determination of the valley-orbit splitting, 4Δ, between the nondegenerate ground state, $1s(A_1)$, and the upper 3-fold degenerate state, $1s(T_2)$. Forbidden optical absorption between the 1s states has been observed in the far infrared at photon energies corresponding to 4Δ for P and As donors [5,6]. Allowed spontaneous Raman scattering between the 1s states has also been studied [7].

2. Experimental

A schematic of the experimental setup for FWMS is shown in Fig. 1. 200 ns CO_2 laser pulses of about 1 kW maximum peak power at frequencies ω_1 and ω_2 are provided simultaneously by using a pair of intercavity mirrors mounted on the opposite sides of a shaft rotating at 150 Hz. Following the Ge beam-

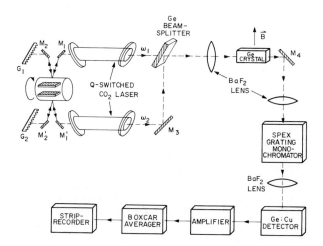

Fig. 1 Schematic of the experimental setup for FWMS in a magnetic field

splitter, the two laser beams propagate collinearly and are focused on a ~0.6 cm long Ge sample at 1.8°K with a BaF_2 lens into ~200 μm spot. Each laser beam incident on the sample has typical peak power of ~200 W, with electric vector $\vec{E}_1 \| \vec{E}_2 \perp \vec{B}$ and propagation vector $\vec{k} \perp \vec{B}$. The ω_4-signal is resolved from the two incident laser beams with a Spex double grating monochromator and is measured with a Cu-doped Ge photodetector. The output of the detector is averaged by a boxcar integrator and displayed on a strip-chart recorder. Wood et al. [1] used a setup similar to that of Fig. 1 to measure dispersion of $\chi^{(3)}$ for As and P donors in Ge at zero magnetic field. Their results, shown in Fig. 2, clearly demonstrate a resonant peak at $\Delta\omega$ corresponding to the $1s(A_1) \rightarrow 1s(T_2)$ valley-orbit transition at $4\Delta(0)$.

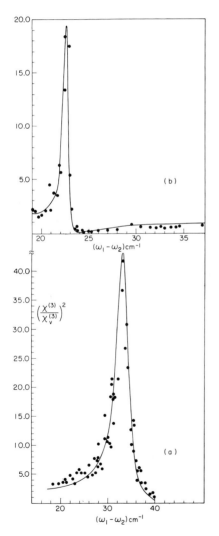

Fig. 2 Normalized dispersion at 11 K and with $\vec{E} // [110]$ of $|\chi^{(3)}(-2\omega_2 + \omega_1, \omega_2, \omega_2 - \omega_1|$ in: (a) Ge:As, $N_D - N_A = 3.1 \times 10^{16}$ cm^{-3} and (b) Ge:P, $N_D - N_A = 9 \times 10^{14}$ cm^{-3}. The solid lines are best-fit calculations from [1].

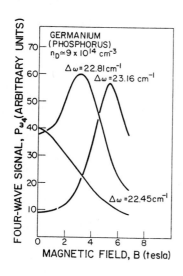

Fig. 3 Four-wave signal, P_{ω_4}, vs. applied magnetic field $\vec{B}//[100]$ in Ge at 1.8°K.

Recently, we [3] have studied the magnetic field dependence of the valley-orbit splitting, $4\Delta(B)$, for P and As donors using FWMS as follows:

For a fixed laser frequency difference $\Delta\omega = \omega_1-\omega_2$, the ω_4-signal, P_{ω_4}, is measured as a function of B. For $\Delta\omega > 4\Delta(0)$, P_{ω_4} shows a maximum at a value of B which satisfies the resonance condition $\hbar\Delta\omega = 4\Delta(B)$. Figure 3 shows the tuning curves obtained for three different values of $\Delta\omega$ for P-doped Ge. Note that no peak is observed for $\Delta\omega$ = 22.45 cm^{-1}, which is less than $4\Delta(0)$ for P donors. A plot of $4\Delta(B)$ vs. B^2 is shown in Fig. 4 for P and As donors. The solid lines in Fig. 4 represent a least-squares straight line fit to the data, giving $4\Delta(B) = 4\Delta(0) + \alpha B^2$ with $4\Delta(0)$ = 22.6 cm^{-1} for P and 34.0 cm^{-1} for As. Values of the coefficient α are given in Table I in the following section.

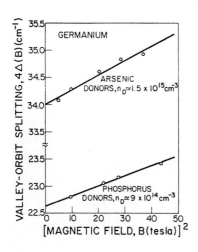

Fig. 4 Valley-splitting in a [100] magnetic field vs. B^2.

3. Theory and Discussion of Results

In the effective mass approximation (EMA), $4\Delta(B) = 4\Delta(0) = 0$ and the $1s(A_1)$ and $1s(T_2)$ states have identical envelope functions. We present arguments indicating that the observed magnetic field dependence of $4\Delta(B)$ discussed in Section 2 is the result of smallness of the spread of the true $1s(A_1)$ envelope functions relative to those of the $1s(T_2)$ levels, which are nearly effective mass like.

A useful donor variational trial envelope function in the EMA for a given valley, i, can be written

$$\chi_i = A \exp\left(-\delta^2 z_i^2 - \kappa\left(\rho_i^2 + \alpha z_i^2\right)^{\frac{1}{2}}\right) \qquad (2)$$

where z_i lies along the axis of cylindrical symmetry of the energy ellipsoid of valley i, and A is a normalization factor. The "best" choice of variational parameters in (2) leads to an EMA energy, $\tilde{R}$, of -2.0884 R, where $R = m_\perp e^4/2\epsilon_0^2\hbar^2$; here we have used $\sigma \equiv m_\perp/m_z = 0.05134$ [8] where $m_\perp$ and m_z are effective masses for motion in valley i perpendicular and parallel, respectively, to z_i. If $\vec{B}$ lies along [100], it makes an angle of $\cos^{-1}(1/\sqrt{3})$ with z_i for all four valleys. In that case the diamagnetic shift is not proportional to $<\rho_i^2>$ as it would be for the ground state of a donor in a spherical band but is given approximately by [9]

$$\delta E = R\left[\left(\gamma_z^2/4 + \sigma(1-\xi)^2\gamma_\perp^2/2\right)<\rho_i^2> + \xi^2\gamma_\perp^2<z_i^2>\right] \qquad (3)$$

where $\xi = \sigma<\rho_i^2>/(\sigma<\rho_i^2> + 2<z_i^2>)$, $<>$ denotes expectation value in χ_i, all lengths are in units of the Bohr radius $\hbar^2\epsilon_0/m_\perp e^2$, and γ_z and $\gamma_\perp$ are dimensionless components of B, along and perpendicular, respectively, to z_i and defined by $\gamma_{z,\perp} = (\hbar e B_{z,\perp}/m_\perp c)/(2R)$. Equation (3) is valid in the EMA for arbitrary direction of $\vec{B}$. Specializing to the case that $\vec{B}$ lies along [100] we find that (3) gives the same diamagnetic shift for all four valleys, which becomes $0.0727\gamma^2 R$; this corresponds to an energy shift of 0.0633 cm^{-1}/T^2. Equation (3) is derived by introducing the gauge

$$\vec{A} = B_z\left(-y_i/2, x_i/2, 0\right) + B_\perp\left(\xi z_i, 0, -(1-\xi)x_i\right) \qquad (4)$$

into the effective mass Schrödinger equation for the donor electron in valley i. In (4) the y_i axis has been implicitly chosen along $\vec{B}_\perp$; the indicated value of ξ to be inserted in (3) was obtained by minimizing δE with respect to ξ.

The true envelope function for valley i in the actual A_1 ground state is not χ_i but a function which is much more strongly concentrated at the donor center than is χ_i. For <u>spherical</u> valleys this function is known to be given by $\phi = W_{n,\frac{1}{2}}(\frac{2r}{n})/r$, where $n = (1/E)^{\frac{1}{2}}$, E is the observed energy of the ground state in units of $\tilde{R}$ (E = 1.31 for P and 1.44 for As in Ge [4,8]) and $W_{n,\frac{1}{2}}$ is a Whittaker function [10]. For nonspherical valleys like those of Ge the appropriate envelope functions are unknown.

To estimate the diamagnetic shift associated with the wave function in the ith valley we employ (2) but reduce the quantities $<\rho_i^2>$ and $<z_i^2>$ appearing there by multiplying them with the scale factor $<\phi|r^2|\phi>/(3<\phi|\phi>)$, where we have used $\int d^3r\ r^2 e^{-2r}/\int d^3r e^{-2r} = 3$. This follows from assuming that the shrinkage of the mean value of r^2 in the true envelope function for given E, relative to the EMA wave function is independent of the mass anisotropy of the valley and that the ratio of mean z_i^2 to mean ρ_i^2 in the true envelope function is $<z_i^2>/<\rho_i^2>$ (i.e. the same as that in the EMA). For the $1s(A_1)$ and $1s(T_2)$ states of Ge:P we obtain scale factors of 0.582 and 0.947, respectively; the corresponding factors for Ge:As are 0.488 and 0.969. From these numbers and the EMA shift of 0.0633 cm^{-1}/T^2 from (3) we obtain the theoretical values in Table I.

Table I Comparison of predicted and observed values of α(cm^{-1}/T^2)

Donor	n_D(cm^{-3})	α Theory	α Experiment
P	9 x 10^{14}	0.023	0.018±.002
P	1 x 10^{16}	---	0.029±.002
As	1.5 x 10^{15}	0.030	0.027±.002

The agreement of experiment and theory indicated in Table 1 supports our contention that the magnetic field dependence of the A_1-T_2 splitting for isolated donors is due primarily to the different orbital diamagnetism of these states. The higher value of α for the 1 x 10^{16} cm^{-3} P-sample is presumably connected with delocalization. However, a real understanding of this effect is lacking at the present time.

Acknowledgement

We are extremely grateful to Dr. David M. Larsen, Prof. Peter Wolff and R. People for useful discussions and help in the preparation of this manuscript.

References

1. R.A. Wood, M.A. Khan, P.A. Wolff, R.L. Aggarwal: Opt. Commun. 21, 154 (1977).

2. M.D. Levenson: Physics Today May, 44 (1977).

3. R.L. Aggarwal, R. People, P.A. Wolff, D.M. Larsen: Proc. of the 15th Intl. Conf. on the Physics of Semiconductors, Kyoto, 1980, to be published.

4. J.H. Reuszer, P. Fisher: Phys. Rev. 135, A1125 (1964).

5. V.V. Buzdin, A.I. Demeshina, Y.A. Kurskii, V.N. Murzin: Sov. Phys.-Semiconductors 6, 1792 (1973).

6. M. Kobayashi, S. Narita: J. Phys. Soc. Japan 43, 1455 (1977).

7. J. Doehler, P.J. Colwell, S.A. Solin: Phys. Rev. Lett. 34, 584 (1975).

8. R.A. Faulkner: Phys. Rev. 184, 713 (1969). See especially Table VIII.

9. D.M. Larsen: unpublished calculations.

10. W. Kohn, J.M. Luttinger: Phys. Rev. 97, 883 (1955).

New Magneto-Optical Transitions in n-InSb: Mid-Gap Deep Defect Level and Three-Phonon Assisted Processes *

D.G. Seiler and M.W. Goodwin

Department of Physics, North Texas State University, Denton, Texas, USA,

W. Zawadzki

Institute of Physics, Polish Academy of Sciences, Warsaw, Poland

The CO_2 laser-induced magnetophotoconductivity in high purity n-InSb exhibits resonant structure which depends upon the magnetic field, lattice temperature, laser intensity, photon energy, and light polarization. Three separate series of resonant structure are resolved and identified as arising from (a) three-LO phonon-assisted cyclotron resonance harmonic (LOCRH) transitions, (b) combined three-LOCRH transitions (CLOCRH) where combined resonance occurs, and (c) transitions originating from a mid-gap deep defect complex to Landau levels or impurity levels in the conduction band.

Figure 1 shows reproductions of the output of the lock-in detector (proportional to the second derivative of the signal) versus inverse magnetic field showing two sets of resonant structure which are periodic in inverse magnetic field. A constant dc electrical current I is applied to a sample while an ac megnetic field of 150 G modulateds the sample conductivity at a Frequency of 43 hz. The laser pulse, 20 μsec wide, produces a photoconductive signal which is processed by simultaneous use of both sampling oscilloscope and lock-in amplifier techniques [1]. All experiments were done in the Voigt configuration with either $\vec{e}|\vec{B}$ or $\vec{e}||\vec{B}$ and with $\vec{I}||\vec{B}||<111>$.

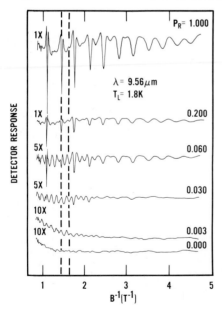

Fig.1 Photoconductive spectra for different relative peak incident CO_2 laser powers P_R and $\vec{e}||\vec{B}$. For $P_R = 1$, the peak incident power is ~1.4 W. Note the gain settings on the left

---

*Work supported in part by the Office of Naval Research and a Faculty Research Grant from North Texas State University.

At low enough laser powers and throughout the spectral range of the CO_2 laser (9.2-10.8μm), spin-conserving and combined (spin-flip) phonon-assisted resonances are observed for both $\vec{e} \parallel \vec{B}$ and $\vec{e} \perp \vec{B}$ polarizations. Figure 2 shows these resonances for the case $\vec{e} \parallel \vec{B}$ and for fields up to 2 T at several photon energies. The envelopes of the observed amplitudes seen in Figs.1 and 2 are a result of the magnetic field modulation method. The actual amplitudes of the resonant structure increase with magnetic field. This fact, along with the periodicity in inverse magnetic field, suggests some type of resonance behavior involving Landau levels. Previous studies [2,3] at lower fields (<1.2 T) did not show the doublet that is clearly resolved at the higher fields. The observed field positions of the resistance minima are plotted in Fig.3 as a function of magnetic field for nine photon energies. To understand this resonance structure, we start with the following resonance condition

$$\hbar\omega = \varepsilon_f - \varepsilon_i + m\hbar\omega_{LO} + \Delta \quad , \tag{1}$$

where the electron in an initial state ε_i absorbs a photon of energy $\hbar\omega$ while successively emitting m LO phonons of energy $\hbar\omega_{LO}$ to arrive at some higher Landau level ε_f. The polar interaction couples electrons only to LO phonons, while the deformation potential interaction with optic phonons is much weaker

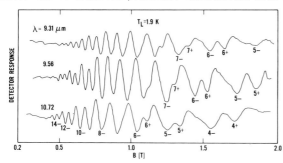

Fig.2 Photoconductive spectra obtained for small laser powers (~30 mW) and $\vec{e} \parallel \vec{B}$. The final state Landau level numbers are assigned from the results obtained in Fig.3.

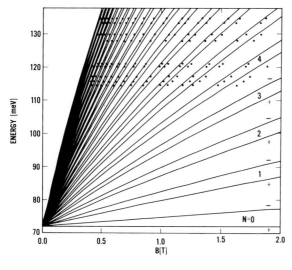

Fig.3 Intraband transition energies versus magnetic field. Theoretical transitions calculated from Eq. (1) are shown as lines

although nonvanishing away from the band edge. The term Δ takes into account possible shifts in resonance positions due to polaron and broadening effects [4]. Landau levels with spin up (+) and spin down (-) will be denoted by $\varepsilon_N^{\pm}$. Several mechanisms must now be considered as possible origins of the observed resonances.

The one-phonon-assisted combined resonance case has already been observed and theoretically discussed [5,6]. In general, we thus expect logarithmic singularities of magneto-absorption for $\vec{e} \parallel \vec{B}$ with $\varepsilon_i = \varepsilon_0^+$ and $\varepsilon_f = \varepsilon_{\bar{N}}^-$ in Eq. (1), with N = 0,1,2,3,... The lines in Fig.3 represent our theoretical calculations based on Eq. (1) with Δ = 0 and using the Landau level model of JOHNSON and DICKEY [7] along with the band parameters g = -51.3 and E_g = 236.7 meV to calculate $\varepsilon_{\bar{N}}^-$ and ε_0^+. Also shown in Fig.3 are the theoretical lines for spin-conserving three-LOCRH transitions seen at higher fields. These can be theoretically described by Eq. (1) if $\varepsilon_f = \varepsilon_N^+$. The lines give a reasonable fit to the data when m = 3, $\hbar\omega_{LO}$ = 24.0 meV and m* = 0.0140 m_0. This value of $\hbar\omega_{LO}$ is lower than the value of 24.4 meV obtained by others, but certainly would be influenced by a finite value of Δ. Alternatively, calculations show that the resonance positions of the high field part of the doublet can also be explained by spin-conserving transitions originating from the ε_0^- Landau level. This upper spin level is populated because of the large current used to enhance the signal-to-noise ratio in our experiments. The large current also rules out transitions from an initial state of an impurity level. Further studies are planned which should enable us to decide the origin of the structure.

Additional studies with $\vec{e} \perp \vec{B}$ show the same resonant structure (positions and amplitudes) as for $\vec{e} \parallel \vec{B}$ given here. This is somewhat surprising, since spin-flip one-phonon-assisted resonances have been predicted [5,6] and observed to be stronger for $\vec{e} \parallel \vec{B}$ than for the $\vec{e} \perp \vec{B}$ polarization. WEILER et al. [8] show that for $\vec{e} \parallel \vec{B}$ combined resonance transitions are allowed, whereas for $\vec{e} \perp \vec{B}$ only the cyclotron resonance transition is allowed for the σ_L polarization component. The slight warping of the conduction band induces a weak transition with ΔN = 3 for $\vec{e} \parallel \vec{B}$ for $\vec{B} \parallel$ <111> [8]. This transition also does not involve k_z and the resulting magneto-absorption will be resonant. However, for $\vec{e} \perp \vec{B}$ a much stronger absorption is possible because of the strong cyclotron resonance transitions with ΔN = 1. Resolution of this conflict requires further investigations that are currently in progress.

At higher laser powers and at photon energies greater than 118 meV a series of sharply defined resonant resistivity minima dominates the photoconductive spectra. Previously we attributed this series to multiple LO phonon emission from the high energy photo-excited electrons [9,10]. However, a two-step process in which an electron is first excited by a photon and then emits 5 optical phonons would require, for its resonant character, separate resonance conditions for the photon absorption and the phonon emission. This would occur only for special values of magnetic field. Thus, we rule out this possible interpretation. Two-photon magneto-absorption processes should be ruled out because of the low intensities used (1-100 W/cm^2) in observing the structure and the linear dependence of the photoconductive amplitudes on intensity. There are no previous two-photon experiments on InSb using a cw CO_2 laser. Recently, we have obtained similar resonant structure at much higher laser powers that are available from a rotating mirror Q-switched laser. No additional structure was observed at intensities up to several hundred kW/cm^2. Consequently, we think that it is highly unlikely (but not entirely improbable) that this resonant structure can be attributed to two-photon processes.

Conclusive proof will come from a detailed comparison of the data with two-photon magneto-absorption calculations.

It is also possible for the structure to arise from transitions originating from a deep level 118 meV below the conduction band edge [2]. However, the nature of the final or initial states involved in the resonant absorption transitions is not at all clear. Several possibilities exist individually or in some combination: (a) conduction band Landau levels, (b) shallow impurities, (c) "exciton-like" states, and (d) excited states of the deep level itself. We report here for the first time that the light polarization strongly affects the spectra as seen in Fig.4 where more structure and a larger amplitude is observed for $\vec{e} \perp \vec{B}$. It appears that the final states must involve more than just the conduction band Landau levels. Further studies are in progress in which we hope to be able to elucidate the nature of these resonances.

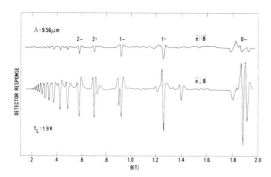

Fig.4 Polarization dependence of the photoconductive spectra at high laser powers (~1 W). The photon energy and magnetic field dependence of the resonance data for $\vec{e} \parallel \vec{B}$ can be fit with a model of a field independent initial state deep in the band gap and conduction band final states as denoted in the figure.

We thank M. H. WEILER for helpful comments.

1. H. Kahlert and D. G. Seiler, Rev. Sci. Instrum. 48, 1017 (1977).
2. D. G. Seiler, M. W. Goodwin, and A. Miller, Phys. Rev. Lett. 44, 807 (1980).
3. M. W. Goodwin, D. G. Seiler, and D. H. Kobe, Solid State Commun. 33, 489 (1980).
4. S. Morita, S. Takano, and H. Kawamura, J. Phys. Soc., Japan 39, 1040 (1975).
5. W. Zawadzki, R. Grisar, H. Wachernig, and G. Bauer, Solid State Commun. 25, 775 (1978).
6. W. Zawadzki, in Proc. of the 13th Int. Conf. on the Physics of Semiconductors, Rome, 1976, edited by F. G. Fumi, (Tipografia Marves, Rome, 1976), p. 1261.
7. E. J. Johnson and D. H. Dickey, Phys. Rev. B 1, 2676 (1970).
8. M. H. Weiler, R. L. Aggarwal, and B. Lax, Phys. Rev. B 17, 3269 (1978).
9. D. G. Seiler and M. W. Goodwin, Proc. Intern. Summer School on Magneto-Optical Properties of Solids, 1979, Ed. J. Devreese, Plenum Press, in print.
10. D. G. Seiler, M. W. Goodwin, and K. Ngai, Optics Commun. 31, 340 (1979).

An Infrared Bolometer for Use in High Magnetic Fields

J.P. Kotthaus and C. Gaus

Institut für Angewandte Physik, Universität Hamburg, Jungiusstr. 11
2000 Hamburg 36, Fed. Rep. of Germany, and

P. Stallhofer

Physik-Department, Technische Universität München
8046 Garching, Fed. Rep. of Germany

A new, sensitive infrared bolometer has been developed that is especially suited for spectroscopy in the presence of high magnetic fields. The bolometer is made from carbon-glass [1], a material successfully used for thermometry in high magnetic fields [2]. The bolometer operates at liquid helium temperatures and exhibits a responsivity comparable to conventional germanium bolometers [3]. Contrary to Ge bolometers it is very insensitive to strong magnetic fields. Therefore it has been found extremely useful for magneto-optical experiments at far infrared frequencies.

As a starting material for the bolometer we use commercially available unencapsulated C-glass thermometers with a resistance at $T = 4.2$ K of $2k\Omega - 10k\Omega$ [4]. The C-glass material is ground to thin platelets, typically $4 \times 1.5 \times 0.2$ mm in size. As ohmic contacts thin gold wires ($\sim 50\mu$m) are ultrasonically soldered with In to the small edge of the platelet. The element is mounted on its lead wires, which also serve as thermal leak, into the center of an integrating cavity into which the infrared radiation enters via a small window (see Fig. 1). The thus obtained bolometer is a small, permanently sealed unit that can be attached to any helium cooled infrared transmission apparatus.

At liquid helium temperatures the element's resistance may be approximated by $R(T) = R_o(T_o/T)^A$ with A increasing from $A \simeq 4.5$ at $T_o = 4.2$ K to $A \simeq 7$ at $T_o = 2$ K. At a bath temperature of $T_o = 4.2$ K a typical bolometer has $R_o \simeq 25k\Omega$ and a time constant $\tau \simeq 10$ msec. The dependence of the low frequency responsivity S on the bias current I, as determined from the dc load curve [1], is shown in Fig. 2. At $T_o = 4,2$ K the maximum responsivity is typically $S_{max} \simeq 5 \times 10^3$ V/W at $I \simeq 40$ µA. At $T_o = 2.1$K we find $R_o \simeq 900k\Omega$, $S_{max} \simeq 7 \times 10^4$ V/W at $I \simeq 3\mu$A and $\tau \simeq 50$ msec. The rms noise voltage, measured with phase sensitive detection at frequency $f \simeq 1/(2\pi\tau)$ and 1 Hz bandwidth is about 100 nV at 4.2 K and 500 nV at 2.1 K. Thus the noise equivalent power (NEP) is typically 10^{-11} W/$\sqrt{Hz}$, which is about two orders of magnitude higher than the theoretical values [1]. At present we cannot decide whether the excessive noise is caused by the C-glass material itself or by imperfect contacts. It should be pointed out that the NEP of commercially available Ge-bolometer [1] is significantly lower and close to the theoretically achievable values.

The attractive feature of the C-glass bolometer is the small variation of the responsivity with magnetic field. Fig. 3 shows

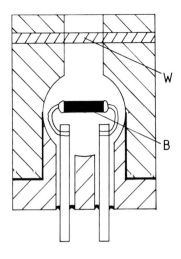

Fig.1: Sketch of the C-glass bolometer detector. The bolometer element (B) is mounted in the center of an integrating cavity, vacuum-sealed by a polyethylene-window (W)

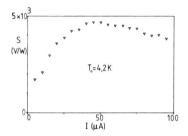

Fig.2: Low frequency responsivity S versus bias current I for a typical bolometer at bath temperature T_o.

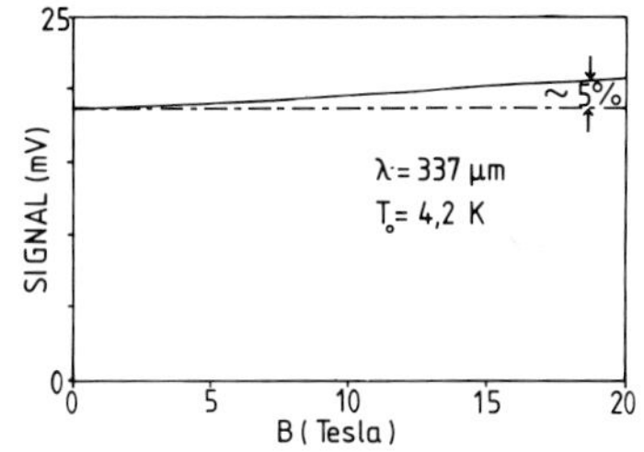

Fig.3: Bolometer signal versus magnetic field B at fixed infrared power of wavelength λ.

the detector signal versus magnetic field B with the incident far infrared laser power kept constant. The responsivity varies with B by less than 5% in magnetic fields up to 20 T. This makes the C-glass bolometer especially useful for spectroscopic investigations at infrared frequencies in the presence of strong magnetic fields, where Ge-bolometers cannot be easily employed.

We have succesfully used C-glass bolometers for a variety of transmission experiments on inversion layers on semiconductors in high magnetic fields and at radiation frequencies between 10 GHz and 10 THz [5]. A typical example are the cyclotron resonance traces of inversion layer electrons on Si (100) under uniaxial compression [6], as shown in Fig. 4. In these experiments the C-glass bolometer could be mounted very close to the samples under investigation. The same experiment had first been tried with a high quality commercial Ge-bolometer which had to be placed about 2 m away from the center of the Bitter coil magnet to sufficiently eliminate the effect of the stray magnetic field on the Ge-bolometer. Though the sensitivity of the Ge-bolometer

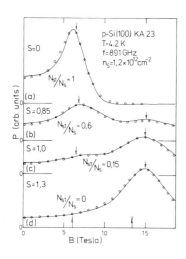

Fig.4: Cyclotron resonance absorption P of inversion layer electrons on Si (100) versus magnetic field B under uniaxial compression S (in kbar) as measured with a C-glass bolometer (from [6]).

had been significantly larger than the one of the C-glass bolometer, the resulting signal to noise ratio was much lower than the one in Fig. 4, because of the considerable loss of radiation intensity between sample and Ge-bolometer.

The above data demonstrate that the C-glass bolometer is very well suited for sensitive detection of infrared radiation in the presence of strong magnetic fields. As of now we have only used the C-glass bolometer at a few discrete frequencies and have not yet measured its spectral response at infrared frequencies. The construction of a composite bolometer [7], in which the C-glass element only serves as thermometer, is under way.

Acknowledgements: We wish to thank the High Field Laboratory of the Max-Planck-Institut für Festkörperforschung in Grenoble for the hospitality extended to two of us during part of the measurements.

References:

1. W. N. Lawless, Rev. Sci. Instrum. 43 (1974) 1743
2. J. M. Swartz, J. R. Gaines and L. G. Rubin, Rev. Sci. Instrum. 46 (1975) 1177
3. F. J. Low, J. Opt. Soc. Am. 51 (1961) 1300
4. Type CGR from Lakeshore Cryotronics, Inc., Columbus, Ohio, U. S. A.
5. see e.g. J. P. Kotthaus, Surface Science 73 (1978) 472
6. P. Stallhofer, G. Abstreiter and J. P. Kotthaus, Solid State Commun. 32 (1979) 655
7. see e.g. N. S. Nishioka, P. L. Richards, and D. P. Woody, Appl. Optics 10 (1978) 1562

Part IV

Impurity States in High Magnetic Fields

D⁻ Centers in Semiconductors in High Magnetic Fields

D.M. Larsen

Francis Bitter National Magnet Laboratory [†],
Massachusetts Institute of Technology
Cambridge, MA 02139, USA

1. Introduction

This paper describes variational studies of bound states of the D⁻ ion, a shallow donor with an extra electron attached, in the presence of a uniform static magnetic field. Of special interest is the calculation of the binding energy of a given state of the D⁻, by which is here meant the least amount of energy required to remove one of the electrons from the D⁻ ion in that state without flipping a spin. Taking away an electron converts the D⁻ to a neutral donor plus a distant electron. For the purposes of our calculations the donor is assumed left in its ground state in the magnetic field, the distant electron in its lowest Landau level.

The first part of this paper treats the case of D⁻ in a single-valley isotropic conduction band, for which the problem is completely analogous to that of the H⁻ ion. Only one bound state exists for the H⁻ ion at zero magnetic field, but there are infinitely many in a finite magnetic field, no matter how small. These latter are very different physically from the state bound at zero field and will be discussed in a separate section. In the second part of the paper the zero-field D⁻ ground state in Si is discussed.

We begin our discussion of the D⁻ ion in simple conduction bands by classifying its various possible states. These states have orbital wave functions which are either symmetric upon interchange of electron coordinates (singlet states) or anti-symmetric (triplet states). The magnetic field, $\vec{B}$, will be taken along the z direction and the D⁻ states will be labeled by M_L, the component of total orbital angular momentum along z. M_L can take on any integer value. Just like the Hamiltonian for the H-atom or the donor in a magnetic field, the D⁻ Hamiltonian is invariant under reflection through a plane perpendicular to $\vec{B}$ and passing through the positive charge center. This means that we should attach a z-parity quantum number, ±1, to the wave functions according to whether they are even or odd upon the above-mentioned reflection. Since no bound states of odd z-parity have been found, we suppress this quantum number and assume it to be +1 for all states considered.

[†]Supported by the National Science Foundation.

For the donor atom in a magnetic field we use the Hamiltonian

$$H(1) = -\nabla_1^2 - 2/r_1 + \gamma^2 \rho_1^2/4, \tag{1}$$

and for the D^- ion, the Hamiltonian

$$H = H(1) + H(2) + 2/r_{12} + \gamma M_L, \tag{2}$$

where $\vec{r}_i$ is the displacement of electron i from the proton, $r_i^2 = \rho_i^2 + z_i^2$, $r_{12} = |\vec{r}_1 - \vec{r}_2|$, and

$$\gamma = \hbar\omega_c/2R, \quad \omega_c = eB/m^*c, \quad R = m^*e^4/2\varepsilon_0^2\hbar. \tag{3}$$

We have ignored the Zeeman spin energy (which doesn't affect our binding energies) and have taken all lengths in units of the hydrogenic Bohr radius, a_0, ($a_0 = \hbar^2\varepsilon_0/m^*e^2$). The Hamiltonians (1) and (2) have been divided through by the hydrogenic Rydberg, R.

If we denote the ground state eigenvalue of (1) by $E_{Hy}(\gamma)$ then, since γ is the lowest energy possible for the free electron in the dimensionless magnetic field γ, the binding energy, $E_B(S,\gamma)$, of the state S of the H^- or D^- ion is given by

$$E_B(S,\gamma) = E_{Hy}(\gamma) + \gamma - E_{H^-}(S,\gamma) \tag{4}$$

where $E_{H^-}(S,\gamma)$ is the eigenvalue of (2) corresponding to state S. State S is bound if $E_B(S,\gamma)$ is positive. We shall use (4) to calculate binding energies for various states S, but since neither E_{Hy} nor E_{H^-} are known exactly we shall insert approximate values for these quantities. For E_{Hy} we use the most accurate values known to us [1], for E_{H^-}, our calculated variational values. Since E_{Hy} will be known in general to much higher accuracy than E_{H^-} and since the variational estimates of E_{H^-} are always higher than the true value, our calculated binding energies using (4) will be lower bounds to the true binding energy. (Thus, in comparing two binding energies obtained via (4) from two variational calculations of $E_{H^-}(S,\gamma)$ the better calculation will be the one giving the higher binding energy.) Negative binding energies correspond to unbound states.

Most previous quantitative calculations have focused upon two states, the $M_L=0$ singlet level which evolves from the ^1S zero-field H^- state when the magnetic field is adiabatically turned on, and the $M_L=-1$ triplet state, which does not bind at zero field but somehow becomes bound in the presence of a magnetic field. For calculating the binding energies of these states a number of authors [2-5] have used variants of the trial function

$$e^{iM_L\phi_1} \rho_1^{|M_L|} e^{-1/4(\gamma_1\rho_1^2 + \gamma_2\rho_2^2)} f(z_1)g(z_2) \pm (1\leftrightarrow 2). \tag{5}$$

These wave functions are simple and physically transparent; however, they are not at all well suited to the low and intermediate field region ($\gamma \lesssim 5$) and, even at very high fields, they do not appear to give very accurate binding energies. Calculations using linear combinations of large numbers of Slater orbitals have been reported [6]. These appear to be excellent for the $M_L=0$ singlet state in "weak" fields ($\gamma \ll 1$) but are considerably less good for the $M_L = -1$ triplet level. Finally, by means of the trial function $\exp(-r_2) \exp(iM_L\phi_1)\rho_1^{|M_L|}\exp(-\gamma\rho_1^2/4)$ it has been shown [7] that for every $M_L < 0$ there exists at least one bound D⁻ state for γ sufficiently small. In the work to be described [1], relatively simple trial functions are introduced which give energies over a wide range of fields nearly as good as or better than any previously obtained and which give physical insight into the structure of the D⁻ ion in a magnetic field.

2. The $M_L=0$ Singlet Ground State
2.1 Low to Moderate Magnetic Field Strengths

Perhaps the simplest of the accurate trial functions proposed for this state, which is the only zero-field bound state of D⁻, in the absence of field is the Chandrasekhar wave function [8], ψ_{CH}, given by

$$\psi_{CH} = \left(1 + C|\vec{r}_1 - \vec{r}_2|\right)\left(e^{-r_1/a_1 - r_2/a_2} + (1 \leftrightarrow 2)\right). \tag{6}$$

Here the electrons are in ground-state hydrogenic orbitals; they are encouraged to stay apart, keeping down their repulsive interaction energy, by the correlation factor $1 + Cr_{12}$, which becomes relatively small as $r_{12} \to 0$. The best values of a_1, a_2 and C are 0.930, 2.092, and 0.312, respectively. With these values the binding energy obtained is 0.0518 R as opposed to the "exact" value of 0.0555 R. Note that the inner orbital in (6), $\exp(-r/a_1)$, is nearly the same as the ground state donor wave function, $\exp(-r)$.

To generalize (6) for the magnetic field case one might well replace the hydrogenic 1S orbitals there by some convenient orbitals appropriate to the H-atom in a magnetic field. However, a simpler approach, and the one adopted here, is to take trial functions of the form

$$\psi^{(k)} = \sum_{n=0}^{k} c_n (z_1^{2n} + z_2^{2n})\, e^{-\delta(r_1^2 + r_2^2)}\, \psi_{CH}. \tag{7}$$

It is well known that in strong magnetic fields the donor 1S state shrinks and becomes roughly ellipsoidal in shape with the major axis of the ellipsoid along the magnetic field. The trial function (7), with δ and the c_n's variational parameters, allows the D⁻ wave function likewise to shrink [via the factor $\exp -\delta(r_1^2+r_2^2)$] and to elongate along the field (via the terms in z_1 and z_2). By calculating the binding energy vs. field curves for various k values, it is possible to extrapolate the binding energy to $k \to \infty$. Fig. 1 compares the binding energies $E_S^{(2)}$ (from (7) with k=2), the extrapolation to large k, the results of [5], and energies obtained from the high-field ansatz (8). A more complete discussion is given in [1].

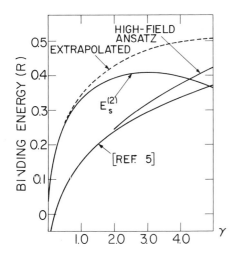

Fig. 1 Comparison of binding energies for the $M_L=0$ singlet ground state of the D^- ion from various trial functions. Solid curves are rigorous upper bounds to the binding energy. $E_s^{(2)}$ comes from (7) with $k=2$, whereas the curve "high-field ansatz" is computed from the trial function (8) with $\kappa=\beta$.

2.2 High-Field Limit

At extremely high fields where the inner electron orbital becomes dominated by magnetic forces ($\gamma \gg 1$), the trial functions $\psi_s^{(k)}$ are not useful. The motion of both electrons in directions perpendicular to z is determined by the magnetic field, and H in (2) becomes approximately separable.

In this limit a trial function of the form

$$\left(1 + C \tanh^2(\varepsilon(z_1-z_2))\right) \left(\exp\left(-\kappa(z_1^2+\alpha^2)^{\frac{1}{2}} -\beta(z_2^2+\alpha^2)^{\frac{1}{2}}\right) + (1\leftrightarrow 2)\right) \exp(-\gamma'(\rho_1^2 + \rho_2^2)/4) \qquad (8)$$

which is similar in concept to the Chandrasekhar wave function (6), is thought to be appropriate. The correlation is provided by the first factor, which goes to 1 when the electron separation in the z direction is small and to 1 + C when large. Although it is undoubtedly possible to improve upon the correlation factor of (8), spherically symmetric correlation, that is, correlation depending upon r_{12}, which is appropriate at zero field, does not seem suitable in the high field limit because such correlation admixes high energy components (energies of order γ) into the x-y motion of the electrons. Put another way, the wave function is expected to be stiff to perturbations in the x-y motion but relatively loose in its z-motion at very large γ.

The distinction between inner and outer orbitals disappears in the high field limit, at least for the trial function (8). Both κ and β when optimized assume a common value, 2.29 at $\gamma = 100$, representing a nearly 30% expansion of the wave function in the z-direction relative to its size when electron-electron repulsion is turned off. On the other hand, for $\gamma = 100$,

γ' in (8) is only 1% smaller with electron-electron repulsion acting than without, demonstrating strikingly the relative stiffness of the wave function to changes in the x-y motion at high fields.

Correlation is very important for obtaining good energies for $\gamma \gg 1$, just as it is at $\gamma = 0$. For $\gamma = 100$ the binding energy found employing (8) and optimizing all six parameters is 1.42 R (assuming 7.58 R for the H-atom binding in this field), but if C is set equal to zero (no correlation) and κ, α, β and α' are then optimized, the best binding energy obtained is only 1.12 R.

3. Bound States Induced by a Magnetic Field

The D⁻ ion has no zero-field bound states for $M_L \neq 0$. For small fields any induced binding would presumably be very weak and the outer electron should travel in an extended orbit around the central donor. The Coulomb field of this outer electron polarizes the donor atom inducing a dipole moment $\vec{p}$, which in turn, produces at the electron an attractive potential of strength $\propto p/r^{-2}$. But, since p itself is proportional to the electric field of the distant electron evaluated at the center of the donor, $(p \propto r^{-2})$ the attractive potential at large distances varies like r^{-4}. The combination of this attractive field and the two-dimensional confinement of the outer electron in the magnetic field produces binding.

The physical picture just presented should apply to all bound states with $M_L \neq 0$ in weak magnetic fields. To give this picture quantitative embodiment in a variational calculation we introduce a trial function of the form

$$\Phi_{M_L}(r_1) \chi(\vec{r}_1, \vec{r}_2) \pm (1 \leftrightarrow 2) \tag{9}$$

where $\Phi_{M_L}(r_1)$ is the orbital for the outer electron and $\chi(\vec{r}_1, \vec{r}_2)$, for large r_1, approximates the ground state wave function in a magnetic field of a hydrogen atom which is polarized by an electron at $\vec{r}_1$. We have chosen χ in the form

$$\chi(r_1, r_2) = (1 + c_1 z_2^2 + P(r_1, r_2) \cos\theta_{12}) e^{-\delta r_2^2 - \kappa r_2} \tag{10}$$

where the polarization coefficient, P, is taken as [1]

$$P(r_1, r_2) = C(r_2^2/2 + r_2)/\left((r_1^2 + b)(\beta r_2^2 + 1)\right).$$

The variational parameters c_1 and δ describe the effect of the magnetic field on the inner orbital while C and b control the strength and cut-off distance, respectively, of the polarization; θ_{12} is the angle between $\vec{r}_1$ and $\vec{r}_2$. A weakness in this choice of χ which may be important at higher fields is that variation of the polarizability of the inner donor with the angle between $\vec{r}_1$ and the magnetic field is not allowed for.

For the outer orbital we have tried to obtain considerable flexibility. Recalling that the lowest lying free particle Landau level wave function of given M_L is $\rho^{|M_L|} \exp(iM_L\phi - \gamma\rho^2/4)$, we have taken

$$\Phi_{M_L}(r_1) = \rho_1^{|M_L|} \exp(iM_L\phi_1 - \gamma'\rho_1^2/4 - \kappa(\rho_1^2 + \alpha z_1^2)^{\frac{1}{2}}) \quad (11)$$

where γ', κ and α are variational parameters and ϕ_1 is the polar angle of $\vec{r}_1$ in circular cylindrical coordinates. The variational calculation shows that Φ_{M_L} is very "Landau-like," with γ' nearly equal to γ at all fields. Binding energy results for $M_L = -1, -2$ and -3 are given in Figs. 2 and 3. Although

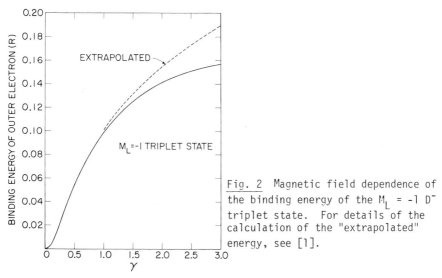

Fig. 2 Magnetic field dependence of the binding energy of the $M_L = -1$ D$^-$ triplet state. For details of the calculation of the "extrapolated" energy, see [1].

arguments have been given [9] which indicate that as $\gamma \to 0$ there exists a bound singlet state at $M_L = -1$, we have not found this state in our computer calculations in the range $0.05 \leq \gamma \leq 3$. This suggests that if such a state exists at $\gamma > 0.05$, its binding energy must remain very small.

The general features displayed in Figs. 2 and 3 have a simple physical interpretation. When the outer orbital keeps the electron far away from the inner atom the binding is relatively weak (due to the rapid diminution of the strength of the attractive potential with distance from the atom) and singlet-triplet splittings, which depend upon overlap of inner and outer orbitals, are small. Since the outer orbital penetrates least at low γ and at high $|M_L|$ we expect in that regime to find weakest binding and smallest singlet-triplet splitting. It is well known that electron-electron repulsion tends to be weaker in anti-symmetric wave functions, which vanish when the electrons come together, than in symmetric states. Thus triplet states have stronger binding than the corresponding singlets (but only for $|M_L| > 0$).

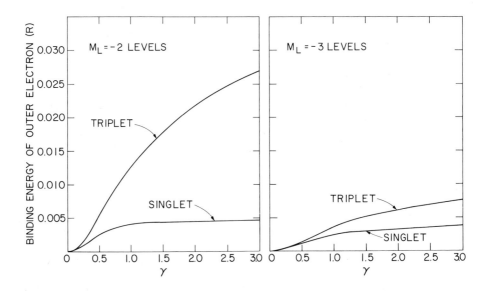

Fig. 3 Estimated D⁻ binding energies of lowest M_L = -2 and -3 levels of singlet or triplet symmetry vs. γ.

The M_L = -1 level is the most penetrating of all the $|M_L| > 0$ states at any field. There, our calculations indicate, the singlet-triplet splitting exceeds or is very nearly equal to the triplet binding energy at all fields investigated, and the singlet fails to bind (or binds only very weakly).

The situation is very different for the M_L = 0 state already described where exchange plays a dominant role in the binding by causing a large increase of charge density near the positive center in the singlet state.

Notice that there is a major qualitative difference at low B between the binding energy curves of Fig. 1 and those of Figs. 2 and 3. The weak-field binding energy of any state with M_L = 0 which is bound at zero field increases like B, because the diamagnetic energy of a bound state is proportional to B^2 at low fields whereas the free-electron energy is proportional to B. On the other hand, states which are bound only in the presence of magnetic field have a Landau-like outer orbital (see (11)), and therefore a diamagnetic energy which increases linearly with B, just like a free particle. Thus, for these states we expect in the limit B → 0 that $\partial E_B(S,\gamma)/\partial \gamma \to 0$. For both kinds of states at very high magnetic fields the orbitals become Landau-like and the binding energy increases relatively slowly with B.

4. Singlet Ground State of D⁻ in Si

Because Si has a multivalley conduction band, anisotropic effective mass associated with each valley and significant valley orbit (VO) interactions in its donor states, D^- centers in Si are much more complicated than the simple D^- ions just discussed. The donor ground state wave function in the effective mass approximation (EMA) can be written $\Sigma \tilde{F}_i(r)u_i(r)$ for substitutional donors, where u_i is the Bloch function for valley i and, for appropriate parameters K, α, and δ, $\tilde{F}_i \cong \exp\left(-K(\rho_i^2 + \alpha z_i^2)^{1/2} - \delta z_i^2\right)$.

Because the substitutional donors have valley orbit splittings large compared to D^- binding energies in Si, we shall use the multivalley donor ground state above as a general form for the inner orbital, ψ, of the D^-; this is also suggested by (6) where the optimal inner orbital found, $\exp(-r/0.93)$, is nearly the same as the donor ground state wave function. However, it is too difficult to use $\tilde{F}_i(r)$ above in a correlated D^- wave function; we employ instead the less accurate but more tractable expanded form $F_i(r) = (1 + Dvz^2)\exp(-Kr)$, where $v = 1/r + \beta$. For the outer orbital, $\psi'(r)$, we try $\Sigma \beta_i F_i'(r)u_i(r)$ where F_i' has the same form as F_i but with different parameters, D' and K'. Our D^- trial function becomes, by analogy with (6)

$$[\psi(r_1)\psi'(r_2) + \psi(r_2)\psi'(r_1)] (1 + Cr_{12}) \tag{12}$$

with K, K', D, D', β, β_i, and C variational parameters. When (12) is expanded in $u_i(r_1)u_j(r_2)$, one finds that all terms appearing can be written in the form

$$\left(F_i'(r_1)F_j(r_2) \pm F_i(r_1)F_j'(r_2)\right)u_i(r_1)u_j(r_2) (1 + Cr_{12}). \tag{13}$$

The energy of such terms in the EMA can be calculated by taking their expectation value in the Hamiltonian $H_{ij} = H_i(r_1) + H_j(r_2) + 2/r_{12}$, where H_i is the EMA donor Hamiltonian for valley i. Terms with the + sign (bonding terms) are relatively low in energy and divide into two types: the intravalley type (valleys i and j are the same or have the same z-axes) and the intervalley type (valleys i and j have different z-axes). Terms with the - sign (antibonding terms) are significantly higher in energy and can be avoided in (12) by an appropriate choice of the β_i coefficients (i.e. $\beta_j = 1$). Note that due to mass anisotropy, terms of the intervalley type are slightly lower in energy than those of the intravalley type, because the electron-electron repulsion is stronger for the latter type [10]. Pure states of the form (13) but with correlation neglected (C set equal to zero) have been carefully studied [10]. It has been found [10] that the binding energy of the intervalley configuration, ε_{ij}, is ~36% greater than ε_{ii}, the binding energy of the intravalley configuration for Si. Our estimates of the effect of correlation reduce this difference to only about 10%.

We have estimated the correlation effect in inter- and intravalley bonding configurations by twice minimizing the energy of the trial function

$$\left\{\left[1 + D'v(r_1)z_{1i}^2 + Dv(r_2)z_{2j}^2\right]e^{-Kr_1 - K'r_2} + \binom{1\leftrightarrow 2}{i\leftrightarrow j}\right\} (1 + Cr_{12}) \tag{14}$$

in H_{ij}, first varying all parameters including C, then varying all parameters with C = 0. In (14) z_{ij} denotes the z coordinate of ith electron in valley j. The correlation corrections found are 0.0347 and 0.0287 for the intra- and intervalley cases, respectively. Using the results of [10] adjusted for the case $m_\perp/m_{||}$ = 0.2079 [11] we obtain ε_{ii} = 0.0752 and ε_{ij} = 0.0831, corresponding, respectively, to 1.50 and 1.66 meV. If we now assume that the energy of the inner electron is lowered due to the short range central cell potential by precisely as much as the donor energy is lowered by that same potential and that this potential has no effect on the outer orbital, which is only weakly penetrating, then to a good approximation the D⁻ binding energy is $(\varepsilon_{ii} + 2\varepsilon_{ij})/3$ or 1.61 meV, in satisfactory agreement with the measured value of 1.7 meV in Si:P [12].

Up to this point we have neglected, in our choice of trial function, the tendency for electrons to avoid the intravalley configuration in favor of the intervalley. For D⁻ centers associated with the interstitial donor Li this kind of correlation can be quite important because of the inverted VO splitting of the Li donor [13]. As an example, the valley-correlated trial function

$$\frac{\left\{\left[F_1(r_1)u_1(r_1) - F_2(r_1)u_2(r_2)\right]\left[F_3'(r_2)u_3(r_2) - F_4'(r_2)u_4(r_2)\right] + \left[F_1'(r_1)u_1(r_1) - F_2'(r_1)u_2(r_1)\right]\left[F_3(r_2)u_3(r_2) - F_4(r_2)u_4(r_2)\right] + (1\leftrightarrow2)\right\}}{(1 + Cr_{12})}$$

gives, in our model, a binding energy equal to ε_{ij}. This corresponds to stronger binding than could be achieved with any trial function lacking valley correlation.

Finally, we note that the binding of D⁻ in a magnetic field can be calculated in much the way indicated for the zero-field case if gauges for the problem are properly chosen. A report on such calculations will be forthcoming shortly.

References

1. D.M. Larsen: Phys. Rev. B20, 5217 (1979).

2. G. Benford, N. Rostoker: Solid State Commun. 6, 705 (1968).

3. R.O. Mueller, A.R.P. Rau, L. Spruch: Phys. Rev. A11, 789 (1975).

4. J. Virtamo: J. Phys. B9, 751 (1976).

5. A. Natori, H. Kamimura: J. Phys. Soc. Jpn. 44, 1216 (1978).

6. R.J.W. Henry, R.F. O'Connell, E.R. Smith, G. Chanmugam, A.K. Rajagopal: Phys. Rev. D9, 329 (1974).

7. J. Avron, I. Herbst, B. Simon: Phys. Rev. Lett. 39, 1068 (1977).

8. H.A. Bethe, E.E. Salpeter:"*Quantum Mechanics of One- and Two-Electron Systems*", in Atoms I, Handbuch der Physik, Vol. 35, ed. by S. Flugge (Springer, Berlin, Göttingen, Heidelberg 1957)

9. B. Simon: private communication.

10. A. Natori, H Kamimura: J. Phys. Soc. Jpn. 43, 1270 (1977).

11. R.A. Faulkner: Phys. Rev. 184, 713 (1969).

12. M. Taniguchi, S. Narita: Solid State Commun. 20, 131 (1976).

13. R.L. Aggarwal, P. Fisher, V. Mourzine, A.K. Ramdas: Phys. Rev. 138, A882 (1965).

High Magnetic Field Zeeman Splitting and Anisotropy of Cr-Related Photo-Luminescence in Semi-Insulating GaAs and GaP

L. Eaves

Department of Physics, University of Nottingham, University Park
Nottingham NG7 2RD, England

Th. Englert, and Ch. Uihlein

Max-Planck-Institut für Festkörperforschung
Hochfeldmagnetlabor Grenoble, B.P. 166 X
38042 Grenoble-Cedex, France

Abstract

The famous 0.84 eV photoluminescence structure arising from recombination at Cr-related defects in GaAs is studied using high resolution Zeeman spectroscopy with magnetic fields up to 10 T. The Zeeman anisotropy for B in the $(0\bar{1}1)$ and (001) planes is measured. From these results the defect is shown to have a dominant trigonal ((111)-axial) perturbation together with a weaker tetragonal Jahn-Teller distortion. The initial and final states of the photoluminescence are described in terms of a simple effective Hamiltonian which can explain all the essential features of the observed Zeeman splitting and anisotropy. In contrast with GaAs, the Zeeman data for the 1.03 eV photoluminescence observed in GaP(Cr) corresponds to defects with (100) axial symmetry. Posssible reasons for this difference are discussed.

1. Introduction

The properties of Cr in III-V semiconductors have attracted great interest in recent years. This is not only because Cr-doping of GaAs leads to the semi-insulating behaviour important for the production of microwave devices, but also because Cr has a host of interesting and complex properties that are amenable to investigation by optical and EPR spectroscopy.

In this paper we describe the results of a high resolution Zeeman spectroscopy study of the intense photoluminescence (PL) features arising from recombination at deep level defects incorporating Cr ions. (Measurements of this type were first reported in References [1] and [2].) In GaAs, this PL spectrum consists of a sharp, prominent multi-component zero phonon structure at 0.84 eV with broader phonon sidebands at lower energy. In GaP, the corresponding zero phonon structure occurs at 1.03 eV. A Cr-associated PL feature is also observed in InP, but this is rather broad and will not be discussed in this paper.

Our results for GaAs show that the PL arises from recombination at defects with C_{3V} (111) axial symmetry. Several different microscopic models for this type of axial defect have been proposed recently [3-6]. One of these models [4], namely that involving a dominant trigonal perturbation of the Cr ion by a nearest-neighbour impurity, together with a weaker tetragonal Jahn-Teller (J-T) distortion will be shown to fit well to our data when it is extended to include the effects of magnetic field.

The Zeeman spectra of the Cr-related PL in GaP also show a pronounced axial symmetry, but rather surprisingly the symmetry axis in this case is (100), in striking contrast with that found for GaAs. We will discuss possible reasons for this difference between the two materials. The piezo- and Zeeman spectroscopy measurements on GaP are described in more detail in the Proceedings of the Kyoto Conference [7].

2. Experimental Details and Material Characteristics

The PL spectra were obtained using a 1 m Jobin Yvon grating monochromator, a cooled intrinsic Ge photodetector and conventional lock-in techniques. The samples were immersed directly in superfluid helium at 2 K and illuminated with 40 to 80 mW of above band-gap laser light. Some improvement in signal band intensity is obtained if the excitation energy does not greatly exceed the band gap. Thus the 647.1 nm line of a Kr laser was used for GaAs and the 514.5 nm line of an Ar laser for GaP. Magnetic fields (B) up to 10 T were generated using the split-coil superconducting magnet at MPI, Grenoble. The samples could be rotated about a vertical axis so that the horizontal field could be swept through various crystalline planes. The details of the samples used in the study are given in Table 1. The two GaAs samples were chosen from a large number of crystals obtained from different sources because they showed the most intense 0.84 eV PL.

Table 1

MATERIAL	Cr CONC.	RESISTIVITY	COMMENTS
GaAs (Sumitomo)	$\sim 10^{17}$ cm^{-3}	$\sim 10^8$ Ω cm	Said to contain Cr and O.
GaAs (Czech)	$\sim 2.10^{17}$ cm^{-3}	$\sim 10^8$ Ω cm	Studied in APR [8].
GaP (MCP)	$\sim 4.10^{17}$ cm^{-3}	$> 10^8$ Ω cm	Studied in photocond. [1].

3. Results for GaAs(Cr)

Figure 1 shows the variation of the Zeeman spectrum with increasing field for B // (100) and B // (111). The spectra are simplest for B // (100) when all the (111) axes are equivalent. The most striking feature is the development of a strong doublet α_1, α_2 at fields above about 5 T whose separation becomes independent of field when these components are clearly resolvable in the high field regime. A similar but weaker doublet β_1, β_2 becomes split-off as the field increases and moves to lower energy with decreasing intensity. The Zeeman splittings for the four principal symmetry axes are plotted in Figure 2.

That the PL arises from recombination at (111)-axial complexes is shown most graphically by plotting the anisotropy of the Zeeman spectrum when a constant magnetic field is swept through one of the principal crystalline planes. The data points in Figures 3 and 4 show the results when B = 10 T is swept through the (0$\bar{1}$1) and (001) planes. Some minor differences are observed between the anisotropies for the Czech and Sumitomo material. (These can be seen by comparing the plot in Figure 3 for the Czech sample with that shown in Figure 2 of Reference [1] which gave the results for the Sumitomo material.) However, all the GaAs(Cr) samples we have studied show essentially the same anisotropies, although differences exist in the sharpness of the features in the spectra. The following features of the anisotropy in the (0$\bar{1}$1) plane and of the Zeeman plots are particularly important.

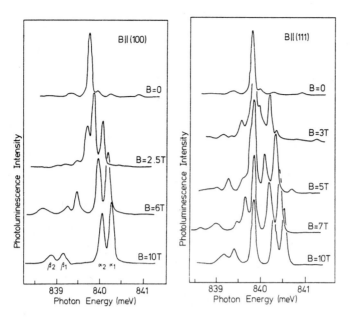

FIGURE 1 *The development of the Zeeman spectrum of the 0.84 eV PL of GaAs (Cr) for: (a) B // (100), (b) B // (111). The spectra were obtained at liquid helium temperatures.*

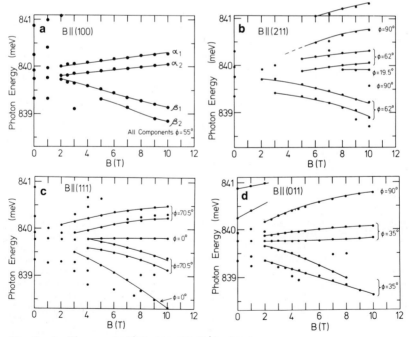

Fig.2a-d. Figure caption see opposite page

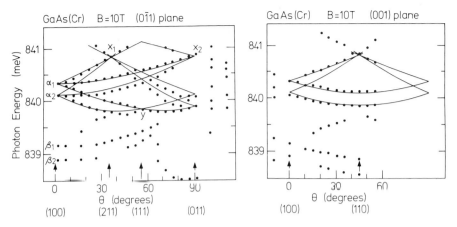

FIGURE 3 The Zeeman anisotropy of the 0.84 eV PL of GaAs(Cr) at 10 T with B rotated in the (0$\bar{1}$1) plane (Czech sample). The data points agree well with the predictions of a (111) axial defect model (solid lines).

FIGURE 4 The Zeeman anisotropy at 10 T with B rotated in the (001) plane (Sumitomo sample). The data points agree well with the predictions of a (111) axial defect model (solid lines).

(1) For B // (100) the spectrum has its simplest form and consists of the two doublets α_1, α_2 (strong) and β_1, β_2 (weak).

(2) As B is rotated away from the (100) direction in the (0$\bar{1}$1) plane each doublet component splits into three sub-components corresponding to axial defects which have now become magnetically non-equivalent. The defects with axes marked a, b, and c in Figure 5 correspond to the higher, central and lower energy sub-components respectively of the split doublets.

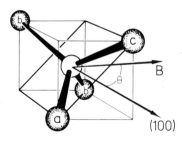

FIGURE 5 Schematic diagram showing the orientation of (111) axial defects as B is rotated in a (0$\bar{1}$1) plane.

FIGURE 2 The magnetic field dependence of the principal components of the 0.84 eV PL of GaAs(Cr) for B along the four principal symmetry axes. The various components correspond to magnetically non-equivalent (111) axial defects. ϕ denotes the angle between B and the axis of the defect.

(3) For B // (211) and (011), crossing points X_1 and X_2 occur in the anisotropy plot at the same energy.

In these configurations B is perpendicular in turn to some of the (111) axial defects, the two type (b) defects for B // (011) and type (a) defect for B // (211).

It will become clear later that whereas most of the crossings in the anisotropy plot are of a configurational type in which some of the different defect orientations become equivalent with respect to the magnetic field direction, those at X_1 and X_2 arise from the degeneracy of two different excited states of a given defect when the defect axis is perpendicular to B. The existence of these crossings is independent of the field strength.

(4) As B is rotated away from (100) the splitting between the two doublet components α_1 and α_2 decreases and is zero, or very close to zero, for B // (111) (crossing point Y). In this configuration the group (c) defects are parallel to the magnetic field.

The full set of anisotropy curves in Figures 3 and 4 are rather complex because they correspond to recombination at axial defects which are differently oriented to the magnetic field. A simpler description of the magnetic behaviour can be obtained by considering the variation of the recombination energy of a *single* axial defect as a function of the angle ϕ between its principal axis and the applied magnetic field. The E-ϕ plot deduced from the data of Figure 3 is shown in Figure 6 for the α lines. The large splitting α - β is essentially independent of ϕ whereas the smaller splitting α_1 - α_2 has zeroes at 0 and $\pi/2$. From Figure 5 the energy of recombination for all the defect orientations can be deduced by calculating the angles between the four non-equivalent defect axes and B. The solid lines in Figures 3 and 4 are the *reconstructed anisotropy plots* resulting from such a calculation and are in quite good agreement with the data points for the α lines.

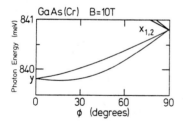

Figure 6 The variation of the recombination energy for the intense, higher energy components α_1, α_2 as a function of ϕ, the angle between B and the principal axis of the defect. The plot is deduced from the data points for GaAs (Cr). This bow-shaped variation can be explained using the energy level scheme in Figure 7. Transitions from the lower branch of the excited state to the split ground states will give the observed ϕ dependence.

(5) The Zeeman plots in Figure 2(a) for B // (100) which are the simplest to describe since all the (111) axes are equivalent in this configuration, show a pronounced non-linear dependence on magnetic field. In particular the energy difference α_1 - α_2 saturates at relatively low field whereas α - β increases linearly with B. Also the *centroid* energy of the α-doublet shows signs of saturation at high fields.

(6) The Zeeman plot in Figure 2(c) for B // (111) shows that the intense component with $\phi = 0$ does not shift with B, suggesting that it arises from a transition between states with the same M_S and g-factors.

All of these features can be used as a fine test of the various microscopic models discussed in the next section and can be fitted remarkably well to one involving a dominant trigonal term and a smaller J-T perturbation.

4. Recent Microscopic Models of the 0.84 eV PL in GaAs

The 0.84 eV PL in GaAs(Cr) was originally thought to arise from an internal transition (^5E-$^5T_2$) of the $Cr^{2+}(3d^4)$ ion substituted on a Ga-site [8], presumably following capture of a photo-excited carrier by neutral $Cr^{3+}(3d^3)$. However, when the fine structure in the zero phonon component was resolved [9,10], it was found that the resulting energy level scheme for the ground state did not agree with that deduced from EPR measurements [11,12] which were interpreted in terms of a Cr^{2+} ion substitutional on a Ga-site (Cr^{2+}_{Ga}), but suffering tetragonal J T distortion. This discrepancy led WHITE [3] to propose that the recombination giving rise to the PL occurred at an *isoelectronic* complex comprising Cr_{Ga} and a second impurity (donor) on a nearest-neighbour As-site. Such a defect would have trigonal (C_{3V}) symmetry and would give rise to the required large number of components in the initial and final states. WHITE favoured a model in which an exciton becomes bound to the complex in an initial state $(Cr^{2+}-D^+)(eh)$. The non-radiative recombination of the hole would give $(Cr^{3+}-D^+)e$ in which the electron bound in a hydrogenic orbit around the complex would finally recombine radiatively to give the original state $(Cr^{2+}-D^+)$ and the observed PL.

More recently, PICOLI et al. [4] have introduced a similar model, but with the important difference that they envisage the recombination as an internal transition of a ^5D ion, either $Cr^{2+}_{Ga}(3d^4)$ or interstitial $Cr^0(3d^6)$ in which the T_d symmetry is also lowered to C_{3V} by the presence of the nearest-neighbour impurity. In their model the ^5E excited state is described in terms of a trigonal axial crystal field, but the ground state requires a vibronic description in trigonal symmetry. Two other models involving a strong J-T distortion and a C_{3V} crystal field treated as a small perturbation of the J-T state have also been proposed recently [5,6]. Both models require a ground state with tetragonal J-T distortion. Our Zeeman anisotropy data appear to rule out such an interpretation for the 0.84 eV PL for GaAs(Cr), but they can be fitted to a model of the type proposed by PICOLI et al. when it is somewhat modified and extended to take account of the Zeeman term.

5. Our Effective Hamiltonian for the (111) Axial Defects: Comparison with Experiment

The non-linearities of the Zeeman plots in Figure 3 suggest a transition from a low-field to a high-field regime at fields around 4 T. At around 10 T we observe that the Zeeman splitting is large compared to the zero-field separation of the various lines. We therefore expect for the high-field regime that M_S is a fairly good quantum number for describing both the initial and the final states where the quantization axis is parallel to the direction of the magnetic field.

Our data are in best agreement with a model where the excited and ground states are both two-fold degenerate (Γ_3 in C_{3V}) excluding electron spin. The interaction of the two-fold degenerate states with the total electron spin S can be described in a very general way by an effective Hamiltonian of the following type in which the Pauli matrix σ describes the 2-fold orbital degeneracy

$$\mathcal{H}_{eff} = D(S_Z^2 - \tfrac{1}{3}S(S+1)) + \lambda'\sigma_Z S_Z + \mu_1[\sigma_-\{S_Z S_+\} + \sigma_+\{S_Z S_-\}]$$
$$+ \mu_2[\sigma_- S_- S_- + \sigma_+ S_+ S_+] + \mu_B \underline{B}\cdot\hat{g}\cdot\underline{S} \quad . \qquad (1)$$

In this equation, spin-orbit interactions are taken into account in first and second order. The Hamiltonian is essentially the type proposed by PICOLI et al. except that it has been modified to take correct account of time reversal symmetry in the μ_1 and μ_2 terms and that it contains the Zeeman term. Although a detailed analysis of the excited and ground states requires the diagonalization of a $2(2S + 1)$-dimension matrix, the required E-ϕ dependence can be obtained assuming a dominant λ' term in the excited state and a dominant μ_1 term in the ground state.

In this limit the Hamiltonian gives the following dependence on the anisotropy angle ϕ for the excited and ground states:

Excited State:
$$E_\pm \sim M_S [(g\mu_B B \cos\phi \pm \lambda')^2 + g^2\mu_B^2 B^2 \sin^2\phi]^{1/2} \qquad (2)$$

Ground State:
(High Field Limit)
$$E_\pm \sim \pm \frac{\mu_1}{4}(S(S+1) - 3M_S^2)\sin 2\phi + M_S g\mu_B B . \qquad (3)$$

Both sets of states and the observed transitions between them are shown in Figure 7. With $\phi = \pi/2$, equations (2) and (3) give the observed crossings at X_1 and X_2 in Figure 3 independent of the strength of magnetic field. Crossing Y corresponds to the zero splitting in the ground state when $\phi = 0$. The small ϕ-dependent ground state splitting gives the observed B-independent splittings $\alpha_1 - \alpha_2$ and $\beta_1 - \beta_2$ when $\phi \neq 0$. Our data require the excited and ground states to have the same g-value (~ 2) thus explaining why the α-line for $\phi = 0$ does not shift with B. This coincidence of g-values follows naturally from the internal transition model of Picoli et al.

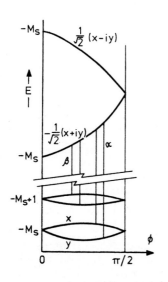

FIGURE 7 Schematic diagram of energy levels deduced from Equations (1), (2) and (3) which gives excellent agreement with the observed anisotropy of the 0.84 eV PL in GaAs(Cr). x and y are the basis functions of the Γ_3 irreducible representation in C_{3V} symmetry. The transitions giving rise to the α and β doublets are indicated. For comparison see Figure 6.

6. Results for GaP(Cr): Comparison with GaAs(Cr)

Our Zeeman and piezo-spectroscopy study of the intense 1.03 eV PL structure in GaP(Cr) is reported in more detail at the Kyoto Conference [7]. Although this PL is superficially rather similar to the 0.84 eV line in GaAs(Cr), the Zeeman anisotropy experiments reveal that it arises from recombination at Cr-related defects with (100)-axial symmetry. This is illustrated in Figure

8 which plots the variation in energy of the principal Zeeman components when B = 10 T is rotated in the (0$\bar{1}$1) plane. The crossing points and the simplification of the spectrum when B is parallel to a (111) axis corresponds to the case when all the defects are equivalently oriented with respect to the magnetic field.

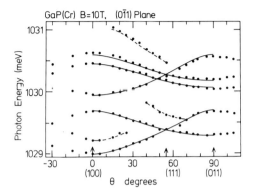
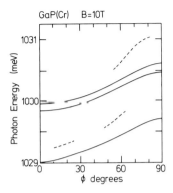

FIGURE 8 The Zeeman anisotropy at 10 T of the 1.03 eV PL of GaP(Cr) with B rotated in the (0$\bar{1}$1) plane. The data points agree well with the predictions of a simple (100) axial defect model (solid lines).

FIGURE 9 The variation of the recombination energy as a function of ϕ, the angle between B and the principal axis of the defect for GaP (Cr). The curves are deduced from the data points in Figure 8.

The sections of Figure 8 corresponding to recombination at the defects with their principal axes in the (0$\bar{1}$1) plane have been used to generate Figure 9, which plots the recombination energy for a given defect orientation as a function of the angle ϕ between B and the principal axis of the defect. The energies of recombination at the remaining defect orientations can then be deduced by calculating the angle between the defect axes and B. The solid lines in Figure 8 are the results of this procedure and are in excellent agreement with the data points.

The (111) axial symmetry for GaAs(Cr) is consistent with a model involving Cr substitutional on a Ga-site and a second impurity on the nearest-neighbour As-site. This is closely analogous to the Cd-O complex found in GaP [13]. Let us consider how a (100) axial defect might arise.

STAUSS et al. [14] have pointed out that (100) axial defects can only be produced by an aligned three-defect complex of the form (X-Cr-X). Such a complex seems extremely implausible. An interstitial defect would give rise to C_{2v} rather than the D_{2d} symmetry suggested by the Zeeman anisotropy data. It is known from EPR measurements that the ground state of substitutional Cr^{2+} in InP [14] and GaAs [11] exhibits a tetragonal J-T distortion and hence has the required axial symmetry to explain our Zeeman PL data. However, care is needed in comparing these types of experiment as they are performed in quite different magnetic field regimes. If a small orthorhombic component existed it could become suppressed in the high magnetic field regime used in the Zeeman measurements. Hence, even a (111) nearest-neighbour axial complex undergoing a strong tetragonal J-T distortion cannot be ruled out as a possible explanation of the Zeeman PL results in GaP(Cr). It is worth mentioning that our Zeeman data in GaAs(Cr) indicate the presence of an additional dynamic J-T distortion of the (111) axial complex.

7. Conclusion

We have shown that Zeeman PL spectroscopy is a powerful means of studying the properties of deep level defects involving paramagnetic ions such as Cr in III-V semiconductors. The method is particularly useful for obtaining information about defect symmetry and the energy level schemes for *both* the initial and final states of the PL transition.

Acknowledgements

One of us (LE) wishes to thank Professor G. Landwehr (MPI, Grenoble) for support and advice during his visits to Grenoble. We also wish to thank H. Krath for expert technical assistance.

References

1. L. Eaves, T. Englert, T. Instone, C. Uihlein and P. J. Williams, Proc. Semi-Insulating III-V Materials Conference (Ed. G. J. Rees, Published by Shiva) 1980.

2. N. Killoran, B. C. Cavanett and W. E. Hagston, Proc. Semi-Insulating III-V Materials Conference (Ed. G. J. Rees, Published by Shiva) 1980.

3. A. M. White, Solid State Commun., $\underline{32}$, 205, 1979.

4. G. Picoli, G. Deveaud and D. Galland, to be published, 1980.

5. C. A. Bates, A. S. Abhvani and S. P. Austen, Proc. Semi-Insulating III-V Materials Conference (Ed. G. J. Rees, Published by Shiva) 1980.

6. F. Voillot, J. Barrau, M. Brousseau and J. C. Brabant, to be published in Lettres au Journal de Physique, 1980.

7. L. Eaves, Th. Englert, Ch. Uihlein and P. J. Williams, Proc. 15th Int. Conference on Physics of Semiconductors, Kyoto, 1980.

8. W. H. Koschel, S. G. Bishop and B. D. McCombe, Proc. 13th Int. Conference on Physics of Semiconductors, Rome, 1065, 1976.

9. E. C. Lightowlers and C. M. Penchina, J. Phys. C: Solid State Physics, $\underline{11}$, L405, 1978.

10. E. C. Lightowlers, M. O. Henry and C. M. Penchina, Proc. 14th Int. Conference on Physics of Semiconductors, Edinburgh, 307, 1978.

11. J. J. Krebs and G. H. Stauss, Phys. Rev. B, $\underline{16}$, 971, 1977.

12. J. J. Krebs and G. H. Stauss, Phys. Rev. B, $\underline{20}$, 795, 1979.

13. C. H. Henry, P. J. Dean and J. D. Cuthbert, Phys. Rev., $\underline{166}$, 754, 1968.

14. G. H. Stauss, J. J. Krebs and R. L. Henry, Phys. Rev. B, $\underline{16}$, 974, 1977.

Impurity States of Tellurium in High Magnetic Fields

K. von Klitzing

Physikalisches Institut der Universität Würzburg
8700 Würzburg, Fed. Rep. of Germany, and

J. Tuchendler

Ecole Normale Superieure
75231 Paris, France

1. Introduction

From infrared magnetotransmission [1-7] and Shubnikov-de Haas [8-10] measurements on tellurium, detailed knowledge about the valence band structure has been obtained. All the experimental results are consistent with a k·p perturbation model, which leads to a camel-back structure of the valence band at the H-point of the Brillouin zone [11,12]. Such a double maximum of the valence band changes the energy spectrum of the acceptors in comparison with the levels of a hydrogenic impurity. Theoretical calculations show that the camel-back structure of the valence band leads to a splitting of the impurity ground state into a bonding and an antibonding state with binding energies of about 0.9 meV and 1.2 meV [13-15]. Without magnetic field the two valleys at the points H and H' in the Brillouin zone of tellurium do not cause a valley-orbit splitting of the impurity levels, since the invariance of the Hamiltonian under the time reversal operation (which transforms the points H and H' into each other) is not broken by the introduction of a Coulomb type impurity potential. However, in strong magnetic fields this degeneracy is removed, and each of the impurity states, especially the ground state, should split into two levels [16]. The absolute value of this splitting is influenced by central cell corrections, since the valley-orbit splitting depends on the impurity envelope function close to the impurity site. Therefore different energies for the impurity levels are expected for P-, As-, Sb-, and Bi-acceptors in tellurium. Up to now, the valley-orbit splitting and the chemical shift of the impurity levels have not been investigated in detail.

Experimentally, impurity lines have been observed in different absorption measurements [1,2,7]. However, in magnetotransmission experiments the impurity lines are mixed up with the intervalence band or cyclotron resonance transitions, and no measurements are possible close to the Reststrahlen region where the transmission becomes very small. Moreover, without magnetic field the impurity absorption lines disappear [17] at carrier concentrations below $4 \cdot 10^{20}$ m^{-3} but are still visible in photoconductivity measurements [18].

In this paper we will present photoconductivity measurements in the energy range 0.3 meV - 20 meV and magnetic fields up to 14.6 Tesla. Undoped samples as well as crystals doped with Bi-, Sb-, As-, and P-impurities were investigated.

2. Experimental Method

Samples with typical dimensions of 2x2x10 mm^3 were cut from Czochralski-grown tellurium single crystals. Four electrical contacts, two current leads, and two potential leads were made by alloying thin gold wires to the specimens.

At a temperature of 2 K the resistance between the potential leads was typically 1000 Ω for undoped samples and 50 Ω for samples doped with Bi, Sb, As or P with an acceptor concentration of about $2 \cdot 10^{20}$ m^{-3}. The measurements were carried out either with a Fourier spectrometer (10 cm^{-1} ... 160 cm^{-1}) and a magnetic field up to 14.5 T or with carcinotrons (80 GHz ... 500 GHz) and a magnetic field up to 6 T.

The photoresponse, i.e. the change in the voltage drop across the potential leads under constant current conditions, originates from optical transitions between discrete impurity states. In addition to the optical excitation, a thermal ionization of the excited impurity states is necessary in order to increase the free carrier concentration and consequently the conductivity [19]. In our experiments the photoconductive signal was largest at T = 1.4 K, the lowest temperature available with our cryostat and most of the results presented in this paper were obtained at this temperature.

3. Experimental Results

3.1 Undoped Samples

Photoconductivity measurements on undoped samples show that, in addition to the "hydrogenlike" impurities with an excitation energy of about 11 cm^{-1} a photoconductive response at about 24 cm^{-1} and 46 cm^{-1} is found [18]. In strong magnetic fields, with the magnetic field direction parallel to the trigonal c-axis, each of these peaks splits mainly into two components. In magnetic fields above 4 Tesla additional structures are visible. The cyclotron resonance is visible as a negative photoconductivity signal.

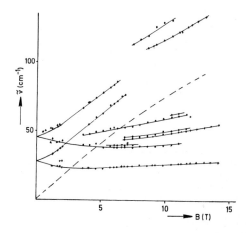

Fig.1 Magnetic field dependence of the energies of the extrema in the photoconductive response for an undoped sample (B $\parallel$ c). The dotted line characterizes the energy of the negative signal in the photoresponse

Experimentally, the photothermal ionization of majority impurities always increases the conductivity, whereas the sign of the cyclotron resonance signal varies for different semiconductors, depending on the energy dependence of the mobility and the energy relaxation process of carriers excited to higher Landau levels [20]. The different sign of the photoresponse for impurity and cyclotron resonance excitation in tellurium simplifies the interpretation of the spectra. A summary of the experimental data is shown in Fig.1. The dotted line characterizes the magnetic field dependence of the negative photosignal. A peculiarity is visible in the magnetic field region of 7 - 8 Tesla. Coming from low magnetic field values, the upper two impurity lines disappear around

B = 7 T but reappear at about B = 8 T at much higher energies. The parallel shift of the lines is about 20 cm^{-1}. A small splitting of about 1.6 cm^{-1} is observed for the lines in the wavenumber region 30 cm^{-1} - 60 cm^{-1} in strong magnetic fields.

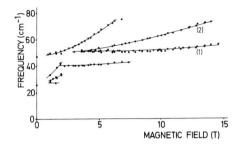

Fig.2 Magnetic field dependence of the energies of the maxima in the photoconductive response for an undoped sample (B⊥c)

For the magnetic field orientation B⊥c, the experimental results for the same crystal as used for the measurements in the orientation B∥c (Fig.1) are summarized in Fig.2. Contrary to the measurements in the orientation B∥c, a photosignal originating from a cyclotron resonance (CR) transition was not visible. Only a small reduction of the photosignal, by less than 10 %, was observed at those magnetic field values where the CR-absorption lines cross the impurity lines. Measurements on another undoped crystal showed additional lines to those shown in Fig.2. One of these additional lines is shifted approximately parallel to the line (2) to higher energies by about 10 cm^{-1} (9.5 cm^{-1} at B = 5 T and 11 cm^{-1} at B = 14 T). Another line appeared in the magnetic field range 5 T - 14 T and is shifted by 20 cm^{-1} to lower energies relative to the line (1) plotted in Fig.2. A systematic error in the energy position of the impurity lines of about 2 cm^{-1} at B = 14 T may result from an estimated misorientation of the sample of maximal ± 5°. In the orientation B∥c, the influence of a misorientation on the energy position of the impurity lines is much smaller than for B⊥c orientations as found from an analysis of the angular dependence of the spectra.

3.2 Doped Samples

For the measurements on doped crystals, samples with Group-V impurities at a doping level of about $2 \cdot 10^{20}$ m^{-3} were used. Within the resolution of 1 cm^{-1} for our measurements with the Fourier spectrometer, no differences in the spectra are observed for the different crystals containing different impurities (Bi, Sb, As, P). The results shown in Fig. 3 (B∥c) are obtained for six different crystals. The dotted line is identical with the dotted line shown in Fig.1, and corresponds to the cyclotron resonance transition. For the strong impurity lines a relatively weak satellite line shifted to higher energies is visible. At B = 14 T the energy difference between these two lines is 7.5 cm^{-1}.

A summary of the experimental results for the magnetic field direction B⊥c is shown in Fig.4. No significant difference is observed for measurements on differently doped samples. As for undoped samples, the CR line is visible only as a small decrease in the impurity signal, which is explained by an attenuation of the infrared radiation within the sample due to CR absorption. The dotted lines correspond to CR transitions obtained from absorption measurements [1].

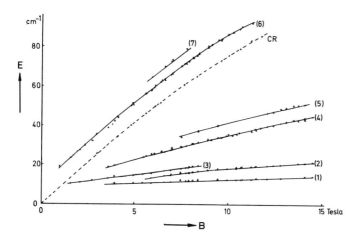

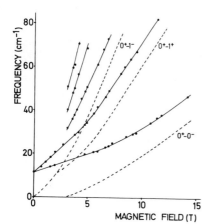

Fig.3 Magnetic field dependence of the energies of the extrema in the photoconductive response for samples doped with Group-V impurities (B ∥ c, T = 1.4 K). The dotted line corresponds to the cyclotron resonance line (negative photosignal)

Fig.4 Magnetic field dependence of the energies of the maxima in the photoconductive response for doped samples (B⊥c, T = 1.4 K). The dotted lines characterize the cyclotron resonance transitions from the lowest Landau level to higher Landau levels obtained from absorption measurements [1]

All the experimental results presented so far do not show significant differences in the energy positions of the impurity lines for different Group-V acceptors. This means that the chemical shift is relatively small. In order to resolve small differences in the impurity excitation energies for different acceptors, in addition to the Fourier spectrometer experiments the photoresponse was measured at a constant carcinotron frequency (80 GHz ... 500 GHz, corresponding to 3 cm^{-1} ... 17 cm^{-1}) as a function of the magnetic field. A typical result for a Bi-doped crystal and the orientation B ∥ c is shown in Fig.5. Three pairs of lines are visible and the comparison with Fig.3 shows, that each line observed with the Fourier spectrometer represents an unresolved double peak. The energy separation of this double peak is largest for Bi-impurities and smallest for P-impurities. A summary of the high resolution experiments on Bi- and P-doped samples is shown in Fig.6 and Fig.7 for the magnetic field orientations B ∥ c and B⊥c, respectively. The impurity lines in Fig.6 and Fig.3 are characterized by the same numbers. The energy separation of the different douple peaks for a Bi-doped sample (B ∥ c) as a function of the magnetic field B is shown in Fig.8. The splitting for line (2) in Fig.6 is

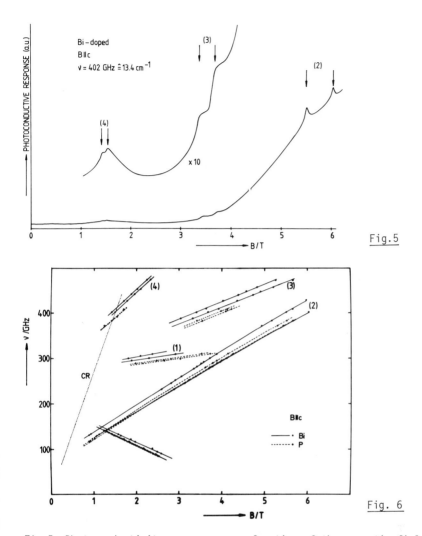

Fig.5 Photoconductivity response as a function of the magnetic field (B ∥ c) for a Bi-doped sample at an excitation energy corresponding to 402 GHz (13.4 cm^{-1})

Fig.6 Magnetic field dependence of the energies of the maxima in the photoconductive response for Bi- and P-doped samples (B ∥ c)

about a factor of two larger than the splittings for the lines (1), (3), and (4). These splittings are reduced by a factor of 0.61, 0.33, and 0.30 for Sb-, As-, and P-impurities, respectively. No fine structure splitting was visible for the orientation B⊥c, but these measurements show clearly, that a parallel shift of the lines appear for different impurities. Relative to the Bi-lines, this shift to lower energies is approximately 6 GHz for Sb-impurities, 11 GHz for As-impurities and 13 GHz for P-impurities.

143

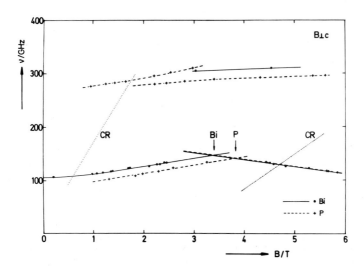

Fig.7 Magnetic field dependence of the energies of the maxima in the photoconductive response for Bi- and P-doped samples (B⊥c)

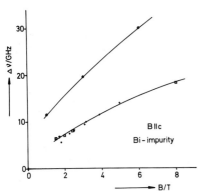

Fig.8 Energy of the fine-structure splitting as a function of the magnetic field B∥c for a Bi-doped sample

4. Discussion

4.1 Undoped Samples

The measurements on undoped single crystals show that in addition to Group-V impurities other impurities are present with an excitation energy of about 3.0 meV and 5.6 meV at B = 0 T (Fig.1). At low magnetic fields B∥c a splitting of each line into two components is observed. The energy of one of these split lines is nearly magnetic field independent, the other one is shifted approximately parallel to the cyclotron resonance line. The agreement of the energy difference between the two impurity lines with the cyclotron energy $\hbar\omega_c$ is better than 10 %. At magnetic fields above B = 4 T a third line can be resolved with an energy position between the split lines discussed before. Such a splitting is typical for a 1s-2p transition of a hydrogenic impurity. In a magnetic field the 2p-state splits into three levels $2p_-$, $2p_0$, and $2p_+$, corresponding to the magnetic quantum numbers m = -1, m = 0, and m = +1. The energy difference $2p_+ - 2p_-$ is equal to $\hbar\omega_c$ for a parabolic band. Since the absolute value of the binding energy of these impurities is much higher than the expected value for a hydrogenlike impurity, attempts were made to explain the spectra on the basis of heliumlike and singly ionized heliumlike impurities [21]. Unfortunately it was not possible to identify heliumlike impurities. For example, measurements on Sn-doped tellurium crystals do not show an increase in the acceptor concentration and therefore we will not discuss in detail the spectra shown in Fig.1 and Fig.2. However, we would like to point out, that the anomalous behaviour of the "$2p_+''$"-states at an energy of about 100 cm^{-1} can be explained as a polaron effect. It is known [22] that electron-LO phonon interaction results in a pinning of the 1s-$2p_+$ transition energy close to the LO-phonon energy $\hbar\omega_{LO}$ and in a line at energies above $\hbar\omega_{LO}$ which is shifted to higher energies relative to a linear extrapolation of the 1s-$2p_+$ line observed at energies well below the LO-phonon energy. The strength of this polaron shift is connected with the polaron coupling constant α which, within the Fröhlich continuum model, depends on the LO-phonon frequency, the high frequency and static dielectric constants, and the effective mass. For tellurium a value of $\alpha = 0.14$ has been calculated [23]. A comparison of the experimental data with calculations on the basis of a simple hydrogenlike impurity [22] indicates that a polaron coupling constant of 0.14 is too small to explain the observed polaron shift. This result is in agreement with an analysis of the magnetic field dependent cyclotron mass [24] which can be explained only if a polaron coupling constant of about $\alpha = 0.35$ is assumed.

4.2 Doped Samples

The experimental data (Fig.7) for $B \perp c$ show, that two different types of transitions are observed - one transition line which remains constant if the doping is changed whereas the other lines show a parallel shift in energy for different impurities. We assume that this shift in energy originates from a chemical shift of the lowest impurity state whereas the transition energy between excited states is not influenced by the chemical nature of the impurity. The level scheme shown in Fig.9 (left side) can explain the experimental results. In this figure all energies are plotted relative to the energy of an excited state (which is not influenced by chemical shifts) and the optical selection rules for a Bi-doped sample are marked by arrows. The two lowest levels represent the bonding and antibonding states and the experimental results show clearly that, as expected, only the bonding state shows a chemical shift because the impurity envelope function at the impurity site is finite for this state only [15].

The energy scheme looks much more complicated for the magnetic field orientation $B \parallel c$ (Fig.9, right side). Both, the bonding state and the antibonding state split into two levels. The observed transitions for a Bi-doped sample are characterized by arrows. The additional splitting of the bonding and antibonding state in a magnetic field $B \parallel c$ agrees very well with theoretical predictions. BIR et al. [16] describe this splitting by the following formula (E = 0 is chosen to be in the middle between the bonding and antibonding state)

$$E_{1,2,3,4} = \pm \left[(\tfrac{1}{2} G \mu_B B)^2 + \Delta^2 \pm G \mu_B B |\delta| \right]^{1/2}$$

2Δ is the energy difference between the bonding and the antibonding state, δ is a constant that determines the mixing of the states at H and H' ($\delta \approx 0.1\Delta$) and G is a constant ($G \approx 5$) which describes the inversion asymmetry splitting of Landau levels in a magnetic field $B \parallel c$ [25]. The change in the splitting between bonding and antibonding state in strong magnetic fields is mainly determined by the term $\pm 0.5 \, G \, \mu_B \, B$. From our experimental data we get a G-factor of $G = 4.8 \pm 0.5$ in good agreement with the value $G = 5 \pm 1$ obtained from Shubnikov-de Haas experiments [25]. A value for δ cannot be deduced from

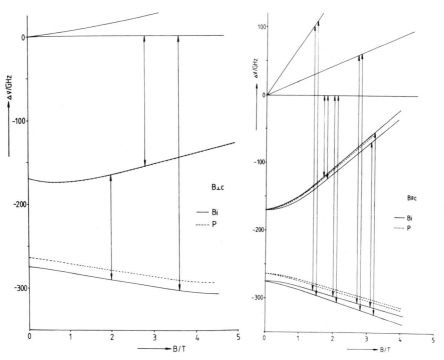

Fig.9 Energy levels of Bi- and P-impurities in tellurium as a function of the magnetic field B for the orientations B⊥c and B∥c

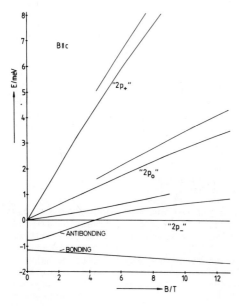

Fig.10 Energy levels for hydrogenic acceptor states in tellurium (B∥c). The fine-structure splitting of the bonding and antibonding state is not shown in this figure

the finestructure splitting since this splitting is strongly enhanced by central cell corrections. The observed energy splitting of about 170 GHz at B = 0 T between antibonding state and the next higher excited state is the same for all impurities investigated and should depend only on parameters of tellurium. However, calculations on the basis of the bandstructure parameters published by COUDER [2] yield a value [21] which is about 10 % smaller than found experimentally.

The interpretation of all impurity lines observed for B ∥c (combination of Fig.3 and Fig.6) leads to an impurity spectrum for hydrogenic acceptors in tellurium as shown in Fig.10. The accurate position of the lowest level (bonding state) is subject to small central cell correction and the fine-structure splittings for the bonding and antibonding state is not shown in this figure. The overall behaviour of the levels may be compared with the levels of a simple hydrogenlike impurity and we introduce tentatively the same symbols $2p_-$, $2p_0$, and $2p_+$ for the lowest excited states. This characterization may be incorrect because preliminary calculations show [21], that the $2p_0$ state should be much lower in energy than found experimentally, whereas the $2p_-$ and $2p_+$ states agree very well with these calculations. A satisfactory explanation cannot be given for the weak satellite lines located about 0.7 meV at B = 10 T above the main impurity lines. The experimentally observed repulsion between the antibonding level and the $2p_-$ satellite line may give more informations about the symmetry and the origin of the satellite lines.

The most interesting result of our impurity measurements is, that the magneto-impurity effect in tellurium [26] can be explained unambiguously. Our measurements show, that the sharp resonances appear at those magnetic field values where the antibonding level is exactly between the bonding level and a higher excited level. Under this condition, the recombination rate for nonequilibrium carriers increases drastically, which can simply be measured as a change in the resistance of the sample. A comparison of the measured impurity levels with the resonance structure observed in the magneto-impurity effect demonstrates, that simple transport measurements on tellurium give informations about impurity levels with a resolution of better than 0.01 meV. A detailed analysis of the magneto-impurity spectra will be published elsewhere.

References

1 R. Yoshizaki and S.Tanaka, J.Phys.Soc.Jpn 30, 1389 (1971)
2 Y.Couder, M.Hulin and H.Thomé, Phys.Rev.B7, 4373 (1973)
3 M.v.Ortenberg, K.J.Button, G.Landwehr and D.Fischer, Phys.Rev.B6, 2100 (1972)
4 M.v.Ortenberg and K.J.Button, Phys.Rev.B16, 2618 (1977)
5 N.Miura, R.Yoshizaki and S.Tanaka, Solid State Commun. 7, 1195 (1969)
6 D.Hardy and C.Rigaux, phys.stat.sol. 38, 799 (1970)
7 D.Hardy, C.Rigaux, J.P.Vieren and Nguyen Hy Hau, phys.stat.sol.(b)47,643(1971)
8 M.S.Bresler, I.I.Farbshtein, D.V.Mashovets, Yu.V.Kosichkin and V.G.Veselago, Phys.Letters 29A, 23 (1969)
9 C.Guthmann and J.M.Thuillier, phys.stat.sol.(b)38, 635 (1970)
10 E.Braun, L.J.Neuringer and G.Landwehr, phys.stat.sol.(b)53, 635 (1972)
11 M.Hulin, Proc. 10th Int. Conf. Physics of Semicond., Cambridge, Mass., 1970 (U.S.AEC, Div.Technical Information, Springfield,Va., 1970) pp. 329-337
12 K.Nakao, T.Doi and H.Kamimura, J. Phys.Soc.Jpn 30, 1400 (1971)
13 L.S.Dubinskaya, Sov.Phys.Semicond. 6, 1267 (1973)
14 D.Thanh, Solid State Commun. 9, 1263 (1973)
15 K.Natori, T.Ando, M.Tsukada, K.Nakao and Y.Uemura, J.Phys.Soc.Jpn 34,1263(1973)
16 G.L.Bir, V.G.Krigel, G.E.Pikus and I.I.Farbshtein, JETP Lett.19-20, 29 (1974)
17 M.Ataka, R.Yoshizaki and S.Tanaka, Solid State Commun. 13, 849 (1973)

18 K.v.Klitzing and C.R.Becker, Solid State Commun. 20, 147 (1976)
19 For a review see, e.g., E.E.Haller, W.L.Hansen and F.S.Goulding, IEEE Transactions on Nuclear Science 22, 127 (1975)
20 P.E.Simmonds, J.M.Chamberlain, R.A.Hoult, R.A.Stradling and C.C.Bradley, J.Phys. C7, 4164 (1974)
21 J.Trylski, (unpublished)
22 D.R.Cohn, D.M.Larsen and B.Lax, Phys.Rev.B6, 1367 (1972)
23 K.J.Button, D.R.Cohn, M.H.Weiler, B.Lax, G.Landwehr, Phys.Lett. 35A, 281(1971)
24 K.v.Klitzing, Phys.Rev.B21, 3349 (1980)
25 V.B.Anzin, Yu.V.Kosichkin, V.G.Veselago, Solid State Commun. 8, 1773 (1970)
26 K.v.Klitzing, Solid-State Electronics 21, 223 (1978).

Part V

Magneto-Transport Phenomena

Anomalous Anisotropies in Rare-Earth Magnetic Compounds and Their Behavior under High Magnetic Field

T. Kasuya, K. Takegahara, M. Kasaya, Y. Isikawa, and H. Takahashi

Department of Physics, Tohoku University
Sendai, Japan, and

T. Sakakibara, and M. Date

Department of Physics, Osaka University, Osaka, Japan

1. Introduction

Recently, the valence fluctuating state in the rare earth compounds is attracting much attention. Among various different characters of valence fluctuation, the dense Kondo-like behavior, which is observed most typically in Ce compounds such as $CeAl_2$, $CeAl_3$ and CeB_6, and the strongly correlated narrow band Fermi liquid like behavior, which is observed typically in Sm compounds such as SmB_6 and SmS under pressure, are most typical and have been studied in detail [1]. Both of them are common in the sense that the Fermi level is in the conduction bands which are made in most cases by the 5d character on the rare earth atom, and the d-f mixing is particularly important in the former case. In most cases, however, the anions are situated in the nearest neighbour sites of the cation and thus the p-f mixing is much larger than the d-f mixing, where valence bands are formed mostly by the p-character on the anion atom. In this sense, the valence fluctuating states in the materials in which the Fermi levels are in the valence bands or near them are very interesting. Ce-monopnictides and Eu-hexaboride, EuB_6 are typical examples for such a category and are attracting considerable attention in the recent years. In this paper, we present some experimental results of these two materials focusing on the high field experiment performed recently by our group and draw several new physical pictures by analyzing the data.

2. Eu-Hexaboride

EuB_6 has the CaB_6 type crystal structure, in which Ca and B_6 molecules form the CsCl type lattice. The similar materials RB_6 in which R is a non magnetic divalent cation such as Ca, Sr, Ba and Yb are known as typical narrow gap semiconductors [2]. The band calculation shows that both the minima of the conduction bands and the maxima of the valence bands are at the three X-points in the k-space and the valence bands are strongly anisotropic with a large effective mass along the Δ axes [3]. In this type of crystal, the main part of the valence bands is formed by bonding orbits of the p-states on borons while the top of the valence bands is formed mostly by non bonding orbits, and a B_6 molecule behaves like a divalent anion. Because borons make a strong network, the crystal is grown usually with some amount of cation vacancies [4]. In the usual sense, cation vacancies induce p-type character. However, the most materials show n-type character. This was explained as follows [5]. Usually each divalent cation vacancy traps two valence band holes. In the present materials, the binding energy for the third hole is estimated to be about 0.2eV. In the usual wide gap semiconductors, the third hole is filled by the valence electrons due to the charge neutrality requirement. In the present narrow-gap semiconductors, the bottoms of the conduction bands are lower than the energy

level of the third hole and then a part of valence electrons occupying the
third holes flows out into the conduction bands. The situation is illustrated
in Fig.1, and this is the origin of the n-type character. In Fig.2 the expected Fermi surfaces of the up spin electrons around the [001] X-point are
shown.

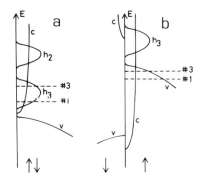
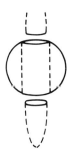

Fig.1 Schematic pictures for density of states of up and down spin, the horizontal axis, as functions of energy E, the vertical axis. a: paramagnetic region and b: ferromagnetic region. The broken lines indicate the expected Fermi levels of #1 and #3 samples. The letter c means the conduction band, v the valence band, h_3 the third hole and h_2 the second hole.

Fig.2 Schematic picture for the Fermi surfaces of up spin electrons, solid lines, and holes, broken lines, around [001] of k-space in the ferromagnetic region. The vertical axis corresponds to the k_z axis.

In EuB$_6$, the 4f level is estimated to be about 1eV below the Fermi energy
E_F [6]. There is strong mixing between the 4f states and the tops of the valence bands [5,7], which pushes up the latter about the order of 0.1eV in the
paramagnetic region and makes the band gap narrower. Actually, from the temperature dependence of the resistivity and the Hall constant, the band gap is
estimated to be of the order of 100K [5]. When the 4f spins on Eu order ferromagnetically, the tops of the down spin valence bands are lowered and those of
up spin are raised. On the other hand, the bottoms of the down spin conduction
bands are raised and those of up spin are lowered by the usual ferromagnetic
d-f Coulomb exchange interaction $-I\vec{S}\vec{\sigma}$, in which $\vec{\sigma}$ is the Pauli spin operator
of the conduction electron and I is estimated to be about 500K [5]. By the
above two mechanisms, the up spin valence and conduction bands overlap by an
amount of order 0.2eV causing an insulator-metal transition. This mechanism
induces a strongly non-linear exchange interaction because the ferromagnetic
energy gain comes from the rearrangement of the up spin valence electrons into
the up spin conduction bands, which causes the first order phase transition to
the ferromagnetic state. Various anomalous magnetic and transport properties
of EuB$_6$ are induced by these mechanisms.

To investigate the above mechanisms and the electronic states in EuB$_6$ more
in detail, magnetic and transport properties under high magnetic field have
been measured both on pure EuB$_6$ and La doped EuB$_6$. Some of the detailed ex-

perimental results is given in another paper [8]. At first, the results on pure EuB_6 are studied in detail. Two kinds of Curie temperatures, T_1 and T_2, are observed in pure EuB_6. The higher Curie temperature $T_1 \sim 14K$ is evaluated to be due to the intrinsic ferromagnetic Heisenberg type exchange interaction and the indirect one through the conduction electrons of a few tenths of percent per Eu atom, which are created by a similar amount of Eu defects existing even in the purest sample through the mechanism mentioned before. The latter exchange interaction is, however, strongly non linear, different from the well known one in the usual metallic materials, due to the smaller Fermi energy compared with the d-f exchange interaction and thus causes quick saturation of the magnetization [9]. Therefore, at T_1, the resistivity shows a sharp peak but there is no observable peak of specific heat staying in a weak ferromagnetism between T_1 and T_2. At $T_2 \sim 9K$, the first order insulator-metal transition described above occurs causing the usual strong ferromagnetism below T_2. Therefore, a large peak of specific heat and the sharpest increase of conductivity are observed at T_2. Note that both T_1 and T_2 are broadened by the giant moments created around each Eu defect [5].

For sample #1 of fairly pure EuB_6 [8], the resistivity decreases from 450 $\mu\Omega cm$ at 30K to 17 $\mu\Omega cm$ at 1.7K while the carrier number n_0, which is estimated from the Hall constant by assuming the single carrier model, increases from 0.0027/Eu to 0.0156/Eu. Because the effective d-f exchange scattering constant I_{dfeff} for the resistivity is estimated to be about 0.036eV from various experiments [5], the resistivity due to the d-f scattering is estimated to be about 125 $\mu\Omega cm$ at 30K. Therefore, the mobility due to the defect scattering increases from 520 $cm^2/vsec$ at 30K to 1830 $cm^2/vsec$ at 1.7K, where 16 $\mu\Omega cm$ was used for the intrinsic residual resistivity as shown in the following. In ferromagnetic materials, we should be careful to analyze the magnetoresistance and the Hall constant data. In Fig.3, these data at 1.7K for the low field region are shown. The effective field H_{eff} acting on the conduction electrons may be given by B, that is $H_{eff} = H_{ext} + (4\pi-D)M$ in which H_{ext} is the external field, D the demagnetization factor and M the magnetic moment. At low temperature, $4\pi M = 11kOe$. In this material, the anomalous Hall effect was checked in several samples and proved to be negligibly small. The initial decreasing in the resistivity ρ seems to be due to the domain effect, from which DM is estimated to be about 8kOe, which is consistent with the sample shape analysis because the magnetic field is applied perpendicular to the flat surface. Note that the effective field can not be smaller than $4\pi M = 11kOe$. The magnetoresistance ratio $\Delta\rho/\rho_M$ thus obtained is plotted in Fig.4, in which ρ_M is the

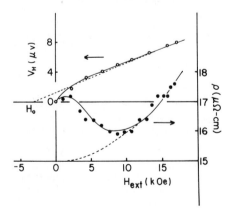

Fig.3 Experimental results for the resistivity ρ and the Hall voltage V_H at 1.7K of sample #1 in the low applied field H_{ext} region. The broken line for V_H represents the extrapolation line from the data for H_{ext} from 20kOe to 40kOe cutting at H_0 = -3kOe. The broken curve for ρ represents a expected extrapolation.

resistivity for H_{eff} = 11kOe and $\Delta\rho = \rho - \rho_M$. The Hall voltage V_H for La-doped samples with larger numbers of carriers shows straight lines as functions of H_{ext} for H_{ext} larger than 10kOe and cut the H_{ext} axis near -2∼-3kOe, consistent with the above considerations. In sample #1, however, V_H is not a straight line. For 20kOe < H_{ext} < 40kOe, V_H is nearly straight and the extrapolated line cuts H_0 = -3kOe as shown in Fig.3 by a broken line. The value n_0 = 0.0156/Eu mentioned before corresponds to this Hall constant R_M, indicating that the Hall constant has a minimum at around H_{ext}∼30kOe. The Hall constant near at H_{eff}∼11kOe, is larger by about 5% than the minimum value. However, there is substantial error. The Hall constant ratio R/R_M thus obtained is also shown in Fig.4. High field magneto-resistance up to 400kOe was measured by using the facility in Osaka University. The result is shown in Fig.5.

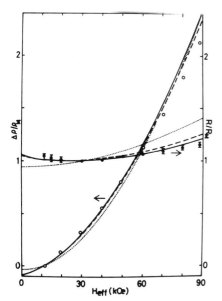

Fig.4 Experimental values for the magneto-resistance ratio $\Delta\rho/\rho_M$ and the Hall constant ratio R/R_M of sample #1 at 1.7K under static field up to 90kOe are shown by ○ and ●, respectively. ρ_M and R_M are values at H_{eff}=11kOe and 30kOe, respectively. The dotted lines show one of best fitted results for the electron-hole two-carrier model. The solid and broken lines are two examples for the three-carrier model. See the text for detailed parameter values.

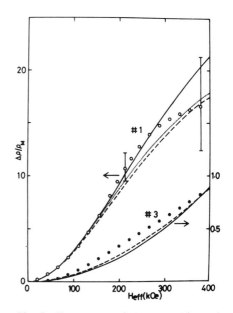

Fig.5 Magneto-resistance ratio $\Delta\rho/\rho_M$ at 1.7K for the high field region up to 400kOe are shown by open circles for sample #1 and by closed circles for sample #3. Note the difference in the scale. The meaning of the three calculated curves for #1 are same as those in Fig.4. Two curves for sample #3 are two examples of the electron-hole two-carrier model.

As shown in Fig.1, it is reasonable to analyze the data in the paramagnetic region by a single carrier model but in the ferromagnetic region both electrons and holes coexist and the Fermi surfaces shown in Fig.2 are expected, casting doubt on the simple single carrier analysis. The analysis was done by using the following semi-classical equations.

$$\rho = \sum_i \frac{\sigma_i}{1+\omega_i^2\tau_i^2} / A , \qquad \frac{\Delta\rho}{\rho_0} = \sum_{(i,j)} \frac{\sigma_i\sigma_j(\omega_i\tau_i-\omega_j\tau_j)^2}{(1+\omega_i^2\tau_i^2)(1+\omega_j^2\tau_j^2)} / A , \qquad (1)$$

$$R = \sum_i \frac{1}{n_i e_i c} \frac{\sigma_i^2}{1+\omega_i^2\tau_i^2} / A , \qquad (2)$$

$$A = \left(\sum_i \frac{\sigma_i}{1+\omega_i^2\tau_i^2} \right)^2 + \left(\sum_i \frac{H}{n_i e_i c} \frac{\sigma_i^2}{1+\omega_i^2\tau_i^2} \right)^2 , \qquad (3)$$

where

$$\sigma_i = \frac{n_i e^2 \tau_i}{m_i}, \quad \omega_i = \frac{e_i H}{m_i c}, \quad \frac{H\sigma_i}{n_i e_i c} = \omega_i \tau_i, \quad \rho_0^{-1} = \sum_i \sigma_i , \qquad (4)$$

and $e_i = \mp e$ for electrons or holes, respectively. At first, the hole contribution is neglected by assuming that the hole relaxation time τ_h is small compared with the electron's τ_e. In the following, we consider the case that the external magnetic field H is along the z-axis and the current along the x-axis. It is clear from Fig.2 that there are at least two kinds of Fermi surface, one shown in Fig.2, which is called electron #1 in the following, in which an electron can circulate freely and others around [100] or [010], which is called electron #2 in the following, in which some electron comes into the valence band in the middle of the orbit and then is scattered rapidly. Therefore, for the latter, the relaxation time becomes the order of ω^{-1} for $\omega\tau > 1$. For fields larger than the gap in Fig.2 due to the d-p mixing, ΔE_{dp}, that is, $\hbar\omega > \Delta E_{dp}$, the magnetic break down due to the quantum effect occurs and then the electron orbital motion is restored. In that case, the d-p mixing energy acts as the scattering mechanism. Therefore, the relaxation time τ_{e2} for electron #2 may be written as

$$\frac{1}{\tau_{e2}} = \frac{1}{\tau_{e1}} + \frac{1}{\tau'_{e2}} \left(\frac{(\omega\tau_{e1})^2}{a^2+(\omega\tau_{e1})^2} \right)^{1/2} . \qquad (5)$$

In the single carrier model used before, the relaxation time, written as τ_0, is estimated to be 2.7×10^{-13} sec. Then $\omega\tau_0$ is unity for H = 51 kOe with the effective mass of the electron 0.25m [5]. Note that for simplicity the value for H_{eff} = 11kOe is used. As shown later, the gap ΔE_{dp} is evaluated to be about 20K. Then the magnetic break down is expected to occur at about H = 40kOe for $\hbar\omega = \Delta E_{dp}$. The above consideration means that in (5) a should be near unity and then $\tau'_{e2}/\tau_{e1} \equiv b$ is also near unity. For the two-carrier model, (1) and (2) are rewritten as

$$\frac{\Delta\rho}{\rho_0} = \frac{\alpha\gamma(1-\nu\alpha)^2\omega_1^2\tau_1^2}{(1+\alpha\gamma)^2 + \alpha^2(1+\nu\gamma)^2\omega_1^2\tau_1^2} , \qquad (6)$$

$$R = \frac{1}{n_1 e_1 c} \frac{1 + \nu\alpha^2\gamma + \nu\alpha^2(\nu+\gamma)\omega_1^2\tau_1^2}{(1+\alpha\gamma)^2 + \alpha^2(1+\nu\gamma)^2\omega_1^2\tau_1^2} , \qquad (7)$$

where

$$\omega_2\tau_2/\omega_1\tau_1 = \nu\alpha, \quad n_2/n_1 = \gamma, \quad \sigma_2/\sigma_1 = \alpha\gamma, \tag{8}$$

and $\nu = \pm 1$ for the same or different charge carriers, respectively. In the present case, $\nu = 1$, $\alpha = \tau_{e2}/\tau_{e1}$ and γ is expected to be around 0.5. Then R has a broad peak at around $\omega_1\tau_1 \sim 1$, not inconsistent with experiment but $\Delta\rho/\rho_0$ is too small compared with experiment and furthermore its H dependence is too strong, H^4, for $\omega\tau \ll 1$. To obtain a larger value of $\Delta\rho/\rho_0$, near cancellation of the Hall currents due to electrons and holes seems to be necessary.

In the next step, the simple electron-hole two-carrier model is studied. In this case, $\Delta\rho/\rho_0$ saturates to the value $\gamma(1+\alpha)^2/\alpha(1-\gamma)^2$, which can be significantly large for a small value of γ. The best fitted curves of $\Delta\rho/\rho_M$ and R/R_M for $\alpha = 0.47$ and $\gamma = 0.67$ are shown in Figs. 4 and 5. R increases monotonically as much as 38% at $H = 80$kOe, more than factor two too large compared with the experimental value of 16%. Furthermore, the H dependence of $\Delta\rho/\rho_0$ in low values of $\omega\tau_0$ is H^2 in the calculation as is clear in (6), but the experiment shows a weaker dependence, about $H^{1.7}$. Nevertheless, the overall agreement is good, indicating clearly the coexistence of valence holes in the ferromagnetic state. Now, using the above determined parameters, we obtain the following quantities; $n_1 \equiv n_e = n_0(1-\alpha^2\gamma)/(1+\alpha\gamma)^2 = 0.439n_0 = 0.0077$/Eu, $n_h = \gamma n_e = 0.0052$/Eu, $n_e - n_h = 0.0025$/Eu, $\tau_e = \tau_0(1+\alpha\gamma)/(1-\alpha^2\gamma) = 1.54\tau_0 = 4.2\times10^{-13}$ sec. In the present model, $n_e - n_h$ gives the number of Eu defects and thus should be larger than the number of the conduction electrons at 30K, 0.0027/Eu. There is a minor disagreement on this point, too. Note that there is more than a 10% allowance for the combination of the parameter values. But the above result indicates clearly that most third hole states are vacant consistent with the earlier analysis [5] of the temperature dependences of the Hall constant and the resistivity that the occupation ratio is about 30%. To improve the above mentioned discrepancies, two types of conduction electrons treated before is introduced. Therefore the three-carrier model is used. To avoid introducing too many adjustable parameters, the following parameters are fixed based on the preceding physical consideration; $n_{e2}/n_{e1} = 0.5$, $\tau'_{e2} = \tau_{e1}$ and $a = 1$. Therefore, there are no adjustable parameters for electron #2. Then the parameters $\alpha = \omega_h\omega_h/\tau_{e1}\tau_{e1}$ and $\gamma = n_h/(n_{e1}+n_{e2})$ are determined from the best fitting procedure. The results for two different combinations of parameters, $\alpha = 0.41$, $\gamma = 0.7$ and $\alpha = 0.42$, $\gamma = 0.66$, are shown in Figs. 4 and 5 by solid and broken lines, respectively. As seen in Fig. 4 the essential improvement appeared in R. R decreases linearly for the initial increase of H making a shallow minimum and increases gradually in good agreement with experiment. This is due to the characteristic form of τ_{e2} given by (5) because, for a small value of $\omega\tau_{e1}$, τ_{e2} decreases linearly with H which dominates the H dependence in low field region because all other H-dependent terms are proportional to H^2. H dependence of $\Delta\rho/\rho_M$ in the low field region is also improved. This is also due to τ_{e2} because the parameter D in the H^2 term, DH^2, now decreases as $D_0/(1+b\omega)$ weakening the H^2 dependence. These facts clearly show that electron #2 with the character described before should exist. Two cases shown in Figs. 4 and 5 may give some hint on the uncertainty of the parameter values. Note that the following physical quantities are obtained for each case; $n_e = 0.0076$/Eu and 0.0077/Eu, $n_h = 0.0053$/Eu and 0.0051/Eu, $n_e - n_h = 0.0023$/Eu and 0.0026/Eu and $\tau_{e1} = 4.5\times10^{-13}$sec. and 4.4×10^{-13}sec.. These are nearly the same as those for the electron-hole two-carrier model. It is clear from Fig. 2 that, for the valence holes, too, there are at least two kinds of holes, one with a small mass and circulating freely and others with a larger effective mass. To see this point, we introduce the second hole and analyse the data by the four-carrier model introducing the additional parameters, $\alpha_2 = \omega_{h2}\tau_{h2}/\omega_{h1}\tau_{h1}$ and $\gamma_2 = n_{h2}/n_{h1}$. However, we could not get any substantial improvement beyond the three-carrier model. Details of the hole state are, therefore, insensitive to the present phenomena and we can not get any more detailed information of the valence holes.

Next the electronic state around each Eu defect is studied more in detail. The effective total cross section σ of each Eu defect for the conduction band electrons is defined by $\sigma = (\ell N_i)^{-1}$, where ℓ is the electron mean free path evaluated from $\hbar k_F \tau_e/m^*$ by using $m^* = 0.25m$ and N_i is the number of Eu defects and is assumed here to be given by 1.3 times the number of carriers at 77K following the preceding analysis. The dimensionless quantity $q \equiv k_F^2\sigma/4\pi$, which is connected more directly to the phase shift ζ_ℓ by $\Sigma_1^\infty \ell\sin^2(\zeta_{\ell-1}-\zeta_\ell)$, then decreases from 0.034 at 30K to 0.01 at 1.7K. This behavior may be interpreted as follows. As shown in Fig.1, in the paramagnetic state, the Fermi level is expected to be in the middle of the third hole levels, the width of which was estimated to be about 2000K from the temperature dependences of the transport coefficients [5]. If this width is caused mostly by the broadening of each level due to the d-p mixing, the resonance scattering should occur which causes q values to be several times of one tenth. The much smaller value obtained above, therefore, indicates that the width is mostly due to the local potential fluctuation and thus the scattering is mostly due to the usual potential scattering. Then the d-p mixing width seems to be less than a few meV, perhaps around 20K, which is consistent with the preceding analysis. The value of q is still smaller than that we expect for the usual non-resonant scattering. This seems to reflect the complicated structure of the Eu defect center. From the above considerations, it is expected that most defect are either neutral with two holes or charged with +1 by holding three holes. The scattering from the former should be small. For the latter, -1 of the screening charge should be provided by the conduction electrons in the paramagnetic region. In the usual charged impurity such as La^{+++} considered later, the main screening occurs on the impurity site, indicating the $\ell = 0$ screening to be the dominant one. Note that in the present band structure $\ell = 0$ means the three 5d(e_g) type characters on the three X-points. In the present defect center, however, the screening begins to occur from the six nearest neighbour sites, where at least one $\ell = 0$, three $\ell = 1$ and two $\ell = 2$ characters for each band are expected to be nearly equally important. For an example, if the screening is made in 40% by $\ell = 0$, 40% by $\ell = 1$ and 20% by $\ell = 2$, the phase shifts are calculated by the Friedel's sum rule to be $\zeta_0 = \pi/15$, $\zeta_1 = \pi/45$ and $\zeta_2 = \pi/60$ which gives q = 0.025 in good agreement with experiment. The smaller value of q in the ferromagnetic state may be explained partly by lack of the resonant scattering and partly by the additional screening by the coexisting valence holes.

Now the behavior of La-doped EuB$_6$ is studied. Here we concentrate on sample #3 in [8]. At 77K, n_0 is 0.0068/Eu. This value corresponds to E_F = 1670K, which is much larger than that in sample #1, 900K, but still smaller than the expected width of the third hole band, and thus means that the Fermi level is in the middle of the third hole band. The total resistivity was observed to be 670 $\mu\Omega$cm at 30K and the resistivity due to the d-f exchange scattering is estimated to be 68 $\mu\Omega$cm leaving the residual resistivity in the paramagnetic region of about 600 $\mu\Omega$cm and the corresponding mobility of 110 cm^2/vsec. This mobility value is much smaller than that of sample #1 and thus suggests that the scattering by a La impurity is much larger than that by a Eu defect. If we neglect the Eu defect scattering and assume the La concentration to be 0.006, the cross section per La is evaluated to q = 0.13, much larger than that of sample #1. This larger value of q is explained as follows. At the La^{+++} impurity, the most dominant screening should be the $\ell = 0$ character as mentioned before. Then, for a case of 80% screening by $\ell = 0$ and 20% by $\ell = 1$, $\zeta_0 = 0.42$ and $\zeta_1 = 0.035$ giving q = 0.142, while for a case of 70% by $\ell = 0$ and 30% by $\ell = 1$, $\zeta_0 = 0.366$ and $\zeta_1 = 0.052$ giving q = 0.1, in reasonable agreement with the experimental value.

The resistivity at 1.7K in the ferromagnetic state, which is obtained by applying more than 10kOe of effective field, is 137 μΩcm with $n_0 = 0.0149$/Eu. The Hall constant is constant within the experimental error up to 80kOe. The magneto-resistance up to 400kOe of pulsed field was also measured and is shown in Fig.5. The resistivity is nearly constant from 20 to 80kOe in agreement with the static measurement and then increases gradually gaining a factor two at 100kOe. To analyze the data, the electron-hole two-carrier model was used. However, we could not obtain satisfactory agreement. At first $\alpha = 0.47$ is used following the result of sample #1. Then $\gamma = 0.8$ fits the data at 400kOe as shown by the solid line in Fig.5. The magnetic field dependence is not so satisfactory. The changes of $\Delta\rho/\rho_M$ and $\Delta R/R_M$ are less than four percent at 80kOe, consistent with experiments within the experimental errors. However, the following values, $n_e = 0.452 n_0 = 0.0067$/Eu, $n_h = 0.0051$/Eu, $n_e - n_h = 0.0016$/Eu and $\tau_e = 1.63 \tau_0 = 5.2 \times 10^{-14}$ sec, are not reasonable. In particular, $n_e - n_h$, which should be the sum of La and Eu defect concentrations and thus expected to be larger than 0.0068/Eu, is unreasonably small. To improve the above point, $\alpha = 0.7$, which seems the maximum value from the physical consideration, is used. Then $\gamma = 0.43$ gives nearly the same degree of agreement with experiments as the former case shown by dotted line in Fig.5. However, the following values, $n_e = 0.0069$/Eu, $n_h = 0.0029$/Eu, $n_e - n_h = 0.004$/Eu and $\tau_e = 5.3 \times 10^{-14}$ sec, are not yet satisfactory. In the present case, introduction of electron #2 causes no measurable effect because of the small τ_{e1} value. We should notice, however, that, in the present case, field dependence of τ_{e1} due to the quantum effect is not negligible and the observed field dependence of $\Delta\rho/\rho_M$ may be due to this effect. Then, there is a considerable probability that the holes are not created in the valence bands in the ferromagnetic region. Then $n_0 = 0.0149$/Eu is the real electron number and is also the sum of the La impurity and the Eu defect. More detailed investigation is necessary to check this problem. The truth seems to be in the middle of the above two extreme cases.

To conclude this section, it has been shown that high field magneto-resistance and Hall effect offer crucial information on the electronic structure of EuB_6 confirming the model we proposed before. Eelctronic structure around La impurity and Eu defect also become clear.

3. Ce-Pnictides

Ce-pnictides crystallize in the NaCl structure. Tetravalent Ce has no 4f electron and trivalent Ce has one 4f electron. Because of a large spin orbit interaction, we consider only the $J = 5/2$ state in the following. In the cubic crystalline field, the $J = 5/2$ multiplet splits into the Γ_8 quartet and the Γ_7 doublet, in which Γ_7 is lower in the point charge model. CeN is known as a possible valence fluctuating material [10]. In other Ce-pnictides, Ce is considered to be trivalence. The trivalent pnictides are known as typical indirect narrow gap or weakly overlapping semimetallic materials [11]. The band calculations show [12] that the band structure is similar to the divalent rare earth chalcogenides, in which the top of the valence bands is at Γ-point and the bottoms of the conduction band made mostly by the $5d(t_{2g})$ states on rare earth atoms are at three X-points. The band gap becomes narrower for heavier pnictogens and is expected to close at about CeAs. Therefore, CeP and CeAs are expected to be very similar to EuB_6 because the 4f level on Ce^{+++} is expected to be about 1eV below the Fermi energy and the f-p mixing is about a factor two larger because of a larger extension of the 4f state in Ce. However, both CeP and CeAs behave like normal trivalent magnetic materials, although the crystalline field splittings are nearly a half of those extrapolated from the values of other pnictides [13]. This seems to be due to the following reason. Both cation and anion vacancies are created very easily in these materials, similar to the trivalent rare earth chalcogenides [14] which always produce conduction electrons in the mechanism described in the preceding section. Actually, even in the purest sample, the number of conduction electrons determined from the Hall constant assuming single carrier is several percent per Ce, which causes too large a Fermi energy to create holes in the valence band even in the ferromagnetic state; therefore the insulator-metal transition described in the preceding section for pure EuB_6 does not occur any more. In CeSb and CeBi, finite overlap is expected even in the paramagnetic state and thus a finite number of holes is expected to exist for relatively pure samples. Thus various unusual magnetic properties have been observed. In the paramagnetic region, the crystal field splittings are more than one order of magnitude smaller than those extrapolated from other monopnictides, 37K in CeSb and 8K in CeBi [15], and furthermore in CeBi the Γ_8 quartet is lower than the Γ_7 doublet in contrast to the point charge model [16]. Therefore they are nearly isotropic in the paramagnetic region. Nevertheless, in the ferromagnetic state, they are strongly anisotropic favoring [100], or two other equivalent directions and behave like an Ising system. The magnetic order is also very complicated [17]. Recently, we proposed a mechanism to explain these anomalies originating from the anisotropic p-f mixing [18]. The treatment so far done is, however, not complete. The more complete description for the mechanism and its effect is shown in the following briefly. More detailed descriptions will be published in separate papers.

We treat the case of an arbitrary Ce concentration. We assume for simplicity that the valence bands are made by the linear combination of the three atomic p-states, x, y and z states, on each pnictogen and introduce the p-f mixing matrices, which are given by two parameters, $(pf\sigma)$ and $(pf\pi)$ [18]. When a detailed numerical value is necessary, we use the following values, $(pf\sigma) = 0.5eV$ and $(pf\pi) = -0.5(pf\sigma)$, which are estimated from the actual band calculation of LaSb [12,18], and the 4f level E_f at 1eV below the Fermi level E_F, or $\Delta E_f \equiv E_F - E_f = 1eV$. Because the number of valence holes is in general small, we use the simplified model at first to replace all the existing hole states by the three states at Γ-point, where we neglect the spin-orbit interaction for simplicity and thus the three states are degenerated. Then it is easily shown that only the quartet can mix with the p-holes at Γ-point. Note that with increasing k vector the mixing of the valence electrons with the doublet appears. We treat the p-f mixing as a perturbation. Then, the lowest order, the second order, gives the shift down of the quartet level and the shift up of the valence band state. For the convenience, we name the quartet as follows; $Q_{\mp 3}$, which are made by $J_z = \pm 5/2$ and $\mp 3/2$, and $Q_{\pm 1}$ made by $J_z = \pm 1/2$, respectively. Then $Q_{\mp 3}$ mix with the p-states of $(x \mp iy)_{\mp}$, or written as $P_{\mp 3}$ corresponding to $j_z = \mp 3/2$ of $j = 3/2$ states for the spin-orbit split p bands, and $Q_{\pm 1}$ mix with $z_\pm$ and $(x \pm iy)_{\mp}$, or $P_{\pm 1}$ correspoinding to $j_z = \pm 1/2$ of $j = 3/2$ states. Then the shift of P_ν, where ν means one of ± 3 and ± 1, is given by

$$\Delta_{p\nu} = \frac{18}{7} \frac{c_0 c_{Q\nu}}{\Delta E_f} [(pf\sigma) - \sqrt{1.5}(pf\pi)]^2 \qquad (9)$$

in which c_0 indicates the concentration of Ce atoms, therefore unity for pure Ce-pnictides, and $c_{Q\nu}$ the population ratio of Q_ν, 1/6 for high enough temperature. Note that $j = 1/2$ states of the valence band have no contribution in the present mechanism. Therefore, there are no essential differences even if the spin-orbit interaction is included in the valence bands [18]. The corresponding shift down of the quartet states, $-\Delta_{Q\nu}$, is obtained from (9) by replacing $n_{p\nu}$ for $c_0 c_{Q\nu}$, in which $n_{p\nu}$ means the number of P_ν holes per pnictogen atom and is divided into two terms, $n^\circ_{p\nu}$ and $\Delta n_{p\nu}$, the former corresponds to the value of the dilute limit and the latter the induced term by $\Delta_{p\nu}$. In the dilute limit of Ce in YSb, for example, the total hole number was evaluated to be about 0.03/Sb. Then $n^\circ_{p\nu}$ may be given by 0.005/Sb and then $-\Delta_{Q\nu}$ is evaluated to be 97K. In CeSb, $\Delta_{p\nu}$ in the paramagnetic state for $c_{Q\nu} = 1/6$ is estimated to be 3230K, which appears as if there is an additional large ℓ-s coupling. Then $n_{p\nu}$ increases to about 0.015/Sb and thus $-\Delta_{Q\nu}$ to 290K, which nearly completely cancels the extrapolated value of about 300K expected from the point charge model, in excellent agreement with experiments. Experimentally, the dilution effect of LaSb into CeSb is much larger than that of YSb [19]. This is also explained naturally because in LaSb the vacant 4f levels are situated several eV above the Fermi energy and they push down the valence bands, in particular at the Γ-point, causing rapid decrease in the number of valence holes. Note that the term for $\Delta n_{p\nu}$ may be interpreted as a kind of multipole-multipole interaction. Note also that the quartet moment of the 4f character decreases through the p-f mixing by about 3%. The experimentally observed reduction is about 10% [16], which is expected to come mostly from the mixing with the conduction bands. The higher order perturbations give scattering and may induce the dense Kondo effect. Experimentally, however, such an effect is not yet observed.

It is clear that, because of importance of the mutual interaction term, the ground state is not the paramagnetic state as described above but the state in which only one of the quartet states, ν_0, is occupied. Then only $\Delta_{p\nu 0}$, which is given by (9) taking $c_{Q\nu 0} = 1$ and thus estimated to be 19380K = 1.67eV, is finite but all other valence states are unchanged, or changed by -3230K from the value in the paramganetic region for the other three states of the quartet. Note, however, that, in such a large shift for $\Delta_{p\nu 0}$, the second order perturbation is definitely not applicable. Then, instead of (9) we should use

$$\Delta'_{p\nu 0} = \frac{\Delta E_f}{2} [(1 + \frac{4\Delta_{p\nu 0}}{\Delta E_f})^{1/2} - 1] \tag{10}$$

which becomes 10270K = 0.886eV, still very large. In our approximation to concentrate on the Γ point, the four kinds of ordered states are degenerated, even though the elementary excitation energies are very large. However, for the shift of 0.886eV, the holes are not localized around the Γ-point but extend far in k-space. In our band used in [18], all the $p_{\pm 3}$ hole states along the Δ_z axis become vacant. In such a situation, the quartet states are no more eigen states because now the mixing between the quartet and the doublet occurs which forces back to the J_z representation. Furthermore, in any spin ordering in the 4f electrons, the d-f Coulomb exchange causes the effective field $I_{df} <\sigma_d>_n$ on the 4f spin at $\vec{R}_n$ similar to the case of EuB_6 studied before, which also causes the J_z representation. Therefore, in ordered states, the local moment is given rather well by $J_z = 5/2$ in agreement with experiment. In this sense, the mutual interaction term now becomes similar, in some sense, to the usual ferromagnetic exchange interaction. Note that the $J_z = \pm 5/2$ states mix only with the p-states on the same [001] plane for the $^2P_{\pm 3}$ valence states of k vector on the Δ_z axis, which causes a strong ferromagnetic like interaction on the x-y plane and rather weak inter-layer interaction as observed in both CeBi and CeSb. To see the strong anisotropy, we should compare the results for the quantization vector on another direction, for example [111]. So far, in [18], we calculated only the n_p^o term and neglected the interaction term, which is the main term. Detailed evaluation of the latter depends sensitively on the valence band structure used. In this sense, the experiment in paramagnetic region under high magnetic field is important because then we can controle both the quantization axis and the amount of induced moment arbitrarily which is favorable to study the present mechanism more in detail. Measurement of the magnetization of CeBi single crystal under the pulsed fields up to 400kOe have been done recently as well as careful static field measurement. However, because of the lack of the space, the detail will be published in a separate paper.

4. Acknowledgments

Various useful discussion and private communication of Profs. A. Yanase and A. Hasegawa on the band structure are highly acknowledged. Comment and private communication of Dr. T. Suzuki on experimental results of CeBi are also highly acknowledged.

References

1. See, for example, Valence Instabilities and Related Narrow Band Phenomena, ed. R. Parks, Plenum Press, N.Y., 1977, and Jour. de Physique C-5, 1979, Physics of Metallic Rare Earth.
2. J.M. Tarascon, J. Etourneau, P. Dordor, P. Hagenmuller, M. Kasaya and J.M.D. Coey, J. Appl. Phys. 51, 574 (1980).
3. A. Hasegawa and A. Yanase, J. Phys. C 12 5431 (1979).
4. J. Etourneau, R. Naslain and S. La Placa, J. Less Common Metals, 24, 183 (1971).
5. T. Kasuya, K. Takegahara, M. Kasaya Y. Isikawa and T. Fujita, Magnetic Semiconductors, Jour. de Physique Colloque C5-161 (1980).
6. Y. Takakuwa, S. Suzuki and T. Sagawa, Japan J. Appl. Phys. 17 (Suppl. 17-2) 284 (1978).
7. A. Hasegawa and A. Yanase, Magnetic Semiconductors, Jour. de Physique Colloque C5-377 (1980).
8. M. Kasaya, Y. Isikawa, K. Takegahara and T. Kasuya, Proc. of Intern. Conf. Semiconductors, held at Kyoto, 1980.
9. M. Umehara and T. Kasuya, J. Phys. Soc. Japan 40, 13 (1976).
10. G.L. Olcese, J. Phys. F 9, 569 (1979).
11. G. Güntherodt, E. Kaldis and P. Wachter, Solid State Commun. 15, 1435 (1974).
12. A. Hasegawa, J. Phys. C (in press)
13. R.J. Birgeneau, E. Bucher, J.P. Maita, L. Passell and K.C. Turberfield, Phys. Rev. B8, 5345 (1973).
14. R. Hauger, E. Kaldis, G. von Schulthess, P. Wachter and Ch. Zürcher, J. Mag. Mag. Mat. 3, 103 (1976).
15. H. Heer, A. Furrer, W. Hälg and O. Vogt, J. Phys. C 12, 5207 (1979).
16. J.X. Boucherle, A. Delapalme, C.J. Howard, J. Rossat-Mignod and O. Vogt, Proc. of Intern. Symp. on Physics of Actinides and Related 4f Materials, held at Zürich, 1980.
17. J. Rossat-Mignod, P. Burlet, J. Villain, H. Bartholin, Wang Tcheng-Si, D. Florence and O. Vogt, Phys. Rev. B 16, 440 (1977).
18. K. Takegahara, A. Yanase and T. Kasuya, Magnetic Semiconductors, Jour. de Physique Colloque C5-327 (1980), K. Takegahara, H. Takahashi, A. Yanase and T. Kasuya, to be submitted to J. Phys. C.
19. B.R. Cooper and O. Vogt, Jour. de Physique 32, C1-1026 (1971).

Anisotropic Scattering by Rare-Earth Impurities: Effect of a High Magnetic Field

J.C. Ousset, S. Askenazy

Laboratoire de Physique des Solides et Service des Champs Intenses, I.N.S.A.
31077 Toulouse-Cedex, France, and

A. Fert

Laboratoire de Physique des Solides
91405 Orsay-Cedex, France

Abstract

Scattering of conduction electrons by rare-earth impurities dilute in a noble metal induce an isotropic (exchange interaction) and an anisotropic (quadrupolar interaction) contribution to the magnetoresistance. Using a high magnetic field to saturate the magnetoresistance, we determine with a good precision the characteristic energies of these coupling processes.

Introduction

Isotropic and anisotropic scattering processes of conduction electrons by rare-earth magnetic ions have been widely studied these last years [1, 2]. But only the application of a high magnetic field (>30 T) which entirely polarizes the magnetic ions, can allow determining with good precision the coupling characteristic parameters:

- classical exchange coupling (V_{exch})
- quadrupolar coupling (V_{qd}) with the 4 f-shell of the rare earth.

In order to determine these parameters, we have done measurements of isotropic and anisotropic magnetoresistance ($\Delta\rho_I/\rho_0 = (\Delta\rho_{//} + 2\Delta\rho_{\perp})/3\rho_0$, $\Delta\rho_A/\rho_0 = (\rho_{//} - \rho_{\perp})/\rho_0$) on silver and gold polycrystals containing about 1 % of heavy rare-earths. First, we present the theoretical models describing the observed effects. Then we give our experimental results and discuss them through these models ; so, we show their limits and the necessity of further theoretical work. Finally we present our results on noble metals-rare-earth (R-E) amorphous alloys ($Ag_{50} Lu_x R-E_{1-x}$) and show the weak influence of the atomic order surrounding the rare-earth ions on the anisotropic scattering due to the 4f-quadrupole.

A. Polycrystalline Alloys

1. Theoretical Models

The interaction potential between conduction electrons and rare-earth impurities includes three terms :

$$V = V_0 + V_{exch} + V_{qd} \qquad (1)$$

V_0 is a spherical coulombian potential ; V_{exch} and V_{qd} are spin exchange and quadrupole terms of the k-f interaction respectively.

Calculations of the anisotropic terms of the k-f interaction based on conduction electrons in plane waves states [3] give values much too small to explain the measured effects. The significant anisotropy found can only be explained by the formation of a 5d virtual bound state (v b s) on the rare-earth impurities, i.e. by the admixture of 5d states into the conduction band. The conduction electrons can strongly feel the anisotropy of the 4f-shell because the admixed 5d states lie closely to the 4f electrons.

1.1. Phenomenological Model

In a monovalent noble metal V_0 attracts about 2 electrons round a trivalent rare-earth ion, mostly in 6s and 5d states. The attraction of screening electrons into 5d states is usually described as the formation of a 5d non-magnetic v b s. Calculations are made using the phase-shift theory and V_{exch} and V_{qd} are treated as perturbations [4]. The k-f interaction with formation of a 5d v b s makes the $\ell = 2$ terms to be predominant [5]:

$$V_{exch} = -4\pi \frac{J^{(2)}}{N} \sum_{m,k,k'} Y^*_{2m}(\Omega_k) Y_{2m}(\Omega_{k'}) \times \left[(a^+_{k'+} a_{k+} - a^+_{k'-} a_{k-}) J_z \right.$$
$$\left. + a^+_{k'+} a_{k-} J^- + a^+_{k'-} a_{k+} J^+ \right] \quad (2)$$

$$V_{qd} = 4\pi \frac{D^{(2)}}{2N} \left[J_z^2 - \frac{J(J+1)}{3} \right] \times \sum_{m k k' \sigma} (m^2 - 2) Y^*_{2m}(\Omega_k) Y_{2m}(\Omega_{k'})$$
$$a^+_{k'\sigma} a_{k\sigma} \quad (3)$$

where $J^{(2)} = (g_J - 1) \Gamma^{(2)}$ and $D^{(2)} = \frac{L(S - 7/4)}{J(J-1/2)} d^{(2)}$ characterize the exchange and quadrupolar interactions.

Under these conditions the anisotropic magnetoresistance for a polycrystal is written, at first order in $D^{(2)}$ (elastic terms) :

$$\frac{\rho_{//} - \rho_\perp}{\rho_0} = 6\pi \; n(E_F) \; \frac{\sin \eta_2 \cdot \cos \eta_2}{\sin^2 \eta_0 + 5 \sin^2 \eta_2} \; D^{(2)} \left[\overline{\langle J_z^2 \rangle} - \frac{J(J+1)}{3} \right] . \quad (4)$$

The symbol $\overline{\langle \rangle}$ represents a thermal average and an average on all the grain orientations. We shall assume values of the phase shifts ($\eta_0 = \pi/2$, $\eta_2 = \pi/10$) corresponding to reasonable values of screening charge ($Z_0 = 1$, $Z_2^2 = 1$) and consistent with the experimental values of residual resistivity. The saturation of the anisotropy by a high field ($J_z^2 = J^2$) allows us to directly determine the parameter $D^{(2)}$. The same calculation for the isotropic contribution (elastic terms) gives a well-known negative behaviour [6] ($\Delta\rho/\rho_0 \sim -(J_z)^2$) towards a saturation. However we have revealed an original positive component that the model is unable to explain. After our measurements FERT and LEVY have developed a new theoretical model [7] which includes the scattering probability terms of second order in $\Gamma^{(2)}$ and $D^{(2)}$ (elastic and inelastic terms).

1.2. The Model of FERT and LEVY

These authors have entirely calculated the k-f interaction and the resulting magnetoresistance. In a first model the v b s is supposed to be split into t_{2g} and e_g v b s by the crystal electric field but the splitting by the t_{2g} spin-orbit interaction is ignored. The best agreement with experimental data is obtained for a pure t_{2g} v b s. In this case they found:

$$\frac{\rho_{//} - \rho_{\perp}}{\rho_0} = \frac{18}{7}\sqrt{\frac{3}{35}} \frac{\cos n_{2t} \sin^3 n_{2t}}{\sin^2 n_0 + 3\sin^2 n_{2t}} \frac{L\ (S-7/4)}{J(J-1/2)} \frac{A_2}{\Delta}\left[\overline{<J_z^2>} - \frac{J(J+1)}{3}\right] \quad (5)$$

$$\frac{\Delta\rho_I}{\rho_0} = \frac{\sin^4 n_{2t}}{\sin^2 n_0 + 3\sin^2 n_{2t}} \left\{ \frac{C_1^2}{\Delta^2}\left[<J - \omega J>_H - <J.\ \omega J>_{H=0}\right] - \frac{C_2^2}{\Delta^2} \right.$$

$$\left. \times \frac{\sin^2 n_{2t}}{\sin^2 n_0 + 3\sin^2 n_{2t}} (<J_z>)^2 \right\} \quad (6)$$

A_2, C_1, C_2 are related to 4f-5d SLATER integrals, Δ is the half-width of the v b's.

$$<J.\ \omega J> = \sum_{II'} P_{(I)}\ (\Delta_{II'}/kT)/(e^{\Delta_{II'}/kT} - 1).\ <I'|J|I><I|J|I'> \quad (7)$$

I and I' being ion states. Eq. (6) includes elastic and inelastic terms. Eq. (5) has been limited to term of first order in $D^{(2)}$ (elastic term) which is predominant for $(\rho_{//} - \rho_{\perp})$.

This model without spin-orbit coupling cannot account for the skew-scattering. LEVY et al. have recently developed a model taking into account the splitting of the t_{2g} v b s in $j = 3/2$ and $j = 1/2$ states ; this splitting is characterized by the parameter $r = Z_{3/2}/Z_{3/2} + Z_{1/2}$ ($r = 2/3$ for an unsplit v b s). Experimental data are well described by a t_{2g} v b s weakly split by the spin-orbit coupling [8] ($r = 0.74$ for Ag-$2g$ rare-earths).

1.3. Crystal-Field Effects

1.3.1. Isotropic Magnetoresistance

The second term in (6) gives the usual negative variation. The first one allows describing the positive contribution we have observed. Indeed if the total momentum is quenched by the crystal field on the ground state and progressively developed by the magnetic field ($<J.\ \omega J>$) increases, the exchange scattering is increased and the magnetoresistance can become positive at high field.

1.3.2. Anisotropy

We have seen that the anisotropy depends on the quadrupolar polarization $J_z^2 - \frac{J(J+1)}{3}$. At high field ($J_z^2 = J^2$) it is independent of the crystal-field but at low field it is very sensitive to the scheme of crystal-field splitting.

163

For example, if the ground state is a quartet or a triplet different values of J_z^2 are associated with the states of the quartet or of the triplet, and the ZEEMAN splitting of the ground state directly results in a contribution to the resistivity anisotropy. This contribution is proportional to $(\mu_B H/kT)^2$ in the low-field limit. Then one expects a partial saturation within the ground state, followed by a slow increase towards the complete saturation, due to the admixture of excited states. If the ground state is a singlet or a KRAMERS doublet, the quadrupolar polarization is only due to the admixture of excited states and the behaviour is very different (no partial saturation).

2. Experimental Results

2.1. Measurements

We have measured the monotonous magnetoresistance of polycrystalline samples of silver and gold containing about 1 % of heavy rare earths. The resistivities are measured by a usual alternating current (f = 100 KHz) four points method. The magnetic field is a pulsed field up to 40 T with a long decreasing time ($\sim$ 200 ms).

2.2. Magnetoresistance from Magnetic Origin

We have to separate the two contributions to the magnetoresistance :
- the first one from magnetic origin is predominant at low temperature and decreases rapidly with increasing temperature.
- the second one, called normal magnetoresistance, obeys KOLHERS' rule ($\Delta\rho/\rho_0 = F(H/\rho_0)$) and is predominant at high temperature.

From measurements at 100 K we have deduced the normal magnetoresistance at 4.2 K and then the magnetoresistance from magnetic origin at low temperature.

So, we define 3 parameters a, b, c characterizing the exchange and quadrupolar scatterings :

$$\frac{\rho_{//} - \rho_{\perp}}{\rho_0} = a \left[\overline{<J_z^2>} - \frac{J(J+1)}{3} \right] \tag{8}$$

$$\frac{\Delta\rho_I}{\rho_0} = \frac{\Delta\rho_{//} + 2\Delta\rho_{\perp}}{3\rho_0} = b \left[<J \cdot \omega J>_H - <J \cdot \omega J>_{H=0} \right] - c \, (<J_z>)^2 \tag{9}$$

Fixing the crystal-field parameters (x, C_4) [9] we have computed $<J_z>$, $<J_z^2>$ and $<J \cdot \omega J>$ and drawn the calculated curves (dashed lines on Fig. 1) giving the best agreement with experimental data (solid lines).

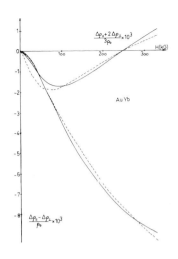

Fig. 1 Anisotropic and isotropic magnetoresistance of a gold based alloy (solid lines) and calculated curves (dashed lines).

2.3. Analysis of the Results

We give in the table the crystal-field parameters x, C_4 and the parameters a, b, c corresponding to the best fits. However we have not been able to fit the isotropic magnetoresistance of many alloys with values of b and c reasonably consistent with the value of $\Gamma(2)$ determined for Gd impurities [4]. Therefore we give b and c for only few alloys.

Table Measured parameters

Rare earth	Matrix	Crystal-field		a	b	c	E_{qd} (eV)
		x	C_4 (K)				
Tb	Ag						
	Au	0.80	-30	$1.02 \cdot 10^{-3}$			0.30
Dy	Ag	0.50	-50	$0.96 \cdot 10^{-3}$	$0.39 \cdot 10^{-3}$	$0.21 \cdot 10^{-3}$	0.45
	Au	0.63	-24	$0.51 \cdot 10^{-3}$			0.24
Ho	Ag	-0.37	-70	$0.15 \cdot 10^{-3}$	$0.15 \cdot 10^{-3}$	$0.07 \cdot 10^{-3}$	0.20
	Au	-0.35	-20	$0.19 \cdot 10^{-3}$			0.10
Er	Ag	-0.30	-50	$-1.25 \cdot 10^{-3}$			-0.58
	Au	-0.40	-24	$-0.23 \cdot 10^{-3}$			-0.11
Tm	Ag	0.56	-70	$-2.85 \cdot 10^{-3}$			-0.84
	Au	0.40	-20	$-1.61 \cdot 10^{-3}$			-0.48
Yb	Ag						
	Au	0.74	-20	$-2.31 \cdot 10^{-3}$	$0.67 \cdot 10^{-3}$	$0.6 \cdot 10^{-4}$	-0.22

From a, we deduce $D^{(2)}$ (see (4) and (8)) or more precisely the characteristic energy of the quadrupolar coupling : $E_{qd} = 5\ J(J - 1/2)\ D^{(2)}$.

On fig. 2 we have plotted the experimental values of E_{qd} for the heavy rare earths in gold. The dashed line gives a variation proportional to the STEVENS factor $\frac{L\ (S - 7/4)}{J(J-1/2)}$ as expected by the theory. The dotted dashed line shows the variation of the exchange energy ($E_{is.exch} = 5\ (g_J - 1)\ \Gamma^{(2)}$) calculated with the value of $\Gamma^{(2)}$ which has been found from the isotropic magnetoresistance of AuGd alloys [4]. (The determination of $\Gamma^{(2)}$ is much more simple and accurate in the S-ion system AuGd than in other non-S alloys).

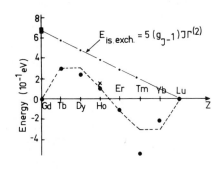

Fig. 2 Experimental results and expected scaling for E_{qd} and $E_{is.exch}$ for heavy rare earths in gold.

3. Discussion of the Results

3.1. Quadrupolar Interaction

The scaling law associated to the variation of the STEVENS factor is well verified in gold. In silver it seems that the quadrupolar coupling increases towards the end of the series for Er and Tm. Only the alloys with Tm have been difficult to study. Indeed with field values up to 30 T we only reach 40 % of the saturation value of the anisotropy ; so it is impossible to deduce E_{qd} with certainty because E_{qd} is very sensitive to the choice of the crystal-field parameters. For the others alloys we can determine E_{qd} with a very good precision (~10 %).

3.2. Exchange Interaction

The saturation of the negative magnetoresistance for alloys with gadolinium is reached at low field (H ~ 30 kG). In this case we can deduce with a good precision $E_{is.exch}$ (~10 %). For the others alloys we suppose an usual scaling law (proportional to DE GENNES factor).

It turns out that the exchange and quadrupolar interactions are of the same order of magnitude. $E_{is.exch}$ is greater than E_{qd} at the beginning of the series of the heavy rare earths while E_{qd} and $E_{is.exch}$ become nearly equal for Er. For Tm E_{qd} is greater than $E_{is.exch}$.

4. Discussion of the Theoretical Model

The model developed by FERT and LEVY expresses the various terms of the k-f interaction with 4f-5d SLATER integrals and the $\ell = 2$ phase shift. The possible choices of the phase shifts compatible with other experimental data (residual resistivity, thermoelectric power) weakly affect the results. The agreement with experiment has been obtained by the choice of 4f-5d integrals. A good agreement is obtained by taking atomic 4f-5d COULOMB integrals. However to interpret the isotropic magnetoresistance of alloys with Gd, one has to reduce the atomic 4f-5d exchange integrals by about a factor 2. It seems that the 5d v b s is more radially extended than a 5d atomic state : this could explain the reduction of the exchange integrals but one would expect a reduction of the direct integrals which depend on the distance between the 4f and 5d shells.

We have interpreted with a good precision the anisotropic magnetoresistance of the studied alloys and the isotropic magnetoresistance of alloys with Gd. The bad results for the interpretation of the isotropic magnetoresistance for non-S ions are probably due to the fact that the model neglects the quadrupolar contribution to the isotropic magnetoresistance. Such a model is going to be developed by LEVY and FERT : an analysis of the upturn of the isotropic magnetoresistance for non-S ions will give interesting informations on the crystal field.

B. Magnetic Amorphous Alloys

Recently we have extended the study of the anisotropic scattering to amorphous metals (silver$_{50}$-rare earth$_{50}$) [10].

Fig. 3 shows the results obtained for two alloys ($Ag_{50} Er_{50}$, $Ag_{50} Lu_{15} Dy_{35}$). We attribute the observed anisotropy to quadrupolar scattering ; indeed it changes its sign between Ho and Er with the STEVENS factor (L (S - 7/4)/J(J-1/2) as in the case of polycrystalline alloys.

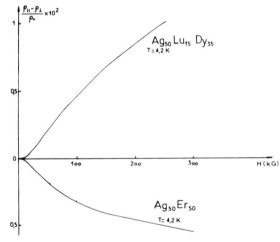

Fig. 3 Anisotropic magnetoresistance of two amorphous alloys.

In the low-field limit, the anisotropy is given by $\frac{d}{dH} (\rho_{\prime\prime} - \rho_{\perp}) = g\mu_B \, A \, \frac{J}{4 D}$. We have deduced A from experimental results, estimated the saturated anisotropy : A (J^2 - J (J + 1)/3) and compared it to the case of a polycrystal for Ag Dy alloys. We found :

$(\rho_{//} - \rho_{\perp})_{sat} = 54.10^{-2}$ μΩ.cm/% of magnetic ions for amorphous alloy
$(\rho_{//} - \rho_{\perp})_{sat} = 24.10^{-2}$ μΩ.cm/% of magnetic ions for polycrystalline alloy.

We conclude that the anisotropy and therefore the quadrupolar interaction is of the same order of magnitude in an amorphous material and in a crystalline one. This is coherent with the formation of virtual bound states well localized around the rare-earth ions.

References

1. A. FRIEDERICH and A. FERT, Phys. Rev. Let., 33, 1214 (1974)
2. G. DE VRIES and J. BIJVOET, J. of Appl. Phys., 39, 797 (1968)
3. T. KASUYA and D.H. LYONS, J. Phys. Soc. Japan, 21, 287 (1966)
4. A. FERT, R. ASOMOZA, D.H. SANCHEZ, D. SPANJAARD and A. FRIEDERICH, Phys. Rev. B, 16, 5040 (1977)
5. N.L. HUANG LIU, J.K. LING and R. ORBACH, Phys. Rev. B, 14, 4087 (1976)
6. T. VAN PESKI-TINBERGEN and R.A. WIENER, Physica, 29, 917 (1969)
7. A. FERT and P.M. LEVY, Phys. Rev. B, 16, 5052 (1977)
8. P.M. LEVY, A. FERT, G. CREUZET and J.C. OUSSET, in preparation
9. K.R. LEA, M.J.M. LEASK and W.P. WOLF, J. Phys. Chem. Solids, 23, 1381 (1962)
10. R. ASOMOZA, J.B. BIERRI, A. FERT, J.C. OUSSET, in preparation
10. J.C. OUSSET, J.P. ULMET, S. ASKENAZY, R. ASOMOZA, J.B. BIERRI, in preparation.

Magnetophonon Effect of Hot Electrons in n-InSb and n-GaAs

C. Hamaguchi, K. Shimomae, and J. Takayama

Department of Electronics, Osaka University, Suita City
Osaka 565, Japan

1. Introduction

The effect of electric field on the magnetophonon structure has been investigated by several workers in n-InSb at 77K with main emphasis on the mechanisms of the shift of the resonance extrema to higher magnetic fields with increasing electric field in the transverse configuration [1-3]. Four mechanisms are proposed for the shift [3], which are (a) repopulation effects of hot electrons at higher Landau levels in the non-parabolic conduction band, (b) electrically distorted band structure (Stark shifted Landau structure), (c) intra-collisional field effects, and (d) distortion of the distribution function produced by hot-electron pile-up close to the LO-phonon emission threshold. At low temperatures (<40K) the magnetophonon oscillations found at weak electric fields disappear as soon as ionized impurity scattering becomes dominant relaxation of the electrons. At a moderately strong electric field magnetophonon oscillations reappear with complex series of peaks at the low temperatures. The reappearance is thought of as arising from an oscillatory variation of the energy relaxation time with magnetic field [4]. These magnetophonon series are identified as arising from a process whereby an electron falls from a Landau level into a bound state of a shallow donor involving an LO phonon emission (impurity series) and arising from the inter-Landau level transition accompanied by the emission of pairs of TA phonons at the zone boundary (X point) (2TA series) [4,5]. Additional series called magneto-impurity resonance are also observed which involve the interaction between a free carrier and a neutral impurity in a magnetic field [6]. Amplitudes of the two series, the impurity and 2TA series, reflect electron population at higher Landau levels and thus the Fourier amplitudes of the two series depend on the applied electric field [7]. These observations predict a possibility to deduce mean temperature of hot electrons [7]. In this paper we report a detailed investigation of the electric field effects on the damping factor in the transverse magnetoresistance at 77K and also an estimation of the hot-electron temperature as a function of applied electric field in the longitudinal magnetic fields at a low temperature (15.5K in GaAs). The samples used are n-GaAs with $n = 3.3 \times 10^{13}$ cm^{-3} and $\mu = 1.74 \times 10^5$ cm^2/V·s and n-InSb with $n = 3.0 \times 10^{13}$ cm^{-3} and $\mu = 4.6 \times 10^5$ cm^2/V·s at 77K.

2. Electric Field Effect on the Damping Factor

Magnetic field modulation technique combined with short pulse electric field application has been developed independently by KAHLERT and SEILER [8] and KASAI et al. [9]. The use of this technique has been shown to provide high resolution measurements of the hot-electron magnetophonon structure which enable us to obtain the damping factor $\bar{\gamma}$ with a good accuracy when our method is applied to the second derivative signals [10,11]. All data were processed digitally using a YHP 9825A personal computer. The magnetophonon structure is represented by the following expression [12]

$$\Delta\sigma_{xx} \sim \exp(-\bar{\gamma}\frac{B_0}{B})\cos(2\pi\frac{B_0}{B}) \tag{1}$$

where B is the applied magnetic field and B_0 is the fundamental field which is given by $\omega_0 m^*/e$ for a parabolic band with effective mass m^*, $\hbar\omega_0$ the LO phonon energy and e the electronic charge. It is shown by simple calculations [10,11] that

$$-B_N^4\left[\frac{\partial^2\Delta\sigma_{xx}}{\partial B^2}\right]_{B=B_N} = C\exp(-\bar{\gamma}N) \equiv I_N \tag{2}$$

where B_N is the magnetic field at the Nth extremum and the integer N is given by the resonance condition $N\hbar\omega_C = \hbar\omega_0$ with ω_C the cyclotron frequency. In the present analysis we take into account the magnetic field dependence of $\bar{\gamma}$ because $\bar{\gamma}$ depends on the magnetic field as

$$\bar{\gamma} = \frac{\sqrt{2}\pi|eEl_0|}{\hbar\omega_C} \tag{3}$$

for the intra-collisional field effect [13], where E and l_0 are the total electric field (the resultant of the applied and the Hall field) and the smallest cyclotron orbit radius at $N=1$, respectively. Using the experimental data of the second derivative conductivity, we calculate I_N and plot $\ln(I_N)$ as a function of N. The slope of the plot gives a value $\bar{\gamma}_D(N)$ defined by

$$\bar{\gamma}_D(N) = -\frac{d\ln(I_N)}{dN} = \bar{\gamma}(N) + N\frac{d\bar{\gamma}(N)}{dN}. \tag{4}$$

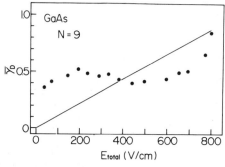

Fig.1 Damping factor $\bar{\gamma}_D$ vs total electric field in n-GaAs (T = 77K).

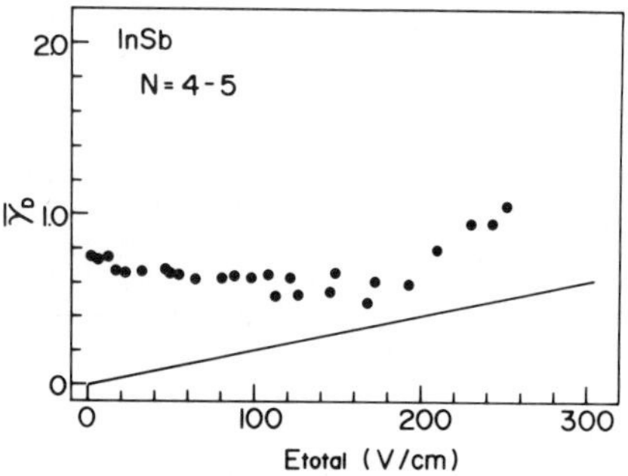

Fig.2 Damping factor $\bar{\gamma}_D$ vs total electric field in n-InSb (T = 77K).

Present results of $\bar{\gamma}_D$ are shown in Fig.1 for n-GaAs and in Fig.2 for n-InSb, where E_{total} is the total electric field (E in (3)). We find in Figs.1 & 2 that the damping factor $\bar{\gamma}_D$ is almost constant at lower electric fields but increases at higher electric fields. These results imply an importance of the intra-collisional field effect at the higher electric fields for the damping processes. The solid curves in Figs.1 & 2 are calculated from (3) and (4), which are in reasonable agreement with the experimental data at higher electric fields. It should be noted here that the damping factor $\bar{\gamma}_D$ obtained experimentally contains various components arising from such processes as the band tailing effect, single impurity scattering, optical and acoustic phonon scattering, and the intra-collisional field effect. The damping factor arising from the band tailing effect which is believed to be dominant at 77K in n-InSb [11,14] and n-GaAs [10,15,16] is independent of the applied electric field and thus the discrepancy between the experimental data and the calculated curves in Figs.1 & 2 indicates an importance of the damping effects besides the intra-collisional field effect at the lower electric fields. The band tailing effect may be most important in this region.

The increase in the damping factor gives rise to a shift of the second-derivative extrema to higher magnetic fields. However, the effect was found to be quite small. The calculated shift at the largest electric field is about 0.6% in InSb for $N = 4 \sim 6$ and about 0.4% in GaAs for $N = 8$ and 9, while the observed shift is about 9% in both samples. The effect of electrically distorted band structure was estimated using the method proposed by BARKER [13], but the effect was found to be negligible for the shift in both InSb and GaAs. The repopulation effects of hot electrons at higher Landau levels are plausible in the case of InSb because of the strong non-parabolicity of the conduction band. However, the large peak shifts, $8 \sim 9\%$ for $N = 6 \sim 8$ in GaAs, cannot be explained by the non-parabolicity. Distortion of the distribution function [3,17,18] gives rise to a peak shift or extremum inversion, but detailed analysis has not been carried out.

3. Hot Electron Temperature

The complex structure of the magnetophonon oscillations at low temperatures was analyzed by means of Fourier transformation. We concentrate our attention on the Fourier amplitudes arising from the impurity series I(imp) and the 2TA phonon series I(2TA) in the longitudinal magnetoresistance. Their resonance conditions are given, respectively, by $\hbar\omega_0 = N'\hbar\omega_c + E_I(B)$ and $2\hbar\omega_{TA} = M\hbar\omega_c$, where $E_I(B)$ is the magnetic field dependent activation energy of the donor, $\hbar\omega_{TA}$ is the TA phonon energy, and N' and M are integers. The present results at 15.5K in n-GaAs are shown in Fig.3. The ratio of the Fourier amplitudes of the 2TA series to the impurity series decreases with increasing

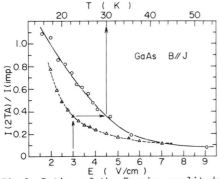
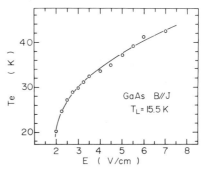

Fig.3 Ratios of the Fourier amplitudes for the 2TA to impurity series as a function of applied electric field E at 15.5K (△) and as a function of lattice temperature at low E (o). (n-GaAs, J∥B)

Fig.4 Electron temperature as a function of applied electric field determined from Fig.3 at 15.5K. (n-GaAs, J∥B)

the electric field, reflecting the relative change in the electron population at the Landau levels involved in the two series. The decrease in the relative amplitudes is interpreted in terms of the increase in the electron temperature since it is proportional to $\exp[-(2\hbar\omega_{TA} - \hbar\omega_0 + E_I)/k_B T_e]$ when we assume Boltzmann distribution for the electrons with mean temperature T_e. These two peaks were also measured in the temperature range from 15.5 to 50K at electric field as low as possible so that the observed oscillations can be understood as arising from scattering of electrons with mean temperature very close to the lattice temperature. The observed ratio of the Fourier amplitudes of the 2TA to the impurity series was found to decrease as the lattice temperature is raised. The present results are shown by open circles in Fig.3. Noting that the occupation number of the LO or TA phonons is much less than unity ($n_q \ll 1$, $n_q + 1 \simeq 1$), the amplitude ratio becomes proportional to $\exp[-(2\hbar\omega_{TA} - \hbar\omega_0 + E_I)/k_B T]$ on the assumption that the energy relaxation of the electrons is limited by the emission of the LO phonons or the pair emission of the TA phonons, where T can be taken to be the lattice temperature at extremely low electric fields. Therefore the ratio decreases as the temperature is raised, which is consistent with the present observations.

The electron temperature at a high electric field can be estimated by the following procedures. We compare the ratios of the Fourier amplitudes for the two independent measurements as shown in Fig.3 in the case of applied electric field E = 3V/cm, where we obtain electron temperature 30K. The electron temperatures thus obtained are shown in Fig.4 for n-GaAs at 15.5K. We find in Fig.4 that the electron temperature increases at high electric fields, reaching 42K at 7V/cm. Another method to deduce electron temperature by comparing the resistivity $\rho(E)$ and $\rho(T)$ [19] was used and we found a good agreement in the case of n-InSb between the two methods.

The present method is based on the following assumptions. The distribution function in the presence of electric and magnetic fields is well represented by a Maxwellian distribution function with an electron temperature T_e and ionized-impurity scattering is the dominant mechanism for the momentum relaxation. The latter assumption seems to be valid in the present experiment as stated earlier and found in the temperature dependence of the electron mobility. The former assumption is believed to be adequate in the case of dominat interelectronic collisions where the energy and momentum relaxation frequency of electron-electron scattering exceeds the frequency of electron-phonon scatterings. The situation becomes favourable when the carrier concentration exceeds a critical value [20]. Although the validity of the assumption has not been proved, theoretical calculations based on the Maxwellian distribution have been extensively carried out [13,21,22]. On the other hand, theoretical analyses have been performed to obtain the distribution functions and have revealed that the distributions are significantly non-Maxwellian [13, 17, 18, 23]. The most significant feature is the cut-off of the distribution function beyond the LO phonon threshold and distortion at the Landau sub-band edges. In spite of the problems stated above several attempts [24] have been made to obtain the electron temperature from the measurements of resistivity [19], Shubnikov-de Haas effect [25], and cyclotron resonance [26]. MATSUDA and OTSUKA [26] have proposed the two electron temperature model to describe distorted distribution function at the sub-bands. They introduced two different electron temperatures: one is the inter sub-band electron temperature T_e^{inter} defined by the relative populations at the Landau sub-bands and the other is the intra sub-band electron temperature T_e^{intra} defined by the electron distribution within each sub-band. They found from an analysis of cyclotron resonance in n-InSb at 4.2K that the difference in lineshape for cyclotron resonance absorption reflects the difference in T_e^{intra}, while the difference in relative intensity of the conduction band cyclotron resonance is related to the difference in T_i^{inter}. If we apply this model to the present results, the obtained electron temperature corresponds to T_i^{inter}. Although the shape of the distribution function is not known, we believe that the present method gives information on electron heating at high electric fields.

References
1. R. C. Curby and D. K. Ferry, Proc. 11th Int. Conf. Phys. Semicond. Warsaw, 1972, p.312. Pol. Scientific Publ. Warsaw, (1972).
2. K. Kasai and C. Hamaguchi, J. Phys. Soc. Jpn. 44, 1749 (1978).
3. H. Kahlert, D.G. Seiler and J.R. Barker, Solid-St. Electron. 21, 229 (1978).
4. R. A. Stradling, L. Eaves, R A Hoult, A.L. Mears and R.A. Wood, Proc. 10th Int. Conf. Phys. Semicond. Cambridge, 1970, p.369. USAEC (1970).
5. R. A. Stradling and R. A. Wood, J. Phys. C: Solid State Phys. 3, 2425 (1970).
6. L. Eaves and J. C. Portal, Application of High Magnetic Fields in Semicond. Phys., Oxford, p.118 (1978) and J. Phys. C:Solid State Phys. 12, 2809 (1979), and references therein.
7. K. Senda, K. Shimomae, K. Kasai and C. Hamaguchi, J. Phys. C: Solid State Phys. 13, 1043 (1980).
8. H. Kahlert and D. G. Seiler, Rev. Sci. Instrum. 48, 1017 (1977).
9. K. Kasai, T. Shirakawa and C. Hamaguchi,Tech. Rept. Osaka Univ. 28, 157 (1978).
10. K. Senda, K. Shimomae, K. Kasai and C. Hamaguchi, J. Phys. Soc. Jpn. 47, 551 (1979).
11. K. Kasai and C. Hamaguchi, J. Phys. Soc. Jpn. 47, 1159 (1979).
12. R. A. Stradling and R. A. Wood, J. Phys. C:Solid State Phys. 1, 1711 (1968).
13. J. R. Barker, Solid-St. Electron. 21, 197 (1978).
14. I. Fujisawa, Jpn. J. Appl. Phys. 17, 667 (1978).
15. L. Eaves, R. A. Hoult, R. A. Stradling, R. J. Tidey, J. C. Portal and S. Askenazy, J. Phys. C:Solid State Phys. 8, 1034 (1975).
16. J. R. Barker, J. Phys. C:Solid State Phys. 5, 1657 (1972).
17. J. R. Barker and B. Mugnusson, Proc. 12th Int. Conf. Phys. Semicond. Stuttgart, 1974, p.811. G. B. Tenbner (1974).
18. E. Yamada, Solid St. Commun. 13, 503 (1974).
19. H. Miyazawa, J. Phys. Soc. Jpn. 26, 700 (1969).
20. R. Stratton, Proc. Roy. Soc. A 246, 406 (1958).
21. R. L. Peterson, Semiconductors and Semimetals, (Edited by R.K. Willardson and A. C. Beer) Vol.10, p.221. Academic , New York (1975).
22. D. C. Rohlfing and E. W. Prohofsky, Phys. Rev. B 12, 3242 (1975).
23. E. Yamada and T. Kurosawa, J. Phys. Soc. Jpn. 34, 603 (1973).
24. G. Bauer:"Determination of Electron Temperatures and of Hot Electron Distribution Functions in Semiconductors", in Solid-State Physics, Springer Tracts in Modern Physics, Vol. 74 (Springer, Berlin, Heidelberg, New York 1974) p.1
25. H. Kahlert and G. Bauer, Phys. Rev. B 7, 2670 (1973).
26 O. Matsuda and E. Otsuka, J. Phys. Chem. Solids 40, 809 and 819 (1979).

Transport Equations Treating Phonon Drag Effect in a Strong Magnetic Field

Y. Ono

Department of Physics, University of Tokyo, Hongo 7-3-1, Bunkyo-ku
Tokyo, Japan

1. Introduction

The first theoretical work on the phonon drag transport in quantizing magnetic fields was made by GUREVICH and NEDLIN [1] who derived a set of coupled transport equations for electron and phonon distribution functions by using a diagramatic method proposed by KONSTANTINOV and PEREL' [2]. Similar equations were derived by FISCHER [3] using a truncation approximation for the equation of motion of Green's functions and by SUGIHARA [4] with a phenomenological argument. These authors all neglected the renormalization of phonon frequency due to the electron-phonon interaction. On the other hand, recent theoretical investigations [5-7] indicate that the electronic density correlation function in strong magnetic fields has singularities as a function of the wave vector because of the one-dimensional character of the electronic motion. This fact means that a phonon with a particular wave vector is expected to be softened due to the electron-phonon interaction. Such a softening of the phonon frequency will lead to an anomalous dependence of the phonon-drag thermopower on the magnetic field and temperature.

The purpose of the present work is to derive a set of coupled transport equations including the phonon drag and the effect of the phonon frequency softening in strong magnetic fields. We use the standard temperature Green's function technique since this technique allows us to take account of the renormalization of quasiparticle energies due to interactions in a rather simple way.

Details of the derivation takes much space and therefore will be published elsewhere. Here we describe only the outline of the formulation.

2. Derivation of Transport Equations

Here we treat a model Hamiltonian of the following form,

$$H = H_e + H_{ph} + H_{e-ph} + H_{e-imp} \tag{1}$$

where H_e and H_{ph} represent the free electron in a magnetic field paralell to the z-axis and the free phonon, respectively, and the couplings of electrons with phonons and static impurities are expressed by the remaining two terms. Other interactions, e.g. electron-electron, phonon-phonon or phonon-defect interactions, are neglected for the sake of simplicity.

To be explicit we consider the linear response of a physical quantity P to a static electric field E. The Kubo formula gives the following expression [8],

$$<P> = -i \sum_j [\frac{\partial}{\partial \omega} Q^R_{Pj}(\omega)]_{\omega=0} E_j \tag{2}$$

where the retarded correlation function $Q^R(\omega)$ is obtained from the thermal correlation function

$$Q_{Pj}(\omega_m) = -e \int_0^\beta d\tau \exp(i\omega_m \tau) <T_\tau P(\tau) v_j> \tag{3}$$

through analytic continuation [9]. Here v_j is the j-th component of the electronic velocity operator, $-e$ the electronic charge, $\omega_m = 2m\pi T$ (m; integer) and $\beta = 1/T$ (T the temperature).

In general the operator P can be written as a sum of three terms, i.e., P_e, P_{ph} and P_{e-ph} where P_e and P_{ph} include only the electron and phonon field operators, respectively, and where P_{e-ph} includes products of electron and phonon operators. (As for the heat current operator, see e.g. [10].) As examples we write the contribution of P_e and P_{ph} to Q,

$$Q^{(e)}(\omega_m) = -T \sum_\mu \sum_\alpha \sum_{\alpha'} P_e^{\alpha\alpha'} G(\varepsilon_\mu,\alpha) G(\varepsilon_\mu+\omega_m,\alpha') \Gamma_{\alpha\alpha'}(\varepsilon_\mu,\varepsilon_\mu+\omega_m), \tag{4}$$

$$Q^{(ph)}(\omega_m) = \frac{T}{2} \sum_\nu \sum_q P_{ph}(q) D(\omega_\nu,q) D(\omega_\nu+\omega_m,q) \Lambda_q(\omega_\nu,\omega_\nu+\omega_m), \tag{5}$$

where G and D are electron and phonon Green's functions, ε_μ and ω_ν the fermion and boson frequencies with integers μ and ν, Γ and Λ the electron and phonon vertex parts, and α (or α') represents the set of quantum numbers of a Landau state. In treating the vertex parts Γ and Λ it is inevitable to use approximations. In order to satisfy the generalized Ward identity, the self-energies of Green's functions and the vertex parts should be calculated within a consistent approximation. In the present calculation, the electron-phonon interaction is treated within the Migdal approximation and the electron-impurity interaction within the self-consistent Born approximation. Correspondingly the vertex parts are calculated within the ladder approximation. It is well-known that these approximations are not sufficiently accurate in strong magnetic fields. However we dare to use them in order to avoid too much complication.

The vertex parts Γ and Λ satisfies integral equations of the Bethe-Salpeter type. Analytic continuation of the equations for Γ and Λ leads to transport equations of the following form,

$$i(\tilde{\xi}_\alpha - \tilde{\xi}_{\alpha'}) \Psi^R_{\alpha\alpha'}(z) = -ev_j^{\alpha\alpha} + I_1(\Psi^R,\Psi^A), \tag{6}$$

$$-i(\tilde{\xi}_\alpha - \tilde{\xi}_{\alpha'}) \Psi^A_{\alpha\alpha'}(z) = -ev_j^{\alpha\alpha} + I_1(\Psi^A,\Psi^R), \tag{7}$$

$$-i(\tilde{\xi}_\alpha - \tilde{\xi}_{\alpha'}) \Phi_{\alpha\alpha'}(z) = -ev_j^{\alpha\alpha} + I_2(\Psi^R,\Psi^A,\Phi,\phi,\chi), \tag{8}$$

$$0 = -ev_j^{\alpha\alpha} + I_3(\Psi^R,\Psi^A,\Phi,\phi,\chi), \tag{9}$$

$$0 = I_4(\Psi^R,\Psi^A,\Phi,\phi,\chi), \tag{10}$$

where $\tilde{\xi}_\alpha$ is the renormalized energy of the state α measured from the Fermi level, and I_1 to I_4 are integral operators of the second order with respect to the electron-phonon or the electron-impurity coupling. The functions Ψ^R, Ψ^A, Φ and ϕ are all related to $\Gamma(\varepsilon_\mu,\varepsilon_\mu+\omega_m)$ through different types of analytic continuation, e.g.,

$$\psi^R_{\alpha\alpha'}(z) = \Gamma^{RR}_{\alpha\alpha'}(z,z)[i(\tilde{\xi}_\alpha - \tilde{\xi}_{\alpha'}) - \gamma(z,\alpha') + \gamma(z,\alpha)]^{-1} , \qquad (11)$$

$$\Phi_{\alpha\alpha'}(z) = \Gamma^{AR}_{\alpha\alpha'}(z,z)[-i(\tilde{\xi}_\alpha - \tilde{\xi}_{\alpha'}) + \gamma(z,\alpha') + \gamma(z,\alpha)]^{-1} , \qquad (12)$$

where, for example, $\Gamma^{AR}(z,z)$ means an analytic continuation from the region, $-\omega_m < \varepsilon_\mu < 0$; $-\gamma(z,\alpha)$ is the imaginary part of the self-energy of the retarded electron Green's function. In the above the static limit, ω (the external frequency) $\to 0$, has been taken, and α' is assumed to be different from α. If we put $\alpha' = \alpha$ in (12) we obtain the expression for $\phi_\alpha(z)$. The function χ is proportional to an analytic continuation of Λ, i.e.,

$$\chi_q(z) = \omega_q \Lambda^{AR}_q(\dot{z},z)/2z\eta_q(z) , \qquad (13)$$

with the bare phonon frequency ω_q and the phonon damping $\eta_q(z)$.

Physically Φ and ϕ are the non-equilibrium parts of the electronic distribution function, off-diagonal and diagonal with respect to the Landau state, respectively, and χ is the deviation of the phonon distribution function from equilibrium. The meaning of ψ^R and ψ^A is not clear but it is found from the calculation of the conductivity tensor that they are related to the diamagnetic current.

3. Iterative Solutions in Strong Magnetic Fields

If the applied magnetic field is strong enough, the above integral equations (6) to (10) can be solved iteratively with $(\omega_c\tau)^{-1}$ as a smallness parameter where ω_c is the cyclotron frequency and τ is the characteristic time scale corresponding to the electronic relaxation time. In the practical calculation we have to include phonon relaxations due to mechanisms other than phonon-electron scatterings. We express the linear transport coefficients connecting the x and y components of the heat current with the x component of the electric field by α_{xx} and α_{yx}, respectively. The zeroth order solutions for Ψ's and Φ are obtained by neglecting I_1 and I_2 in (6) to (8). Substituting these solutions into (9) and (10) we obtain integral equations for $\phi^{(0)}$ and $\chi^{(0)}$ which can be solved easily in the case of an isotropic system [1]. As a result we have an expression for the zeroth order phonon contribution to α_{yx},

$$\alpha^{ph}_{yx} = -\frac{e}{m\omega_c} \sum_q \tilde{\omega}_q (\partial\tilde{\omega}_q/\partial q_y) q_y [1+\tau_{e-p}(q)/\tau_f(q)]^{-1} \frac{1}{4T} \text{cosech}^2(\tilde{\omega}_q/2T) \qquad (14)$$

where m is the effective mass of an electron, $\tilde{\omega}_q$ the renormalized phonon frequency, $\tau_{e-p}(q)$ and $\tau_f(q)$ the phonon relaxation times due to electron-phonon scattering and other scattering mechanisms, respectively. This expression is the same as that obtained by the previous authors except for the replacement of the bare phonon frequency by the renormalized one. The singularity due to the phonon softening ($\tilde{\omega}_q \to 0$) is found to be cancelled because of the $\tilde{\omega}_q$-dependence of $\tau_{e-p}(q)$. In the zeroth order, $\alpha_{xx} = 0$ from the symmetry. Going one step in the iteration, the first order solution for χ is obtained. It yields non-zero contribution to α_{xx} and zero contribution to α_{yx}. The expression for α_{xx} is omitted here since it is too lengthy. An important point is that the singularity due to the phonon softening does not disappear in this case. The previous authors did not give the expression for the phonon contribution to α_{xx}. However α_{xx} will be much more impotant than α_{yx} in some cases, e.g., in a very pure Bi crystal with complete compensation between electrons and holes (see [4] where the case of incomplete compensation is treated).

4. Remarks

As concluding remarks we mention the advatages of the present formulation compared to the previous ones.

(i) Since we used the diagram technique, it is rather easy to discuss what is neglected and what is included.

(ii) The diagram method used by GUREVICH and NEDLIN [1] is much more complicated than the present one because they treated the interactions included in the Hamiltonian expressing the time dependence of operators and in the equilibrium density matrix in different ways.

(iii) In the derivation of the transport equations in [1], the contribution of Ψ's giving rise to the inhomogeneous terms of the integral equations for Φ, ϕ and χ had to be calculated from complicated diagrams in every order of iteration. In the present formulation, however Ψ's can be obtained by solving (6) and (7) iteratively, and substituting the result into (8) to (10), we have the integral equations for Φ, ϕ and χ. Because of this simplification it became much easier to calculate higher order terms.

References

1. L. E. Gurevich and G. M. Nedlin, Soviet Phys. -JETP, 1961, 13, 568.
2. O. V. Konstantinov and V. I. Perel', Soviet Phys. -JETP, 1961, 12, 142.
3. S. Fischer, Z. Phys., 1965, 184, 325.
4. K. Sugihara, J. Phys. Soc. Jpn., 1969, 27, 356.
5. H. Fukuyama, Solid State Commun., 1978, 26, 783.
6. R. Gerhardts and P. Schlottmann, Z. Phys. B, 1979, 34, 349.
7. U. Paulus and J. Hajdu, Solid State Commun., 1976, 20, 687.
8. R. Kubo, J. Phys. Soc. Jpn., 1957, 12, 570.
9. A. A. Abrikosov, L. P. Gorkov and I. Ye. Dzyaloshinskii, Quantum Field Theoretical Method in Statistical Physics, 1965, Pergamon.
10. Y. Ono, J. Phys. Soc. Jpn., 1973, 35, 1280.

Part VI

Excitons and Magneto-Optics

Exciton and Shallow Impurity States of Semiconductors in an Arbitrary Magnetic Field

N.O. Lipari[*]

IBM T.J. Watson Research Center, P.O. Box 218
Yorktown Heights, NY 10598, USA, and

M. Altarelli

Max-Planck-Institut für Festkörperforschung, Heisenbergstraße 1
7000 Stuttgart 80, Fed. Rep. of Germany

1. Introduction

A considerable amount of theoretical and experimental investigations has been devoted to the understanding of the electronic states of solids in the presence of an external magnetic field, over the last few decades [1]. The theoretical problem of Landau levels of cubic semiconductors was first satisfactorily treated by Kohn and Luttinger [2] and their solution was later refined by several investigators [3]. This Landau-level analysis, although successful in the interpretation of cyclotron resonance experiments, is not adequate to describe magnetoabsorption near the fundamental edge. Excitons represent the fundamental electronic excitations of a semiconductor crystal. The importance of excitons was first clearly shown the measurements of Edwards and Lazazzera [4] on the direct edge of Ge, where the evident nonlinear dependence on magnetic field is in direct contrast with the predictions of the Landau-level theory. In addition, the fact that the lowest peak extrapolates as the magnetic field goes to zero, to the energy of the direct-exciton ground state is a definite indication of the importance of the electron-hole coulomb interaction. Since then, magneto-spectroscopy of excitons has received great attention because it can provide precise information on various fundamental parameters, such as effective masses, g-values, k-linear terms, exchange interaction, etc. [5]. It has also become very clear that, to gain such information, realistic exciton models and their accurate solutions in the presence of an external magnetic field are necessary [6].

Any theory that aims at interpreting experiments and extracting from them useful and accurate information must incorporate the complicated valence band structure in addition to the electron-hole coulomb interaction. The importance of the coulomb interaction was theoretically supported long ago [7] and the relevance of the complexity of the valence band structure is evident from the observed structure which shows a great number of peaks, unevenly spaced and of different intensities [8]. Efforts to include these effects have been numerous over the last several years [9,10,11]. The main difficulty of the problem lies in the simultaneous presence of the coulomb interaction and of the localization along the magnetic field lines. When one of the two effects dominates, it is possible to consider the other as a perturbation and to include it by a suitable approximation. It has been shown that, in the low field region, where the cyclotron energy is small compared to the binding energy of the exciton, a perturbation approach can be applied [6,9,12] This method is valid only for $\gamma \leq 0.4$ (γ is the dimensionless magnetic field, as defined in the next section). For most materials, however, the applied magnetic fields reach values of $\gamma \approx 5 \sim 10$. For very high fields, one can assume that the magnetic localization dominates over the coulomb interaction; this has allowed an adiabatic method to be introduced and applied with good success to a variety of materials [11,13]. For most materials, however, the highest values of γ attainable are not high enough to warrant the accurate use of the adiabatic approximation, as shown in Table I. Furthermore, even for those materials for which high values of γ are achievable, the available experimental data sweep all values of γ, and interesting effects occur at intermediate fields. Such effects cannot be explained by the perturbative method, nor by the adiabatic approximation. Finally, the two methods, perturbative in the limit of very low, or adiabatic for very large magnetic fields, employ such different basis states that it is impossible to predict how the various levels are connected at intermmediate fields and how their oscillator strengths vary. Recently [14], Ekardt has introduced a method which allows an exciton treatement for intermediate values of the magnetic field but is not very accurate for small fields. More recently [15], Bajaj and Aldrich have also considered the case of direct excitons for arbitrary magnetic fields.

Table I Magnetic field H (kG) corresponding to $\gamma = 0.4$ (see definition of γ in text) for direct excitons and acceptors in various semiconductors. The values of ϵ and γ_1 are taken from ref. 20; the values of m_e^* are taken from P. Lawaetz, Phys. Rev. B4, 3460 (1971).

Substance	ϵ	m_e^*	γ_1	Excitons H	Acceptors H
AlSb	12.0	0.018	4.15	2.70	381.
GaP	10.75	0.170	4.20	80.3	463.
GaAs	12.56	0.067	7.65	11.7	102.
GaSb	15.7	0.045	11.80	3.31	27.5
InP	12.4	0.080	6.28	17.4	156.
InAs	14.6	0.023	19.67	1.11	11.4
InSb	17.9	0.014	35.08	.260	2.39
Ge	15.36	0.038	13.35	2.50	22.4
ZnS	8.1	0.28	2.54	385.	2230.
ZnSe	9.1	0.14	3.77	96.0	802.
ZnSe	16.46	104.			

For impurity states, the values of the effective magnetic field γ may be considerably different than that for excitons, since one of the two particles, the electron for acceptors and the hole for donors, is missing. Here we shall discuss mainly the acceptor problem because it is intrinsically more complicated, even though the analysis can be applied to donors as well. For acceptors, the effective field γ is about an order of magnitude smaller than the corresponding one for excitons, as shown in table I. As a result, one is very often in the perturbative region for the magnetic field. It is however important to note that, for acceptors, the Hamiltonian in the absence of external field is completely different than that of a simple hydrogen atom. It is, in fact, well known that acceptors have spectra considerably more complicated than the simple hydrogen atom even in the absence of external fields. A proper inclusion of the magnetic field, therefore, even in the perturbative region, is very difficult and any treatment that assumes hydrogenic states for the unperturbated problem is inaccurate.

The purpose of the present paper is to describe a method which is able to treat accurately the exciton and shallow impurity problem for all values of the magnetic field. The method is based on the tensor operator formalism introduced previously for the case of impurities [16]. It is completely general and does not use any explicit matrix representation, thus making it simple and efficient. In section II, the method is introduced and described. In section III, the general solution is presented. In section IV, the accuracy of the method is tested and compared with other available methods. Finally, in the last section, we summarize the main results and we discuss possible lines of future developments.

II. Formulation of the problem

In the presence of an external magnetic field $\vec{H}$, the wave function of an exciton at rest in the crystal can be written as [17]:

$$\Omega_n(\vec{r}_e,\vec{r}_h) = V^{-\frac{1}{2}} \exp\left(\frac{ie}{2\hbar c}\vec{H}\cdot(\vec{r}_e\times\vec{r}_h)\right)$$

$$\sum_{i=1}^{r}\sum_{j=1}^{s} \Psi_{n,ij}(\vec{r}_e-\vec{r}_h)\, \phi_{ci}(\vec{r}_e)\, \phi_{vj}(\vec{r}_h) \qquad (1)$$

where V is the volume of the crystal, $\phi_{ci}(\vec{r}_e)$ and $\phi_{vj}(\vec{r}_h)$ are the Block functions of the ith conduction band and the jth valence band at the Γ point, respectively, and $\vec{r}_e$ and $\vec{r}_h$ denote the electron and the hole position, respectively. The summation over i, in the cases of interest here, i.e. in cubic semiconductors, includes two terms, corresponding to spin up or down in the conduction band; the sum over j includes four terms corresponding to the upper multiplet of the spin-orbit-split valence band. The index n labels all the quantum numbers associated with the relative electron-hole motion. The function $\Psi_{n,ij}(r)$, where $\vec{r} = \vec{r}_e - \vec{r}_h$ satisfy the effective mass equation:

$$H_{ex}\Psi = E\Psi \qquad (2)$$

where, using the gauge $\vec{A} = 1/2(\vec{H}\times\vec{r})$

$$H_{ex} = H_e\left(-i\vec{\nabla} + \frac{e}{2\hbar c}(\vec{H}\times\vec{r})\right)$$

$$- H_h\left(i\vec{\nabla} + \frac{e}{2\hbar c}(\vec{H}\times\vec{r})\right) + H_{exch} - \frac{e^2}{\varepsilon r} \qquad (3)$$

$\vec{A}$ is the vector potential describing the field $\vec{H}$ and ε is the static dielectric constant. Since $\Psi_{n,ij}$ (i=1,2,; j=1,2,3,4) is an eight-component wave function, Eg. (2) corresponds to an eight-by-eight system of differential equations. H_e represents the kinetic energy near the minimum of the conduction band, H_h is the corresponding quantity for the valence band, and H_{exch} represents the electron hole exchange coupling. The explicit expressions for the conduction and valence band hamiltonians are:

$$H_e(p) = \frac{\hbar^2 p_3^2}{2m_l} + \frac{\hbar^2(p_1^2 + p_2^2)}{2m_t} + \bar{\mu}^* \vec{\sigma}\cdot\vec{H} \qquad (4)$$

and

$$-H_h(p) = \frac{\hbar^2}{2m_o}\left[(\gamma_1 + \frac{5}{2}\gamma_2)\frac{1}{2}p^2 - \gamma_2(p_x^2 J_x^2 + p_y^2 J_y^2 + p_z^2 J_z^2)\right.$$

$$-2\gamma_3(\{p_x p_y\}\{J_x J_y\} + \{p_x p_z\}\{J_x J_z\} + \{p_y p_z\}\{J_y J_z\})$$

$$\left.-\frac{e}{c}\kappa\vec{J}\cdot\vec{H} - \frac{e}{c}q(J_x^3 H_x + J_y^3 H_y + J_z^3 H_z)\right] \quad (5)$$

respectively, where γ_1, γ_2, γ_3, κ and q are the five valence band parameter introduced by Luttinger [18]. In Equation 4, m_t and m_l are the transverse and longitudinal conduction effective masses and $\bar{\mu}^*$ the effective magnetic moment of the conduction band. In the present paper, we assume that the electron ellipsoidal axes 1,2 and 3 are the same as the crystal cubic axes x, y and z, and that the magnetic field H is in the z direction. More general situations can be treated by a simple extension of the present analysis. In Equation 5, m_o is the free-electron mass, J_x, J_y and J_z are the 4x4 angular momentum matrices corresponding to a spin 3/2 state (4 fold degeneracy of the top of the valence band). The exchange hamiltonian is approximated as [19]:

$$H_{exch} = \Delta_1 \delta(r)\, \vec{J}\cdot\vec{\sigma} \quad (6)$$

In a series of papers [16,20,21], it has been shown that a deeper understanding and simplification of the problem can be obtained if one reformulates the effective mass equations in terms of new operators. In absence of external fields, we have introduced the second-rank cartesian tensor operators.

$$P_{ik} = 3p_i p_k - \delta_{ik}\, p^2 \quad (7)$$

and

$$J_{ik} = \frac{3}{2}(J_i J_k + J_k J_i) - \delta_{ik}\, J^2 \quad (8)$$

and the corresponding irreducible components $P_q^{(2)}$ and $J_q^{(2)}$ (q = -2,-1,0,1,2). In presence of the magnetic field one needs to define also the following second-rank tensor:

$$R_{ik} = 3x_i x_k - \delta_{ik}\, r^2 \quad (9)$$

$$T_{ik} = \frac{3}{2}(x_i p_k + x_k p_i) - \delta_{ik}(\vec{x}\cdot\vec{p}) \quad (10)$$

and their irreducible components. Also to be defined are the first order tensor:

$$\vec{L} = \vec{r}\times\vec{p} \quad (11)$$

and the third order tensor

$$J_0^{(3)} = J_x\left(2J_z^2 - J_x^2 - J_y^2\right) \tag{12}$$

In the following, we use the effective Rydberg energy:

$$R_0 = \frac{e^4 \mu_0}{2\hbar^2 \varepsilon^2}, \tag{13}$$

the effective Bohr radius

$$a_0 = \frac{\hbar^2 \varepsilon}{\mu_0 e^2} \tag{14}$$

and the dimensionless quantity

$$\gamma = \frac{e\hbar H}{2\mu_0 c R_0} \tag{15}$$

as units of energy, length and magnetic field respectively. Here

$$\frac{1}{\mu_0} = \frac{1}{\mu_{0e}} + \gamma_1 \tag{16a}$$

$$\frac{1}{\mu_{0e}} = \frac{1}{3}\left(\frac{2}{m_t} + \frac{1}{m_l}\right). \tag{16b}$$

$$\frac{1}{\mu_{1e}} = \frac{1}{3}\left(\frac{1}{m_t} - \frac{1}{m_l}\right) \tag{16c}$$

After tedious but straightforward calculations [22], eq. (3) can be written as a linear combination of products of the above defined terms.

$$H = H_0(\mu,\delta,\mu_{1e}) + \gamma H_1(\mu,\delta,\mu_{0e},\mu_{1e},\gamma_1) + \gamma^2 H_2(\mu,\delta,\mu_{0e},\mu_{1e},\gamma_1) \tag{17}$$

where we have defined

$$\mu = \frac{6\gamma_3 + 4\gamma_2}{5\mu_0} \tag{18}$$

$$\delta = \frac{\gamma_3 - \gamma_2}{\mu_0} \tag{19}$$

The Hamiltonian (17) in the new formulation contains terms which are independent of the magnetic field, some which are linear in γ, and others quadratic in γ. The first 5 terms, describe the exciton in the absence of magnetic field. The next nine terms are linear in H; they describe the magnetic moment effect of the electron and of the hole, together with terms linear in the gradient operator. Finally, the last six terms are quadratic in the magnetic field. It is worth noting that in the absence of degenerate and anisotropic valence bands and of spin, the Hamiltonian for excitons with anisotropic electron masses becomes:

$$H_{exc}^{hydr} = p^2 - \frac{2}{r} - \sqrt{\frac{2}{3}\frac{\mu_0}{\mu_{1e}}} P_0^{(2)} + \mu_0 \left(\frac{1}{\mu_{0e}} + \frac{1}{\mu_{1e}} - \gamma_1\right)\gamma L_0^1$$

$$+ \mu_0\left(\frac{1}{\mu_{0e}} + \frac{1}{\mu_{1e}} + \gamma_1\right)\left(\frac{\gamma^2}{6}r^2 - \frac{1}{12}\sqrt{\frac{2}{3}}\,\gamma^2 R_0^{(2)}\right) \qquad (20)$$

For the case of isotropic masses and states with m = 0 (m being the angular momentum component in the direction of the field) hamiltonian (20) can be written as [23]:

$$H_{exc}^{hydr} = p^2 - \frac{2}{r} + \frac{\gamma^2}{6}r^2 - \frac{1}{12}\sqrt{\frac{2}{3}}\,\gamma^2 R_0^{(2)} = p^2 - \frac{2}{r} + \frac{\gamma^2}{4}(x^2 + y^2) \qquad (21)$$

This hamiltonian however, is completely inappropriate for excitons in cubic semiconductors and is written here only for purpose of comparison with the appropriate one (17). The method presented in this paper allows the solution of hamiltonian (17) to the same accuracy as the best calculations for (21), for all values of the magnetic field.

III. Method of solution

In the previous section, we have formulated the problem of excitons and shallow impurities in a magnetic field for a semiconductor with degenerate and anisotropic bands in terms of products of tensor operators. In analogy to the case of excitons and impurities in the absence of external fields, we write the eigensolutions of hamiltonian (17) in the form:

$$\Psi_i(\vec{r}) = \sum_{L,F,F_z} f_{i,L,F,F_z}(r)\, |L,(J,\sigma)I,F,F_z\rangle \qquad (22)$$

where i plays the role of a principal quantum number, L runs over all even or odd integers (as parity is also a good quantum number); $I = J + \sigma$ can assume the values 2 and 1 thus giving rise to the eight-fold structure of the ground state; also $F = I + L$. Using expression (22) we can eliminate analytically the angular part of the hamiltonian. This is accomplished by using standard angular momentum theory to calculate the matrix elements of (22) with hamiltonian (17), in terms of 3J-,6J-, and 9J-symbols and of the reduced matrix elements [24]

The problem has now been reduced to a system of radial differential equations of order N, where N is the number of terms in expansion (22). This system of radial differential equations is solved by expanding each radial component into a number of exponentials. The problem is therefore finally transformed into the diagonalzation of a secular determinant of order N x M, where M is the number of exponentials in the expansion for the radial functions. Previous calculations [25] have shown that 10 exponentials are always sufficient to attain good numerical accuracy, and therefore the matrices to be diagonalized are always well within the range of any computer.

IV. Results

In this section we discuss some numerical results and show the advantages of the present method over others available ones. In Table II, we show the results obtained for hamiltonian (21), i.e. for the hydrogen atom using only terms up to L = 8 in expansion (22). This implies N = 5, i.e. matrices of order 50. In the table we also provide, for comparison, the exact results of Cabib et al. [26] We see that our results are very accurate well up to $\gamma = 10$, where the adiabatic method starts being valid. We see, therefore, that the same method of solution is able to describe accurately the low, intermediate and large field regions.

Table II. Energy of the hydrogen ground state as a function of the magnetic field for different values of L_{Max}. Comparison with Cabib et al. is also shown.

γ	L=2	L=4	L=6	L=8	Ref. 19
0.0	-1.000	-1.000	-1.000	-1.000	-1.000
0.1	-0.995	-0.995	-0.995	-0.995	-0.995
0.2	-0.981	-0.981	-0.981	-0.981	-0.981
0.4	-0.929	-0.929	-0.929	-0.929	-0.929
0.6	-0.855	-0.855	-0.855	-0.855	-0.855
0.8	-0.764	-0.765	-0.765	-0.765	-0.765
1.0	-0.662	-0.662	-0.662	-0.662	-0.662
1.5	-0.369	-0.371	-0.371	-0.371	-0.371
2.0	-0.040	-0.044	-0.044	-0.044	-0.044
3.0	0.684	0.673	0.671	0.671	0.671
4.0	1.464	1.443	1.439	1.439	1.439
5.0	2.281	2.247	2.241	2.240	2.239

We now discuss, as examples, two realistic cases in order to illustrate the efficiency and accuracy of the method. We first consider the case in which the bands are degenerate and anisotropic. In Table III we present the results for a typical semiconductor whose parameters are given in the table and correspond to those of InP, using terms up to $L_{Max} = 4$ in eq. (22). In this calculation we have set to zero the exchange interaction in order to compare the present results with those obtained using the adiabatic method. It is clear that the present results are always more accurate than the adiabatic over in the range of γ shown. For large γ, the two methods agree well with each other. In the intermediate range, however, $0.4 < \gamma < 5$, where the adiabatic method fails, the present results represent the only accurate method.

In Figure 1, we show the same results to illustrate an important point. The ground state multiplets undergo crossing in the intermediate magnetic field region. This clearly reveals that the level correspondence between low and large field region is not trivial and that only a proper calculation, such as the one reported here, allows a smooth transition from the first to the latter region.

We consider briefly now the case of excitons in anisotropic, polar materials, such as PbI_2. In these types of materials, in addition to the anisotropic masses indicated above, one has to take into account the exciton-phonon interaction [27]. This latter interaction can be described

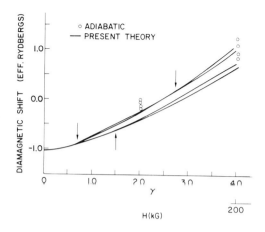

Fig.1. Diamagnetic shift of the exciton ground state multiplet for a typical semiconductor. The parameters are the same as those given in table III.

Table III. Diamagnetic shifts of the ground state multiplet for a typical semiconductor. The parameters used are $\gamma_1 = 4.80$, $\gamma_2 = 1.45$, $\gamma_3 = 2.57$, $m_e^* = 0.0803$, $\varepsilon = 12.57$, $\kappa = 1$, $q = 0.$, $\bar{\mu} = 0.0$ and $\Delta_1 = 0.0$. In parenthesis, the results obtained using the adiabatic method are shown.

γ	-3/2	-1/2	1/2	3/2
0.0	-1.060	-1.060	-1.060	-1.060
0.1	-1.063	-1.062	-1.053	-1.051
0.2	-1.057	-1.058	-1.041	-1.035
0.4	-1.025	-1.032	-1.005	-0.987
0.6	-0.972	-0.988	-0.954	-0.923
0.8	-0.903	-0.930	-0.893	-0.849
1.0	-0.822	-0.861	-0.824	-0.766
2.0	-0.311	-0.414	-0.394	-0.269
	(-0.048)	(-0.143)	(-0.144)	(-0.028)
4.0	0.980	0.745	0.690	0.956
	(1.183)	(0.953)	(0.866)	(1.107)
6.0	2.405	2.044	1.890	2.304
	(2.585)	(2.222)	(2.027)	(2.392)
8.0	3.916	3.423	3.173	3.702
	(4.070)	(3.576)	(3.263)	(3.751)
10.0	5.487	4.867	4.528	5.214
	(5.608)	(4.982)	(4.544)	(5.156)

by a spherically symmetric, exciton-radius dependent screening of the electron-hole coulomb interaction, which can be straightforwardly incorporated our formalism [28]. In Fig. 2 we show the influence of the anisotropy alone on the shift of the exciton ground state in a magnetic field. It can be seen that the anisotropy quenches considerably the diamagnetic shift [29]. The effect of the exciton-phonon interaction is also to quench the diamagnetic shift and the two effects combine in PbI_2 to produce,for an applied field of 180kGauss a diamagnetic shift of only 0.3 meV, which is below the experimental resolution and explains why no shift could be detected, while the hydrogenic description in would have predicted a shift of 1.5 meV, which is above the experimental resolution ,thus leaving the puzzle regarding the lack of diamagnetic shift [30].

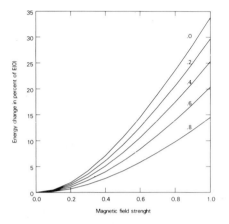

Fig. 2. Diamagnetic shift of the exciton groung state as a function of the magnetic field for various values of the anisotropy parameter .

V. Conclusions

In this paper, we have discussed a theory valid for excitons and shallow impurities at all values of an external magnetic field. The method is general, simple and efficient, and allows an unified description for excitons,donors and acceptors. This theory is able to accurately describe the exciton ground state multiplets in cubic semiconductors, by using the tensor operator formulation previously introduced in the absence of external fields. The necessity of a proper description of the intermediate field region has been demonstrated by the complexity and crossing of the exciton levels, thus rendering any simple interpolation between the two regions invalid. The present method applies as well to anisotropic excitons,as illustrated for the case of PbI_2. Recently [31], it has also been applied to the calculation of the polarizability of donors in Si, and can be used for the analysis of acceptor impurities. Finally, it can be employed to extract band parameters for those materials for which they are not accurately known.

*   Supported in part by the ONR under contract No. N00014-80-C-0679.
1   See, for example, J.G. Mavroides,in *Optical properties of Solids*, edited by F. Abeles (North-Holland, Amsterdam,1972).
2   W. Kohn and J.M. Luttinger, Phys. Rev. **96** ,529(1954).
3   R.R. Goodman, Phys. Rev. **122** ,397 (1961); E. Kazmarek, Phys. Rev. **132** ,1929 (1963); C.R. Pidgeon and R.N. Brown, Phys. Rev. **146** ,575(1966).
4   D.F. Edwards and V.J. Lazazzera, Phys. Rev. **120** ,420 (1960).
5   For a recent review see, for example, D. Bimberg *Advances in Solid State Physics* (Pergamon-Vieweg, 1977), Vol XVII.

6   M. Altarelli and N.O. Lipari, Phys. Rev. B **7**, 3798 (1973).
7   R.J. Elliott and R. Loudon, J. Phys. Chem. Solids **8**, 382 (1959).
8   L.M. Roth, B. Lax and S. Zwerdling, Phys. Rev. **114**, 90 (1959).
9   K. Cho, S. Suga, W. Dreybrodt and F. Willmann, Phys. Rev. B **11**, 1512 (1975).
10 G.J. Rees, J. Phys. C **4**, 2822 (1971), ibid **5**, 549 (1972).
11 N.O. Lipari and M. Altarelli, Solid Stat. Commun. **13**, 1791 (1973); M. Altarelli and N.O. Lipari, Phys. Rev. B. **9**, 1733 (1974).
12 C.H. Uihlein and S. Feierabend, Phys. Stat. Solidi (b) **94**, 157 (1979).
13 K. Hess, D. Bimberg, N.O. Lipari, J.V. Fishbach and M. Altarelli, Proc. XIII Int. Conf. Phys. Semiconductors, Rome, 1976, p.142.
14 W. Ekardt, K. Losch and D. Bimberg, Phys. Rev. B **20**, 3033 (1979).
15 K.K. Bajaj and C.H. Aldrich, Sol. Stat. Commun. **35**, 163 (1980).
16 N.O. Lipari and A. Baldereschi, Phys. Rev. Letters **25**, 1660 (1970).
17 J.O. Dimmock, in *Semiconductors and Semimetal*, edited by R.K. Willardson and A.C. Been (Academic, New York, 1972), Vol. 3, p. 259.
18 J.M. Luttinger, Phys. Rev. **102**, 1030 (1956).
19 K. Cho, Inst. Phys. Conf. Ser. **43**, 841 (1979).
20 A. Baldereschi and N.O. Lipari, Phys. Rev. B **8**, 2677 (1973); ibid B **9**, 1525 (1974).
21 N.O. Lipari and A. Baldereschi, Sol. Stat. Commun. **25** 605 (1978).
22 N.O. Lipari and M. Altarelli, to be published.
23 F. Bassani and A. Baldereschi, Surface Science **37**, 304 (1973).
24 A.R. Edmonds, *Angular Momentum in Quantum Mechanics*, University Press, Princeton, N.J. 1960.
25 N.O. Lipari and M. Altarelli, Solid Stat. Commun. **33**, 47 (1980).
26 D. Cabib, E. Fabri, and G. Fiorio, Il Nuovo Cimento **10B**, 185 (1972).
27 G. Harbeke and E. Tosatti, J. Phys. Chem. **37**, 304 (1976).
28 J. Pollmann and N.O. Lipari, Sol. Sta. Commun. **20**, 203 (1978).
29 N. O. Lipari and J. Pollmann, Inst. Phys. Conf. Ser. **43**, 1097 (1979).
30 M. Skolnik, Le Chi Tanh, F. Levy and G. Harbeke, Physica **89B**, 143 (1977).
31 D. Dexter and N.O. Lipari, Inst. Phys. Conf. Ser. **43**, 969 (1979).

Magnetic Field Induced LT Mixing of the Multicomponent Polaritons in CdTe

K. Cho [1]
Faculty of Engineering Science, Osaka University, 560 Toyonaka, Japan, and

S. Suga [1]
Institute for Solid State Physics, The University of Tokyo, Roppongi, Minato-ku, Tokyo, Japan, and

W. Dreybrodt [1]
Universität Bremen, Fachbereich Physik, 2800 Bremen, Fed. Rep. of Germany

Recently there has been increasing interest in the internal structure of excitons which arises from various spin configurations of electron-hole pair states [1]. This feature is important especially (A) in the presence of external perturbations which mix the sublevels, and/or (B) for finite translational wave vector K, for which the level scheme is determined by K-dependent interactions among the sublevels. The exciton states consist of multi-branch levels whose energies and transition dipole moments are both K-dependent. The polaritons derived from these complicated (and realistic) exciton states play essential roles in many optical measurements, such as absorption, reflection, scattering, luminescence, etc. In order to treat general cases of such systems, conventional polariton theory has been improved with respect to multicomponent character [2] and the possible mixing of longitudinal (L) - transverse (T) modes [3]. The use of the polarizability tensor α defined for pure external field (without depolarizing field) simplifies the treatment of the LT mixed modes, leading to a simple dispersion equation of the form

$$\det \begin{pmatrix} 1+4\pi\alpha_{\xi\xi} - \kappa^2, & 4\pi\alpha_{\xi\eta} \\ 4\pi\alpha_{\eta\xi}, & 1+4\pi\alpha_{\eta\eta} - \kappa^2 \end{pmatrix} = 0 \quad (1)$$

where the ζ-axis of the cartesian coordinates (ξ,η,ζ) is parallel to $\vec{K}$, $\kappa = cK/\omega$, and the (K,ω) dependence of α should be understood. Note that $1+4\pi\alpha \neq \varepsilon$ (dielectric function) for systems with LT mixing. For the calculation of α from linear response theory, we have to include the long range part of the Coulomb interaction, which causes the depolarization field, in the Hamiltonian of the unperturbed system. This is equivalent to including the non-analytic part of the electron-hole exchange interaction, H_{exch}. For long wavelengths, this term has the following matrix element between two arbitrary exciton states with common K_* [4]:

$$(i| H_{exch} |j)_{NA} = 4\pi M_\zeta(i)^* M_\zeta(j) / \Omega \quad (2)$$

where $M_\zeta(j)$ is the ζ-component of the transition dipole moment of the state $|j)$ and Ω the unit cell volume. The rest of the Hamiltonian consists of kinetic (K-dependent) energies, external perturbations if any, and the analytic part of the exchange interaction.

In this note we apply this formalism to the analysis of the magnetoreflectance measurement of the 1s excitons in CdTe, especially in Voigt geometry.

---

[1] Former address: Max-Planck-Institut für Festkörperforschung, Stuttgart, Fed. Rep. of Germany

Among various zincblende type semiconductors, this system is particularly interesting because of the remarkable effect of the K-linear term: The zero field reflectance anomaly and the existence of three components in each of the σ_{+1} and σ_{-1} Faraday spectra can be well explained by the K-linear term. This was explicitly demonstrated for Faraday geometry by calculating the magneto-reflectance spectra, which excellently simulate the measured curves in each of the spectral details [5]. A further theoretical study in Voigt geometry is of interests because LT mixed modes appear in σ-polarization, which has not yet been fully investigated as a polariton problem, and because the full cubic anisotropy of the K-linear term emerges from comparison between the various components of Faraday and Voigt spectra.

The effective Hamiltonian for the eight-fold 1s exciton states at finite K and magnetic field H can be deduced from the general group theoretical scheme proposed earlier [4]. For $\vec{K}//\langle 110\rangle$ and $\vec{H}//\langle 1\bar{1}0\rangle$, the 8 x 8 matrix splits into two block-diagonals, corresponding to σ and π polarizations. In terms of the basis used in [4], the K-dependent (suffix K) and independent (suffix 0) parts of the submatrices are given as follows:

$$H_K^\sigma = \bar{\gamma}_1 + \begin{pmatrix} \bar{\gamma}_2 & -\sqrt{3}\phi & \phi & -\sqrt{3}\bar{\gamma}_2 \\ & -2\bar{\gamma}_2 & 0 & -2\phi \\ (\text{h.c.}) & & 2\bar{\gamma}_2 & 0 \\ & & & -\bar{\gamma}_2 \end{pmatrix} \qquad \begin{array}{c}|2+\rangle \\ |y''\rangle \\ |x'\rangle \\ i|z\rangle\end{array} \qquad (3)$$

$$H_0^\sigma = E_0 + 2\bar{C} + \begin{pmatrix} -\bar{C}_2 & \bar{g}_1 - \bar{g}_0 & \sqrt{3}(\bar{g}_2 + \bar{g}_0) & -\sqrt{3}\bar{C}_2 \\ & -\bar{C}_2 & \sqrt{3}\bar{C}_2 & -\sqrt{3}(\bar{g}_4 + \bar{g}_0) \\ (\text{h.c.}) & & \bar{C}_2 + \Delta_{Tt} + \Delta_{LT} & -\bar{g}_5 - \bar{g}_0 \\ & & & \bar{C}_2 + \Delta_{Tt}\end{pmatrix} \qquad (4)$$

$$H_K^\pi = \bar{\gamma}_1 + \begin{pmatrix} -\bar{\gamma}_2 & \phi & \sqrt{3}\bar{\gamma}_2 & \sqrt{3}\phi \\ & \bar{\gamma}_2 & 0 & -\sqrt{3}\bar{\gamma}_2 \\ (\text{h.c.}) & & \bar{\gamma}_2 & -2\phi \\ & & & -\bar{\gamma}_2\end{pmatrix} \qquad \begin{array}{c}|2,0\rangle \\ |x''\rangle \\ i|2-\rangle \\ |y'\rangle\end{array} \qquad (5)$$

$$H_0^\pi = E_0 + 2\bar{C} + \begin{pmatrix} -\bar{C}_2 & \sqrt{3}(\bar{g}_1 - \bar{g}_0) & -\sqrt{3}\bar{C}_2 & -\bar{g}_2 - \bar{g}_0 \\ & -2\bar{C}_2 & \bar{g}_3 - \bar{g}_0 & 0 \\ (\text{h.c.}) & & \bar{C}_2 & \sqrt{3}(\bar{g}_4 + \bar{g}_0) \\ & & & 2\bar{C}_2 + \Delta_{Tt}\end{pmatrix} \qquad (6)$$

where

$$\bar{\gamma}_j = \tilde{\gamma}_j \hbar^2 K^2/2m_0 \; (j=1,2), \quad (\bar{C},\bar{C}_2) = (\tilde{C},\tilde{C}_2)\mu_B^2 H^2/4R_y, \quad \phi = -2K_\ell K/\sqrt{27},$$
$$\Delta_{Tt} = \text{Transverse-triplet splitting}, \quad \Delta_{LT} = \text{LT splitting}$$
$$\bar{g}_j = g_j \mu_B H; \quad g_0 = \tilde{g}_c/4, \quad (g_1,g_2,g_3,g_4,g_5) = \frac{1}{8}(\tilde{\kappa},\tilde{q})\begin{pmatrix} 12, & 4, 12, & 4, 20 \\ 27, & 1, 15, 13, 41 \end{pmatrix} \qquad (7)$$
$$\begin{array}{c}|x''\rangle \\ |y''\rangle\end{array} \} = [|1+\rangle \pm |1-\rangle]/\sqrt{2}, \quad \begin{array}{c}|x'\rangle \\ |y'\rangle\end{array} \} = [|x\rangle \pm |y\rangle]/\sqrt{2}$$

The interaction mechanisms considered above are as follows: H_K represents the light-heavy mass splitting ($\tilde{\gamma}_2$) as well as the usual spatial dispersion ($\tilde{\gamma}_1$), and the K-linear effect (K_ℓ). H_0 includes the Zeeman terms ($\tilde{g}_c$, $\tilde{\kappa}$, $\tilde{q}$), the diamagnetic effects ($\tilde{C}$ and $\tilde{C}_2$), and the exchange interaction (Δ_{Tt} and Δ_{LT}).

For the calculation of reflectance spectrum $R(\omega)$, we need to know the polariton dispersion $\{K_s(\omega); s=1,2,3,4,5$ for each polarization$\}$ as functions of frequency ω. According to the method in [2], which is equivalent to solving (1), the dispersion relation is obtained from the equation

$$\det \begin{pmatrix} \begin{pmatrix} H_0 + H_K - \hbar\omega \end{pmatrix} & \begin{matrix} 0 \\ 0 \\ 0 \\ M\sqrt{4\pi/\Omega} \end{matrix} \\ 0, \; 0, \; 0, \; M\sqrt{4\pi/\Omega}, & \kappa^2 - \varepsilon_b \end{pmatrix} = 0 , \qquad (8)$$

$$(\Delta_{LT} = 4\pi M^2/\Omega\varepsilon_b = 2\pi\beta E_0/\varepsilon_b)$$

where M is the transition dipole moment of $|z\rangle$ or $|y'\rangle$, which takes a common value in this case, and ε_b the transverse background dielectric constant. (The longitudinal component is implicitly included in Δ_{LT} [3].) Numerical solution of (8) for a given value of ω is a standard computational problem. Once the set $\{K_s(\omega)\}$ is known, one can calculate $R(\omega)$ by imposing appropriate ABC (additional boundary conditions). The same problem has been treated in Faraday configuration [5], and the use of a modified "Hopfield-Pekar" type ABC [2] together with an exciton dead layer was shown to work nicely in fitting the measured reflectance curves. The ABC can be stated as follows: " The amplitude of the j-th exciton should, when summed over all the polariton

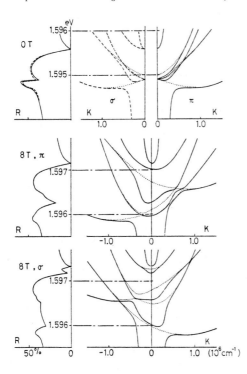

Fig.1 Calculated dispersion curves and reflectance spectra with a common energy axis (see text)

branches, vanish at the interface between bulk and exciton dead layer, and this is required for all j." The same procedure has been followed in this calculation. Figure 1 shows the calculated results for reflectance and the corresponding dispersion curves for H = 0 and 8 Tesla. The parameter values are same as in Faraday configuration [5]: $E_0 = 1.5949$ eV, $\Delta_{Tt} = 0.1$ meV, $\tilde{\gamma}_1 = 2.17$, $\tilde{\gamma}_2 = 0.07$, $K_\ell = 0.5 \times 10^{-9}$ eV cm, $4\pi\beta = 0.0066$ ($\Delta_{LT} = 0.58$ meV), $\varepsilon_b = 9.0$, $\tilde{g}_c = -1.59$, $\tilde{\kappa} = 0.11$, $\tilde{q} = 0.00$, $\tilde{C} = 1.94 \times 10^5$, $\tilde{C}_2 = 0.13 \times 10^5$. For $R(\omega)$ in this figure, we took vanishing values for life time broadening (Γ) and dead layer thickness (d). The dotted lines show the exciton dispersion. For "H=0 T, σ" and "H=8 T, π", the second highest branch overlaps with the exciton dispersion.

Because of the cubic anisotropy of the K-linear term and of the LT splitting, there is a difference, for H = 0 T, between the σ (dashed lines) and π (solid lines) polarizations. At finite fields, the dispersion is not symmetric for +K and -K, in contrast with the Faraday geometry. Figure 2 shows the reflectance spectra calculated for σ-polarization with finite Γ and d (Γ = 0.05 meV and d = 100 Å). They should be compared with the measured curves in Fig.3, which were taken at 6 K for normal incidence on the (110) cleaved surface. The change in spectral shape is well reproduced by the present theory. Certain disagreements at H = 12 T may be ascribed to the inadequacy of the low field scheme adopted here. The overall agreement lends strong support to the present theoretical framework. Though reflectance spectroscopy is not a direct method to determine polariton dispersion, the consistent results for all the Faraday and Voigt geometries indicate that the dispersion curves of the multicoponent polaritons in Fig.1 (and also those in [5]) would be very near to reality, which could be determined later by more direct experimental methods.

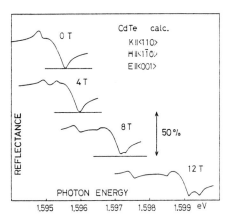

Fig.2 Calculated reflectance Zero level is shifted for each curve.

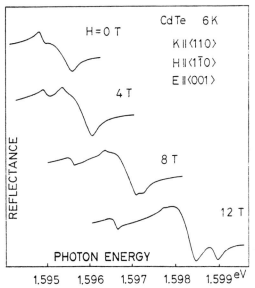

Fig.3 Measured reflectance in arbitrary units

Multicomponent systems provide not only complexity but also certain situations that cannot be realized in usual two-branch polaritons. For example, two "photon-like" branches may coexist for a given range of ω (e.g., ℏω ∼ 1.5967 eV, K > 0, for H = 8 T, σ-polarization in Fig.1). This is interesting in connection with the ABC problem: An incident picosecond pulse with this carrier wave frequency will split into two pulses with comparable intensities whose ratio would sensitively reflect the ABC to be used. The measurement of this ratio would give a conclusive result as to the choice of the correct ABC. For usual two-branch polaritons, there is little hope to generate two pulses with comparable intensities [6]. The ±K asymmetry of dispersion curves for H ≠ 0 is another interesting feature. If we invert the direction of H, we get dispersion curves which are mirror images of Fig.1. This system has apparently different character from that for +H.

The dispersion relations of multicomponent polaritons are interesting objects containing much useful information, and are expected to be further studied from both experimental and theoretical points of view.

References

1. K.Cho: "Internal Structure of Excitons", in *Excitons*, ed. by K.Cho, Topics in Current Physics, Vol. 14 (Springer, Berlin, Heidelberg, New York 1979) Chapt. 2
2. K.Cho: Solid State Commun. $\underline{27}$, 305 (1978)
3. K.Cho: Solid State Commun. $\underline{33}$, 911 (1980)
4. K.Cho: Phys. Rev. $\underline{B14}$, 4463 (1976); $\underline{B11}$, 1512 (1975); $\underline{B12}$(E), 1608 (1975)
5. W.Dreybrodt, K.Cho, S.Suga, F.Willmann, and Y.Niji: Phys. Rev. $\underline{B21}$, 4692 (1980)
6. M.Yamane and K.Cho: unpublished

Magneto-Optical Effects in III-VI Layer Compounds

Y. Sasaki, N. Kuroda, and Y. Nishina

The Research Institute for Iron, Steel and Other Metals, Tohoku University
Sendai 980, Japan, and

H. Hori, M. Shinoda, and M. Date

Department of Physics, Faculty of Science, Osaka University
Toyonaka 560, Japan

1. Introduction

The magneto-optical spectra of III-VI layer compounds GaSe and InSe have been measured up to the maximum pulsed field of 50 T at 4.2 K. The direct absorption edge in GaSe (2.130 eV) and InSe (1.353 eV) corresponds to the transition from the B-valence band to the conduction band, and the next higher absorption edge (3.41 eV in GaSe and 2.562 eV in InSe) from the A-valence band in the conventional notation for the wurtzite materials [1]. The excitons associated with these band edges show the hydrogenic n-2 series [1,2]. Neither the top of the valence bands nor the bottom of the conduction band are degenerate. We give experimental analysis on the magneto-absorption of the B-exciton in GaSe and of the A-exciton in InSe in the Faraday and Voigt configurations. In the analysis of the data, particular attention is paid to the level connection problem from the hydrogenic Rydberg series to the adiabatic limit in the extremely high magnetic field [3,4,5]. The maximum of our field falls in the intermediate region so that $\gamma \equiv \hbar\omega_C/2Ry \simeq 1$ (ω_C: cyclotron frequency and Ry: effective Rydberg constant). Since the electric field of the light is normal to the c-axis, in the present experiments, the spin state of the exciton observed here is triplet mixed with a weak singlet component for the B-exciton [2], whereas it is singlet for the A-exciton.

2. Experimental Arrangements

Figure 1 shows the block diagram of the experimental arrangements with tran-

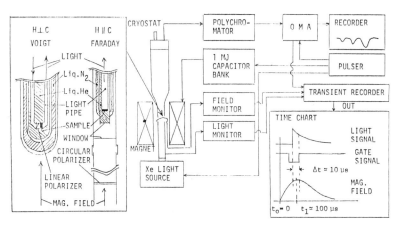

Fig.1 Experimental arrangements

sient signals. The pulser serves to synchronize the field generation, Xe-lamp triggering and gating of the OMA ISIT detector. The light transmitted through the sample is dispersed by a polychromator (Spex 1701) for the ISIT detection. The resolution of this spectrometer system is about 0.6 meV, which is mainly determined by the entrance slit width of the polychromator. A double helix maraging steel magnet of High Magnetic Field Laboratory, Osaka University, with its inner diameter of 20 mm can produce the maximum field of 60 T. It is used for the measurements in the Faraday configuration, $H \parallel c$. A single helix magnet with its inner diameter of 60 mm is used for the Voigt configuration, $H \perp c$. The field is calibrated in terms of the submillimeter spin resonance [6]. The magnetic field inhomogeneity over the sample size and the field variation during the gating time do not affect our spectral line shapes. Other details of experiments are described in [7]. The single crystals of GaSe and InSe have been grown by the Bridgman method. The X-ray analyses show that they are ε- and γ-polytypes for GaSe and InSe, respectively. Thin cleaved samples are $\simeq 10~\mu$ thick for GaSe and $\simeq 1~\mu$ for InSe.

3. Results and Discussions

Figure 2 shows an example of the recorder traces of the transmission spectra along with the Ne-He mixed gas calibration lines. The transmission spectra of Fig.2 have been measured for both RCP and LCP in the Faraday configuration and for σ and π polarizations in the Voigt configuration. Figure 3 shows the plots of the positions of the transmission minima (absorption peaks), A through E, for both RCP and LCP data. The dotted lines in the figure show the low field (H < 10 T) magneto-absorption data [8]. The energies of the N = 0 and 1 Landau levels are given by the dashed lines for the reduced mass of $\mu_\perp = 0.14~m_0$. The data in the Voigt configuration are analysed only for the A line.

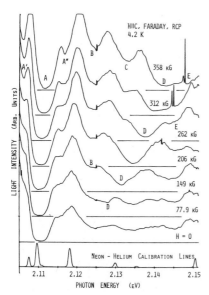
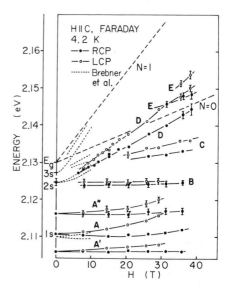

Fig.2 Transmission spectra of the direct exciton of GaSe at 4.2 K in the Faraday configuration

Fig.3 Magnetic field dependence of absorption peaks of GaSe in the Faraday configuration at 4.2 K

Table 1 Exciton parameters of GaSe and InSe for the 1s state

Material	GaSe		InSe
Ry [meV]	20		15
Configuration	Faraday (H‖c)	Voigt (H⊥c)	Faraday (H‖c)
σ_{1s} [10^{-3}meV/T^2]	3.2 ± 0.2^a, 4.3^b, 2.5^c	3.6 ± 0.3^a, 4.3^b	7.9 ± 0.4^a
$\|g_e + g_h\|$	$2.7 \pm 0.2^{a,b}$, 2.9^c	1.8 ± 0.15^a, 1.9 ± 0.15^b	
$\|g_e - g_h\|$		$< 0.2^a$	
$\mu_\perp$ [m_0]	0.15 ± 0.07^d, 0.13^e, 0.14 ± 0.05^f		0.119 ± 0.007^d
$\mu_\|/\mu_\perp$	0.9 ± 0.2^d, 1 ± 0.2^e		

a) present work (H < 50 T) b) after [2] (H < 10 T) c) after [9] (50 < H < 160 T)
d) present work, from σ_{1s} e) after [2], from σ_{1s} f) after [2], from Ry

3.1 1s State

The A line corresponds to the 1s state [2]. The strain induced by the sample cleavage may be responsible for the small deviation from [8] in the 1s exciton energy at H = 0. Table 1 summarizes the diamagnetic shift, $\sigma_{1s}H^2$, effective g-value of the exciton, $|g_e \pm g_h|$, and reduced masses of the 1s exciton estimated in the present work in comparison with the results of MOOSER and SCHLÜTER [2] measured in the lower field (H < 10 T) and those of KIDO et al. [9] in the higher field (50 < H < 160 T). Apparent difference in σ_{1s} with increase in H comes from the breakdown of the simple $\sigma_{1s}H^2$ relation in higher fields. The value of $|g_e - g_h|$ may be deduced from the σ-polarized spectrum in the Voigt configuration in GaSe. The experiment up to 50 T suggests that $|g_e - g_h|\mu_B H < \Delta_{ex}$, where $\Delta_{ex} = 1.8$ meV [2] is the exchange splitting energy between the singlet and triplet states, even at the maximum field. $|g_e - g_h|$ is, therefore, estimated to be smaller than 0.2 for H⊥c. The A' and A" lines possibly originate from the excitons trapped by the stacking fault.

3.2 2s and/or 3rd State

The D line is located between the 2s and $3d_0$ line of the low field spectrum for H < 10 T [4,8], but these two lines are not resolved probably due to the strain-induced broadening of the exciton states. Since the oscillator strength of the $3d_0$ line is about 20 % of the 2s line at about 9 T [8], the D line consists primarily of the 2s exciton at H < 10 T. The $3d_0$ component of this line may increase with H as expected from the low field data. LEE, LARSEN and LAX [5] have shown by the variational calculation that the states originating from 1s to 4s, 3d and 4d are found below the N = 0 Landau level for any strength of magnetic field. On the other hand, the nodal surface conservation rule predicts that the state originating from the 2s [3] or $3d_0$ [4] state at H = 0 crosses the N = 0 Landau level at some intermediate field. Figure 3 shows that the D line pair starts from the low field 2s peak [8] and it crosses the N = 0 Landau level at H ≃ 30 T ($\gamma \approx 0.65$). Even if the D line is a composite of 2s--, 3s--, $3d_0$--, like states as pointed out in [5], it crosses the N = 0 Landau level. The line may be extrapolated to the absorption peak observed just below the N = 1 Landau level in the ultra high field region (50 < H < 160 T) [9]. The intensity of the D absorption line increases with increase in H as shown in Fig.2. Such magnetic field dependence of the line intensity is in qualitative agreement with the character of (100^+) line in the high field formalism of TANAKA and SHIMADA [10]. They claim that the oscillator strength of the (100^+) state increases linearly with H and it exceeds the oscillator strength of (001^+) at $\gamma \simeq 0.5$. The crossing of the D line with the $N=0$ Landau level as well as the magnetic field dependence of the absorption intensity suggests that our experimental results are consistent with the theoretical model on the basis of the nodal surface conservation rule rather than that based on the non-crossing rule. The D line and the $N=0$ Landau level cross at $\gamma \simeq 0.65$ for the B-exciton of GaSe while at $\gamma \simeq 0.33$ for the A-exciton of InSe, which are close to the calculated value of [10]($\gamma \simeq 0.35$). This qualitative aspect of the magneto-optical data disagrees with that of TlCl measured by NAKAHARA [13], who claims that there is no indication of the level crossing between the $N=0$ Landau level and $2s/3d_0$ level. At the same time he finds the level crossing between the $N=0$ Landau level and the 3s-like level. These material dependence in the aspects of level crossing suggests that the exciton levels in magnetic fields may not be calculated in a common formalism. The effective g-value deduced from the splitting of the D line pair in Fig.3 agrees with that of the 1s (A) line for $H < 25$ T but it tends to deviate from 1s at a higher field. This deviation is probably due to the crossing or anticrossing of the 2s and $3d_0$ levels and to the appearance of the E line which seems to show a repulsion against D around 30 T.

3.3 p States

Since the $2s/3d_0$ level, or the D line, crosses the $N=0$ Landau level at $\gamma \simeq 0.65$ in GaSe, the weak B and C lines which appear between the 1s and D lines may be assigned to p states. Similarity of the result in Fig.3 with the magneto-spectroscopic data of shallow donor of GaAs [5,11,12] suggests that the B and C lines represent the $2p_-$ and $3p_-$ (or $2p_0$) like states, respectively.

4. Conclusions

The magneto-optical studies of GaSe and InSe up to 50 T show that the state originating from the 2s or $3d_0$ exciton crosses the $N=0$ Landau level at $\gamma \simeq 0.5$. Among the existing models of the exciton levels in the magnetic field, the present results are in favor of the level scheme deduced from the nodal surface conservation rule rather than that from the non-crossing rule.

This work was supported in part by the Grant-in-Aid for Scientific Research from the Ministry of Education in Japan.

1. N. Kuroda, I. Munakata and Y. Nishina, Solid State Commun. 33, 687 (1980).
2. E. Mooser and M. Schlüter, Nuovo Cimento 18B, 164 (1973).
3. W.H. Kleiner, Lincoln Lab. Prog. Rep., Feb. (1958).
4. M. Shinada, O. Akimoto, H. Hasegawa and K. Tanaka, J. Phys. Soc. Jpn. 28, 975 (1970).
5. N. Lee, D.M. Larsen and B. Lax, J. Phys. Chem. Solids 34, 1059 (1973).
6. M. Date, M. Motokawa, A. Seki, S. Kuroda, K. Matsui and H. Mollymoto, J. Phys. Soc. Jpn. 39, 898 (1975).
7. H. Hori, H. Mollymoto and M. Date, J. Phys. Soc. Jpn. 46, 908 (1979).
8. J.L. Brebner, J. Halpern and E. Mooser, manuscript for oral presentation at Symposium on Anisotropy in Layer Structures, Taormina, 1968.
9. G. Kido, H. Katayama, N. Miura and S. Chikazumi, to be published.
10. K. Tanaka and M. Shinada, J. Phys. Soc. Jpn. 34, 108 (1973).
11. R. Kaplan, M.A. Kinch and W.C. Scott, Solid State Commun. 7, 883 (1969).
12. G.E. Stillman, C.M. Wolfe and J.O. Dimmock, Solid State Commun. 7, 921 (1969).
13. J. Nakahara, Solid State Commun. 29, 115 (1979).

Interband Faraday and Kerr Effects in Semiconductors: An Analysis by Means of Equivalent Modulated Magneto-Optical Spectra of Ge and Si

J. Metzdorf

Institut B für Physik und Hochmagnetfeldanlage der Technischen Universität Braunschweig, 3300 Braunschweig, Fed. Rep of Germany

1. Introduction

The magneto-optical (mo) properties of a (nonmagnetic) material in the Faraday configuration are all contained in either the complex Faraday effect or the complex (polar mo) Kerr effect. Each of these differential effects involves two complete spectra, one of the rotation Θ_F or Θ_K and one of the ellipticity Ψ_F or Ψ_K. The theoretical relations, including sign, between the Faraday and the Kerr effects and the off-diagonal component $\tilde{\varepsilon}_{xy}$ of the complex dielectric tensor can be written as $(B||z)$

$$\tilde{\varepsilon}_{xy} = -i \cdot \tilde{n} \cdot \Delta\tilde{n} = (\Theta_K + i\Psi_K) \cdot (1-\tilde{n}^2) \cdot \tilde{n} = (\Psi_F - i \cdot \Theta_F) \cdot h \cdot c \cdot \tilde{n}/\pi \cdot E \cdot d,$$

where E is the photon energy, d is the thickness of the sample, $\tilde{n} = n - i \cdot k$ is the complex index of refraction (at zero field), and $\Delta\tilde{n} = \tilde{n}_r - \tilde{n}_\ell \ll \tilde{n}$; the subscripts r, ℓ refer to right- and left-circularly polarized waves.

In studying interband transitions there are two serious limitations. Firstly, there is an upper limit of the "transparent" energy range for measuring Faraday effects due to (i) the thickness of thin films (d≳1 μm) that can be prepared self-supporting and free from strain and (ii) the enhancement of surface effects with decreasing thickness. Secondly, Kerr effects due to indirect transitions are usually too small to be measured. Therefore, to overcome these drawbacks, we have performed overlapping measurements of the interband Faraday and Kerr effects (B≤15T) in Ge and Si up to 4.6eV photon energy. The effects of the boundary surface (neglecting inhomogeneous surface layers) and internal multiple reflections have been taken into account where necessary on the basis of the Fresnel reflection and transmission coefficients.

In measuring both the angle of rotation and the mo change of ellipticity of the transmitted and reflected light absolutely and continuously, related linear and circular polarization modulation techniques [1] have been modified and improved. The measurement sensitivity limit of $\Delta I/\bar{I}$ is $3 \cdot 10^{-6}$ (TC = 4s) under the most favourable conditions.

The main subject of this paper is a summary of aspects of the utilization of independently measured interband mo rotation and ellipticity of the reflected and/or transmitted light using our experimental results of Ge. Moreover, the identification of the second indirect gap in Si will be quoted as an illustration of the sensitivity of the interband Faraday ellipticity (IFE) to indirect transitions.

2. Results and Discussion

(i) Comparing magneto-reflection and -transmission spectra (Fig.1), the Kerr effects (Θ_K and Ψ_K) yield the correct bulk interband mo properties except for the first peak of the E_0 exciton transitions in Ge. This result is supported by the fact that the nonresonant interband Faraday rotation (IFR) can be analyzed quantitatively by means of Kerr effect data as discussed below. Thus, both the transparent and opaque regions of interband transitions can be covered.

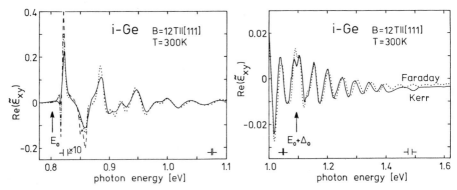

Fig.1 Comparison of the real part of $\tilde{\varepsilon}_{xy}$ of Ge obtained from Kerr effects (full lines) with that obtained from Faraday effects (dotted lines: d = 1 μm; dashed line: d = 2.3 μm). The difference at 1.6eV corresponds to a Kerr rotation of 0.0015°.

(ii) Considering the thickness dependence of the Faraday effects in Fig.1 and the only failure of the Kramers-Kronig relations (KKR) at the same exciton transitions [2,9], this discrepancy gives indication of a dead layer where bulk excitons are damped or screened. The spectral range of effectiveness of such a surface layer, however, is essentially limited to that of the discrete or bound exciton states where the Kerr effects of intrinsic Ge are equivalent to the Faraday effects of doped Ge (of a distinct free carrier concentration of $3 \cdot 10^{16}$ cm^{-3}) where the screening of excitons is extended to the whole volume. Though commonly assumed to be responsible for similar features in other cases, the effect of spatial dispersion is negligible here because of strong damping in comparison with very small longitudinal-transverse splitting.

(iii) The exact tensorial KKR or dispersion relations between the real and imaginary parts of $\tilde{\varepsilon}_{xy}$ have been applied successfully to check the self-consistency of our spectra, a procedure which is similar to the use of sum rules. This is discussed in detail in a separate paper [2]. Contrary to other authors [3], we have not found it possible to derive KKR between the directly measurable quantities themselves (rotation and ellipticity) without restrictive approximations that are satisfied only in case of weak absorption.

(iv) As the IFR is dispersive, positive and/or negative contributions from different indirect and direct transitions have to be taken into account at each wavelength in order to explain the nonresonant IFR which has been a long-standing problem. Using the Kramers-Kronig transform of the absorptive real part of $\tilde{\varepsilon}_{xy}$ determined from IFE and Kerr effects, respectively, we have assigned the contributions of the different interband transitions up to 4.6eV to the nonresonant FR in Ge (Fig.2) and Si quantitatively for the first time. As illustrated in Fig.2 and summarized in Table 1, the nonresonant change in sign of rotation in Ge is due to a competition mainly between negative contributions from E_0, $E_1+\Delta_1$, and E_0' and a positive contribution from E_1 transitions. Different theoretical approaches, as listed in Table 1 and reviewed by RAY [4], have been proposed to account for this change in sign; the effective gap E_{eff} introduced by RAY [4] is analogous to the Penn gap.

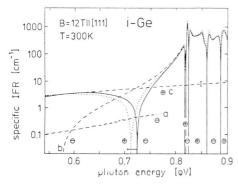

Fig.2 Analysis of the specific IFR of Ge (signs are represented by ⊕ and ⊖). Full line: experimental curve; dashed lines: contributions from indirect (a) and direct transitions below (b) and above (c) 1.4eV; dotted line: resultant Kramers-Kronig transform from all transitions below (4±0.7)eV

Table 1 Signs of the contributions to the nonresonant IFR (<0.6eV) in Ge from various interband transitions (contributions of less than 10% in parentheses; E_{eff} = 4.02eV [4])

transitions (energy gap)	this work	theoretical approaches [5]	[6]	[7]	[4]
indirect	(-)	+	0	0	0
E_0 (light hole)	-	-	-	-	-
E_0 (heavy hole)				+	
$E_0 + \Delta_0$	(-)			(-)	(-)
E_1	+		+		
$E_1 + \Delta_1$	-		+		
E_0'	-				+(E_{eff})
(E_2)	(+)				

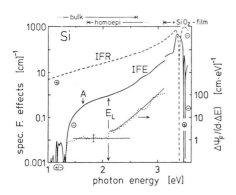

Fig.3 Specific IFR, IFE, and numerical derivative of IFE of various single-crystal Si. E_L and A indicate the positions of the second indirect gap and the Fraunhofer line, respectively (B=12T; T=300K)

(v) As the IFE (or MCD) is absorptive and therefore its spectra are selective to different transitions, we were able to identify the "sensitive" second indirect transition $\Gamma'_{25} \to L_1$ of Si with an energy gap of 2.04eV. Although there is only a weak break of slope in the original curves at E_L in Fig.3 (even at 1.He temperatures), the numerical derivative of the IFE with respect to photon energy shows that the change of slope is at the beginning of a persistent additional increase due to transitions into the second conduction band. A weak structure detectable at 1.63eV (position A in Fig.3) could be identified as due to the Fraunhofer A line from the oxygen in the air. Therefore, an interpretation of a structure at this particular energy as the second indirect gap [8] may not be maintained.

I should like to thank Prof. F.R. Keßler for his kind interest in this work and for continuous support.

References:

1. J. Metzdorf, F.R. Keßler: Phys. Status Solidi B71, 237 (1975)
2. J. Metzdorf, R. Nies, F.R. Keßler: to be published
3. S.O. Sari: Phys. Rev. B6, 2304 (1972);
   D.Y. Smith: J. Opt. Soc. Am. 66, 454 and 547 (1976)
4. B.K. Ray: Indian J. Phys. 50, 818 (1976); 53A, 496 (1979)
5. B. Lax, Y. Nishina: Phys. Rev. Lett. 6, 464 (1961)
6. I.M. Boswarva, A.B. Lidiard: Proc. Int. Conf. Phys. Semicond., Exeter 1962, p. 308;
   C.R. Pidgeon, C.J. Summers, T. Arai, S.D. Smith: Proc. Int. Conf. Phys. Semicond., Paris 1964, p. 289;
   F.R. Keßler: Physica 89B, 7 (1977)
7. I.M. Boswarva, A.B. Lidiard: Proc. R. Soc. London A278, 588 (1964)
8. R.A. Forman, W.R. Thurber, D.E. Aspnes: Solid State Commun. 14, 1007 (1974)
9. C. Jacobsen, T. Skettrup, I. Filinski: Surf. Sci. 37, 568 (1973)

Magnetoplasma Modes at the Interface between a Semiconductor and a Metallic Screen in the Faraday Geometry

P. Halevi

Instituto de Ciencias de la Universidad Autonoma de Puebla, Apdo. Post. J-48 Puebla, Pue., Mexico

In this article we study electromagnetic waves and polaritons which may propagate at the interface between a semiconductor and a highly conducting metallic screen. We assume that a static magnetic field $\vec{B}_0$ is parallel to the plane of the interface and consider only propagation in the direction of the applied field, i.e. the wave vector Re $\vec{q}$ is parallel to $\vec{B}_0$ (Faraday geometry). The polariton solutions for the Voigt geometry, with Re $\vec{q} \perp \vec{B}_0$, were recently published by HALEVI and GUERRA - VELA [1]. In the present work we shall discuss both the low-frequency or electromagnetic wave solutions and the high-frequency or polariton solutions in the Faraday configuration.

The semiconductor magnetoplasma is characterized by the following, gyrotropic dielectric tensor:

$$\epsilon_{xx} = \epsilon_{yy} = \epsilon_\infty - (\epsilon_\infty \omega_p^2 \tilde{\omega}/\omega)(\tilde{\omega}^2 - \omega_c^2)^{-1}$$
$$\epsilon_{yx} = -\epsilon_{xy} = -i(\epsilon_\infty \omega_p^2 \omega_c/\omega)(\tilde{\omega}^2 - \omega_c^2)^{-1} \qquad (1)$$
$$\epsilon_{zz} = \epsilon_\infty - \epsilon_\infty \omega_p^2/(\omega\tilde{\omega}), \quad \tilde{\omega} = \omega + i\nu.$$

The z axis is parallel to $\vec{B}_0$, ϵ_∞ is the high-frequency dielectric constant, and ω, ν, ω_p, and ω_c are the applied, collision, plasma, and cyclotron frequencies. We have not taken into account the effect of phonons. As for the metallic screen, we assume that its plasma frequency is much greater than ω_p, and so we may take $|\epsilon_{ij}| \to \infty$ for all the dielectric tensor elements.

There is no penetration of the fields into the metallic screen and therefore all components of E and B vanish in this half-space. In the semiconductor we must take a superposition of two plane-wave solutions with different decay constants α_1 and α_2:

$$\vec{E}(y,z) = (\vec{E}_1 e^{-\alpha_1 y} + \vec{E}_2 e^{-\alpha_2 y})e^{i(q_z z - \omega t)} \qquad (2)$$

We have chosen the y axis normal to the interface. The α_i satisfy a biquadratic equation given by WALLIS et al. [2] for a free semiconductor surface; see also a recent review article by HALEVI [3].

The boundary conditions at the interface $y = 0$ imply that

$$E_{1x} + E_{2x} = E_{1z} + E_{2z} = 0 \tag{3}$$

We find [4] the following result for the wave vector q_z of the interface modes:

$$q_z = \frac{\omega}{c}\left(\epsilon_{xx} \pm \epsilon_{yx}\sqrt{\frac{\epsilon_{xx}}{\epsilon_{zz} - \epsilon_{xx}}}\right)^{1/2} \tag{4}$$

This formula was first given - without proof - by DAVYDOV and ZAKHAROV [5].

Low-Frequency Solutions ($\omega \ll \nu \ll \omega_c$)

With the substitution of (1) in (4) we find the following, approximate solutions:

$$q_z = \frac{\omega}{c}\frac{\sqrt{\epsilon_\infty}}{2} , \quad q_z = (1+i)\frac{\omega_p}{\omega_c}\frac{\sqrt{\omega \nu}}{c} . \tag{5}$$

At this point we wish to comment on the low-frequency, slow magnetoplasma modes observed by BAIBAKOV and DATSKO [6] at n-type InSb surfaces in the Faraday geometry. It has been suggested by DAVYDOV and ZAKHAROV [5] that a metallic screen which was present in the experiments (at a certain distance from the surface) might have affected the characteristics of the observed "surface helicon" mode. ZAKHAROV [5] claims that (4)"was found to be in good agreement with the experimental results throughout the range of magnetic fields employed by BAIBAKOV and DATSKO [6]". Unfortunately, the solutions given by (5) are not acceptable. The solution on the left side corresponds to a fast wave, whose phase velocity is of the order of c. The solution on the right side of (5) does not describe a bonafide mode. While it does give a parabolic dispersion ($\omega \propto q_z^2$) characteristic of a volume helicon, the dependence on the magnetic field is incorrect. Moreover, it is difficult to accept that Re q_z depends on the collision frequency ν. Although DAVYDOV and ZAKHAROV [5] allowed for the presence of holes (which we neglected), we expect that their principal effect is to increase the damping, rather than convert the ν-dependent solution of (5) into a bona fide mode, which must have a real dispersion relation for $\nu \to 0$.

The effect of a finite gap between the surface of the semiconductor and the metallic screen was studied by YI, QUINN, and HALEVI [7]. It was concluded that the presence of the screen is rather unimportant in the way of explaining the experimental findings of BAIBAKOV and DATSKO [6]. It has been pointed out by HALEVI and QUINN [8] that sofar all the theoretical attempts to provide a basis to understanding these experiments have failed. Unfortunately, as far as we are aware, the results of BAIBAKOV and DATSKO [6] have not been confirmed experimentally by other authors. We also wish to comment that the three-media-geometry (semiconductor-vacuum-screen) has been studied in a different context by other authors [10].

High - Frequency Solutions ($\omega \gg \nu$)

We substitute again (1) in (4), this time with the assumption that $\nu = 0$. The result is

$$q_z = \frac{\omega}{c}\left[\epsilon_\infty \frac{\omega_H^2 - \omega^2 \pm \omega_p\sqrt{\omega_H^2 - \omega^2}}{\omega_c^2 - \omega^2}\right]^{1/2} \tag{6}$$

where $\omega_H = (\omega_p^2 + \omega_c^2)^{1/2}$. Numerical dispersion relations for the polariton modes given by (6) are shown in Fig.1. Two values of

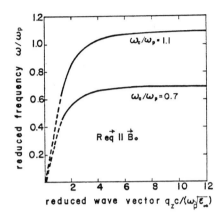

Fig.1 Dispersion relations for polaritons at the interface between a semiconductor and a metallic screen in the Faraday geometry for two values of the parameter ω_c/ω_p (computational work by Mr. K.S.Yi)

the parameter ω_c/ω_p have been considered. For these, as well as other values [4], the graph rises linearly from the origin at low frequencies and exhibits a resonance at the cyclotron frequency ω_c. Interestingly, this is just the frequency range for volume helicon waves.

It is important to investigate the nature of the decay constants α_1 and α_2, see (2). These must be real quantities for pure interface polaritons. The solid sections of the dispersion curves in Fig.1 indeed correspond to such modes. The lower portions of the dispersion curves (broken lines), on the other hand, correspond to generalized modes [2] whose decay constants are complex conjugates ($\alpha_2^* = \alpha_1$). The transition from generalized to pure modes takes place at a point defined by the intersection of the $\omega(q_z)$ dispersion curve and the curve $\alpha_1(\omega, q_z) = \alpha_2(\omega, q_z)$. The latter curve forms a loop; the generalized modes are located inside the loop and the pure modes outside.

We expect that the pure interface modes predicted here might be observable by ATR spectroscopy [9]. It seems that the best geometry would be sandwiching a metallic film (any metal) between a high-index prism and a semiconductor such as InSb. The desireable thickness of the film would be c/ω_p' (where ω_p' is the plasma frequency of the metal), i.e. a few hundred Å. Otherwise, if one sandwiches the semiconductor between the prism and the metal, the optimum thickness of the film would depend strongly on the frequency. As for the generalized modes, we have doubts [4] concerning their observability.

1. P.Halevi and C.Guerra-Vela, Phys.Rev.B$\underline{18}$, 5248 (1978).
2. R.F.Wallis, J.J.Brion, E.Burstein, and A.Hartstein, Phys.Rev. B$\underline{9}$, 3424 (1974).
3. P.Halevi, in <u>Electromagnetic Surface Modes</u>, ed. A.D.Boardman, J.Wiley and Sons, to be published.
4. P.Halevi, to be published.
5. A.B.Davydov and V.A.Zakharov, Sov.Phys.Solid State $\underline{17}$, 117 (1975); V.A.Zakharov, ibid.$\underline{18}$, 670 (1976).
6. V.I.Baibakov and V.N.Datsko, Jetp Lett. $\underline{15}$,135 (1972); Sov. Phys.Solid State $\underline{15}$, 1084 (1973); Sov.Phys.Semicond. $\underline{12}$, 855 (1978); ibid. $\underline{13}$, 472 (1979).
7. K.S.Yi, J.J.Quinn, and P.Halevi, to be published.
8. P.Halevi and J.J.Quinn, Solid State Commun. $\underline{33}$, 467 (1980).
9. E.D.Palik, R.Kaplan, R.W.Gammon, H.Kaplan, R.F.Wallis, and J.J.Quinn, Phys.Rev.B$\underline{13}$, 2497 (1976).
10. M.Nakayama and M.Tsuji, J.Phys.Soc.Japan $\underline{43}$, 164(1977) and references therein.

Part VII

Electron-Hole Drops and Semimetals

Electron-Hole Liquid in Ge in High Magnetic Field

M.S. Skolnick

Royal Signals and Radar Establishment, St. Andrews Road
Great Malvern, Worcestershire, United Kingdom

1. Introduction

The existence of the electron-hole liquid (EHL), in a number of indirect gap semiconductors (Ge, Si, GaP, SiC) has now been firmly established [1-3]. The Fermi liquid nature of the degenerate electron-hole system is very clearly demonstrated in magnetic field experiments. Quantum, de Haas-van Alphen-like oscillations are observed in a variety of EHL properties eg luminescence intensity [4], lifetime [5], ultrasonic absorption [6] and infrared absorption [7], [8], [9] whenever an electron or hole Landau level crosses its respective Fermi energy. The observation of such Landau oscillations forms part of the subject of the present paper. Hole oscillations in the EHL luminescence intensity are observed here for the first time at high fields between 5 and 18 T by the use of sensitive field modulation techniques. These fields are high enough to reach the extreme quantum limit for electrons where only the lowest spin level of the lowest Landau level is populated. This spin polarization may have important consequences for the non-radiative, Auger recombination probability in the EHL and is discussed in section 4.

EHL luminescence spectra in magnetic fields up to 10 T were first studied by Alekseev et al. [17]. This was followed by the detailed measurements of Störmer and Martin [12] up to 19 T. The magnetic field variation of the EHL ground state energy and density were obtained from these measurements in addition to a value for the mass renormalization of carriers in the EHL.

In the present paper attention is focused on experimental studies of the total luminescence intensity I_{tot} and lifetime τ of the EHL in Ge from 0 to 18 T. The density variation of the EHL with magnetic field is deduced by two independent methods from the experimental results. Efficient electron thermalization between magnetic field split valleys is deduced from the field positions of the Landau oscillations and luminescence line shape studies for $\vec{H}//<111>$.

2. Experimental Results

The variation of I_{tot} and τ with magnetic field from 0 to 18 T for $\vec{H}//<001>$ and <111> are shown in Figs.1, 2 and 3. Similar results are obtained for $\vec{H}//<110>$ but are not shown here [5]. High field hole oscillations are clearly visible in the derivative traces $\frac{dI_{tot}}{dH}$ in Figs.2, 3 obtained using magnetic field modulation. Well-resolved electron Landau oscillations are seen in the range 0 to 5 T.

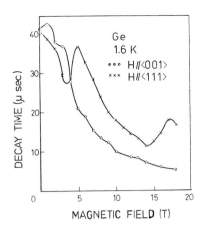

Fig.1 Low temperature EHL lifetime in Ge against magnetic field for $\vec{H}//<001>$, $<111>$

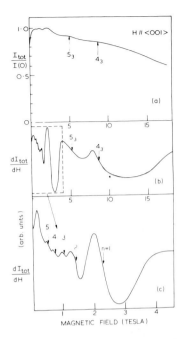

Fig.2 (a) EHL luminescence intensity I_{tot} at 1.6 K against magnetic field $\vec{H}//<001>$ from 0 to 18 T (b) Magnetic field derivative of EHL luminescence intensity $\frac{dI_{tot}}{dH}$. 3_1 and 4_3 indicate positions of hole oscillations (c) Expanded region of 2(b) from 0 to 4.5 T. n = 1 to 5 electron oscillations are indicated.

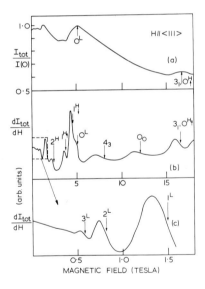

Fig.3 (a) EHL luminescnece intensity I_{tot} for $\vec{H}//<111>$ from 0 to 18 T. (b)(dI_{tot}/dH) 4_3, 0_0 and 3_1 are hole oscillations. 1^H, 2^H and 0^L are electron oscillations. H and L stand for heavy and light electron Landau levels. (c) Expanded region of 3(b) from 0 to 1.5 T. Light electron oscillations are indicated

3. Density Variation with Magnetic Field

Minima in the EHL density occur each time a Landau level crosses its respective Fermi energy. The dependence of I_{tot} and τ on the EHL density n is given by the relations

$$I_{tot} = \alpha N n \tau B \qquad (1)$$

$$\frac{1}{\tau} = Bn + Cn^2 \qquad (2)$$

where N is the total number of carriers in the condensed phase, α is a constant of proportionality, and B and C are the coefficients of radiative and three particle Auger recombination respectively. Density minima correspond to maxima in I_{tot} and τ (and zeroes in dI_{tot}/dH) at each Landau level crossing. The hole oscillations in Figs.2 and 3 are identified using the notation of HENSEL and SUZUKI [10] and the electron oscillations by the corresponding electron Landau level number. Once the oscillations have been firmly identified (for further details see ref 5) the EHL density variation with magnetic field can be determined in a straightforward and accurate manner. For the analysis of the hole oscillations, the accurate hole density of states tabulations of HENSEL and SUZUKI [11] were employed. From the known field position of an identified oscillation, the EHL Fermi level at that field is easily calculated from the Landau level energy dependence on magnetic field. A simple integration over the carrier density of states up to the Fermi energy then gives the EHL density at that particular field. In this way, the density values, at various magnetic fields, given by the crosses on Figs. 4 and 5, were calculated.

A renormalization of the transverse electron mass and of the hole masses of + 10% was assumed in the density of states and Fermi level calculations, in accordance with the findings of STÖRMER and MARTIN [12].

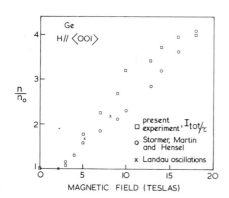

Fig.4 Normalized EHL density against magnetic field for $\vec{H}//<001>$. The squares are obtained from the I_{tot}/τ ratio, the crosses from the field ositions of the Landau oscillations and the circles from the luminescence lineshape fits of Störmer et al.(ref. 12)

From (1) it can be seen that the density variation can also be calculated from the ratio I_{tot}/τ divided by its zero magnetic field value for normalization. The results of such a determination are given on Figs.4, 5 together with the Landau oscillation values and the independent values of STÖRMER and MARTIN [12] deduced from high field magneto-luminescence lineshape fits.

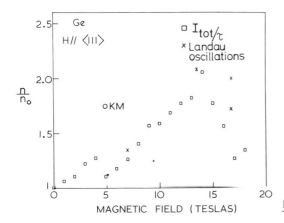

Fig.5 As Fig.4 but for $\vec{H}//<111>$.

Good agreement between the values obtained by the different methods is found. The maximum discrepancy between the I_{tot}/τ values and the precise Landau oscillation results is about 20%. From (1) this implies that the radiative coefficient B does not change by more than 20% from 0 to 18 T. The variation of the density n with magnetic field has also been determined by YAMANAKA ET AL from analysis of far infrared (FIR) EHL magnetoplasma resonances [9]. At 7 T they deduce an EHL density of $\sim 8 \times 10^{17}$ cm^{-3}, about a factor of two higher than the average value given in Fig.4. The discrepancy between the density deduced from the luminescence and FIR measurements may arise from inaccuracies in the more complex fitting procedure used in the analysis of the FIR results.

For $\vec{H}//<001>$ a factor of 4 increase in the EHL density from 0 to 18 T is found. For $\vec{H}//<111>$ the density increase is smaller, the ratio $n_{<100>}/n_{<111>}$ at 15 T being ~ 2. The density increases with magnetic field can be understood qualitatively as arising from the increase in carrier densities of states (d.o.s.) with H. The electron d.o.s. for one particular Landau level increases linearly with H whilst for holes the d.o.s. increases as $H^{0.3}$ to $H^{0.6}$ depending on the field direction [5]. It has been shown [12] that the sum of the exchange and correlation many body contributions to the e - h pair energy is approximately magnetic field independent. The increase of the densities of states leads to a reduction in carrier kinetic energy, at constant Fermi energy, with magnetic field and thus a stabilization and density increase of the EHL. In the high field limit the electron and hole densities of states are 2 and 1.5 times higher for $\vec{H}//<001>$ than for $\vec{H}//<111>$, thus explaining the higher EHL densities observed for $\vec{H}//<001>$.

4. Field Dependence of Auger Coefficient

Having determined n from the I_{tot}/τ ratio, $\frac{1}{\tau n}$ is now plotted against n in Fig.6. From (2)

$$\frac{1}{\tau n} = B + Cn \qquad (3)$$

and so the slope in Fig.6 should be given by the Auger coefficient C and the intercept on the $1/\tau n$ axis by B. However, good straight line behaviour for the $1/\tau n$ against n plots is not found in Fig.6. Furthermore a marked angular dependence between $\vec{H}//<001>$ and $<111>$ is observed. B was shown to be almost field independent for $\vec{H}//<001>$ and $<111>$ in the earlier analysis in section 3. Thus the non-linearity and angular dependence in Fig.6 can be attributed to changes in C rather than B.

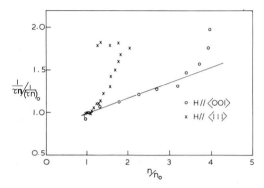

Fig.6 Normalized plot of $1/\tau n$ against n, circles $\vec{H}//<001>$, crosses $\vec{H}//<111>$. n values are obtained from Figs.4, 5 and τ values from Fig.1.

The Auger recombination is a complicated three particle, phonon assisted process but its stronger field dependence can nevertheless be understood qualitatively. At fields above ~ 10 T the electrons become increasingly spin polarized as the extreme quantum limit is reached. For $\vec{H}//<001>$ only the n = 0 spin up electron Landau level is occupied above 18 T. The spin polarization results in a strong exchange repulsion between the electrons. This leads in turn to a marked field dependence of the two-electron, one hole Auger process which depends sensitively on the degree of electron-overlap [13]. Furthermore since the conduction band of Ge is composed of four <111> valleys, at fields > 3 T for $\vec{H}//<111>$ the valley parallel to the magnetic field is depopulated whereas for $\vec{H}//<001>$ all four valleys remain equally populated. Thus a larger Auger coefficient for $\vec{H}//<111>$ than <001> would be expected at the same EHL density, due to the greater number of electrons on average in each populated valley. It can also be remarked that the electron spin polarization will give rise to a decrease in the exchange contribution to the electron-hole pair energy in the condensed phase [14]. This should have a marked effect on the EHL ground state energy in very high fields and would be an interesting problem for future investigation.

5. Thermalization Between Magnetic Field Split Valleys

There have been recent experimental reports of the observation of EHL luminescence from "hot" and "cold" electrons with separate Fermi energies and densities in valleys split apart by uniaxial stress [15] and magnetic field [16]. It is shown in this section, from analysis of the electron Landau oscillations and from luminescence lineshape studies that hot electrons do not exist in the EHL in Ge for magnetic fields $\vec{H}//<111>$. For $\vec{H}//<111>$ one ("light electron") conduction band valley parallel to the magnetic field has cyclotron mass $m_c = 0.090\ m_0$ and the other three ("heavy") valleys have $m_c = 0.224\ m_0$. As stated in section 4, with increasing magnetic field the light valley is split away to higher energy, the splitting being eg 6.2 meV at 15 T.

From section 3, maxima in I_{tot} are expected for light electrons wherever

$$E_F^{el} + \tfrac{1}{2}\hbar\omega_{heavy} - \tfrac{1}{2}g_{heavy}\mu_B H = \left(n + \tfrac{1}{2}\right)\hbar\omega_{light} \pm g_{light}\mu_B H \qquad (4)$$

and for heavy electrons when

$$E_F^{el} - \tfrac{1}{2}g_{heavy}\mu_B H = n\hbar\omega_{heavy} \pm \tfrac{1}{2}g_{heavy}\mu_B H \quad . \qquad (5)$$

The Fermi energies are taken equal to one another in (4), (5) and are measured from the lowest spin state of the lowest heavy electron Landau level. The fan diagram in Fig.7 shows the Landau level energies plotted against magnetic field calculated from (4) and (5). The positions of the experimentally observed electron oscillations in Fig.3 are indicated by horizontal bars and are seen to be in good agreement with the fields predicted for a constant $E_F^{el} = 2.43$ meV represented by the dashed-dotted line drawn parallel to the lowest Landau level. This shows clearly that both heavy and light electrons have the same Fermi energy measured relative to the 0^H level, the main assumption in (4) and (5). Thus it can be concluded that electrons in both types of valley form part of the same equilibrium system with common Fermi energy and uniform density. Rapid thermalization of electrons between the different valleys, in times much less than the EHL lifetime (Fig.1), is seen to occur. The observation of both hot and cold EHL's under uniaxial stress [15] probably arose because of a drastic shortening of the EHL lifetime due to stress inhomogeneities leading to enhanced surface recombination. EHL lifetimes of 1 to 3 μsec were measured in these experiments [15], times of the order of the electron intervalley thermalization time as compared to EHL lifetimes of several hundred microseconds expected for uniform uniaxial stresses. Inhomogeneities of the perturbing field leading to a reduction of τ cannot arise in the magnetic field, the minimum value of τ being $\sim$ 12 μsecs at 14 T for $\vec{H}//<111>$.

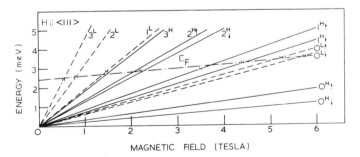

Fig.7 Fan diagram of heavy and light electron Landau levels for $\vec{H}//<111>$. The field positions of the experimentally observed maxima are shown by horizontal bars. The dashed-dotted line represents a field independent Fermi energy of 2.43 meV drawn parallel to the lowest heavy electron level.

The above conclusions are supported by the EHL luminescence spectra for $\vec{H}//<111>$ shown in Fig.8. No splitting of the EHL band is observed at any magnetic field even though the light-heavy valley splitting is 6.2 meV at 15 T. Once again, this demonstrates convincingly the occurrence of efficient intervalley thermalization between the valleys split apart by high magnetic field.

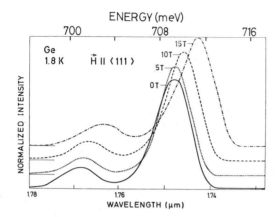

Fig.8 EHL luminescence spectra at 0, 5, 10 and 15 T for $\vec{H}//<111>$. LA and TO phonon replicas are shown.

Acknowledgements

The experimental work described in this paper was carried out at the Hochfeld Magnetlabor of the Max Planck Institut, Grenoble together with D.Bimberg. The author wishes to express his sincere thanks to him for a very enjoyable and successful collaboration. He is also very grateful to K.Dransfeld and G.Landwehr for their valuable support of this work during his stay in Grenoble.

References

1. See the comprehensive review by J C Hensel, T G Phillips and G A Thomas in Solid State Physics Vol 32, H Ehrenreich, F Seitz and D Turnbull editors, Academic Press, New York 1977.

2. D Bimberg, M S Skolnick and L M Sander, Phys Rev B19, 2231, 1979.

3. D Bimberg, M S Skolnick and W J Choyke, Phys Rev Lett 40, 56, 1978.

4. V S Bagaev, T K Galkina, N A Penin, V B Stopachinskii and M N Churaeva, Pisma Zh. Eksp. Teor. Fiz 16, 120, 1972 (JETP Lett 16, 83, 1972).
    K Betzler, B G Zhurkin, A L Kurazskii and B M Balter, J Lumin 12/13, 651, 1976.

5. M S Skolnick and D Bimberg, Phys Rev B21, 4624, 1980.

6. T Ohyama, A D A Hansen and J L Turney, Solid State Commun 19, 1083, 1976.

7. V N Murzin, V A Zayats and V L Kononenko, Fiz Tverd Tela 15, 3634, 1973. (Sov. Phys. Solid State 15, 2421, 1974)

8. H Nakata and E Otsuka, Proceedings of the 15th International Conference on the Physics of Semiconductors, Kyoto 1980.

9. M Yamanaka, K Muro, S Narita, J. Phys Soc Japan 47, 1168, 1979.

10. J C Hensel and K Suzuki, Phys Rev B9, 4148, 1974.

11. J C Hensel, private communication 1979.

12. H L Störmer and R W Martin, Phys Rev B20, 4213, 1979.

13. V N Abakumov and I N Yassievich, Fiz. Tech Poluprovodn 11, 1302, 1977 (Sov. Phys. Semicond 11, 766, 1977).

14. R Zimmerman and M Rosler, Solid State Commun 25, 651, 1978.

15. H H Chou, G K Wong and B J Feldman, Phys Rev Lett 39, 959, 1977.

16. J Vitins, R L Aggarval and B Lax, Solid State Commun 30, 103, 1979.

17. A S Alekseev, V S Bagaev, T I Galkina, O V Gogolin, N A Penin, A N Semonov and V G Stopachinskii, Pisma. Zh. Eksp. Teor. Fiz 12, 203, 1970 (JETP Lett 12, 140, 1970).

Far-Infrared Magneto-Plasma Absorptions of Electron-Hole Drops in Germanium

S. Narita, K. Muro, and M. Yamanaka

Department of Material Physics, Faculty of Engineering Science
Osaka University, Toyonaka
Osaka 560, Japan

1. Introduction

The far-infrared (FIR) magneto-absorption measurement of electron-hole drops (EHD) has been a valuable tool for investigating the properties of the plasma sphere of EHD: carrier density, mass-renormalization, collision frequency of the carrier, dimension of EHD, etc. However, in previous reports [1-3] the absorption measurements have been done in relatively limited conditions: in a magnetic field parallel to the [111]-axis of Ge in the Faraday configuration.

Our measurements have been done in magnetic fields not only parallel to [111], but also to [100] and [110] directions both in the Faraday and Voigt configurations up to 8T at quasi-continuous wavelengths from 250μm to 1.73mm with FIR lasers pumped by a CO_2 laser in a time-resolved scheme [4]. The laser lines used for the present experiment were more than 30.

The merits of such wide range measurements in a time resolved scheme are as follows: We can observe not only the transverse mode of the magneto-plasma resonance but also the longitudinal mode. So, we can rightly assign the observed peaks by comparing both the modes with the theoretical results of peak intensity and the order of the appearance with increasing the magnetic field. Otherwise it becomes difficult to distinguish the kind of peaks, including the electric dipole resonance (EDR) and the magnetic dipole resonance (MDR).

Moreover, in the process of the mass-renormalizations, we took the method of the best fits of the many spectral peaks in the calculation to the experimental ones with common renormalized carrier masses, so we could obtain more reliable values for the mass-renormalizations. In addition, with the time resolved method we can hold the EHD in a same condition in spite of changing the laser frequency and the magnetic field, namely 2μsec after the pulse excitation, so that we could obtain the magnetic field dependence of the carrier density and radius of EHD at a new standpoint.

2. Theoretical considerations

The complex dielectric tensor of EHD is assumed to have the following form in a static magnetic field (defined to be parallel to the z-axis) applied along the [111], [100] or [110] crystallographic direction.

$$\tilde{\varepsilon} = \begin{pmatrix} \varepsilon_{xx} & \varepsilon_{xy} & 0 \\ -\varepsilon_{xy} & \varepsilon_{yy} & 0 \\ 0 & 0 & \varepsilon_{zz} \end{pmatrix} . \qquad (1)$$

In the cases of H∥ [100] and H∥ [111], the above tensor can be rewritten as follows, because $\varepsilon_{xx} = \varepsilon_{yy}$:

$$\tilde{\varepsilon} = \begin{pmatrix} \varepsilon_+ & 0 & 0 \\ 0 & \varepsilon_- & 0 \\ 0 & 0 & \varepsilon_\parallel \end{pmatrix} \qquad (2)$$

where $\varepsilon_\pm = \varepsilon_{xx} \pm i\varepsilon_{xy}$ and $\varepsilon_\parallel = \varepsilon_{zz}$. Thus the dielectric tensor is diagonalized into three linearly independent modes (denoted by m), which consist of right and left circular polarizations (transverse)(m = ±) to the static magnetic field H (Faraday configuration) and of a longitudinal polarization (m = ∥) where the radiation field is parallel to H (in the Voigt configuration).

The dielectric tensor elements $\varepsilon_\pm$ and $\varepsilon_\parallel$ are calculated by the formula given in some text books (e.g.[5]). However in the case of H∥ [110], we must use the following formula:

$$\varepsilon_{xx} = 1 - \frac{4\pi e^2}{m_0 \varepsilon_0 \omega} \left\{ \frac{\alpha_T N_A(\omega + i\Gamma)}{(\omega + i\Gamma)^2 - \omega_A^2} + \left(\frac{2\alpha_L + \alpha_T}{3}\right) \frac{N_B(\omega + i\Gamma)}{(\omega + i\Gamma)^2 - \omega_B^2} \right\}$$
$$- \frac{4\pi e^2 N_l}{m_l \varepsilon_0 \omega} \frac{\omega + i\Gamma}{(\omega + i\Gamma)^2 - \omega_l^2} - \frac{4\pi e^2 N_h}{m_h \varepsilon_0 \omega} \frac{\omega + i\Gamma}{(\omega + i\Gamma)^2 - \omega_h^2} , \qquad (3\text{-a})$$

$$\varepsilon_{yy} = 1 - \frac{4\pi e^2}{m_0 \varepsilon_0 \omega} \left(\frac{\alpha_L + 2\alpha_T}{3}\right) \left\{ \frac{N_A(\omega + i\Gamma)}{(\omega + i\Gamma)^2 - \omega_A^2} + \frac{N_B(\omega + i\Gamma)}{(\omega + i\Gamma)^2 - \omega_B^2} \right\}$$
$$- \frac{4\pi e^2 N_l}{m_l \varepsilon_0 \omega} \frac{\omega + i\Gamma}{(\omega + i\Gamma)^2 - \omega_l^2} - \frac{4\pi e^2 N_h}{m_h \varepsilon_0 \omega} \frac{\omega + i\Gamma}{(\omega + i\Gamma)^2 - \omega_h^2} , \qquad (3\text{-b})$$

$$\varepsilon_{xy} = i\frac{4\pi e^2}{m_0 \varepsilon_0 \omega} \left[\left\{ \frac{\alpha_T(\alpha_L + 2\alpha_T)}{3} \right\}^{\frac{1}{2}} \frac{N_A \omega_A}{(\omega + i\Gamma)^2 - \omega_A^2} + (\alpha_L \alpha_T)^{\frac{1}{2}} \frac{N_B \omega_B}{(\omega + i\Gamma)^2 - \omega_B^2} \right]$$
$$- i\frac{4\pi e^2 N_l}{m_l \varepsilon_0 \omega} \frac{\omega_l}{(\omega + i\Gamma)^2 - \omega_l^2} - i\frac{4\pi e^2 N_h}{m_h \varepsilon_0 \omega} \frac{\omega_h}{(\omega + i\Gamma)^2 - \omega_h^2} , \qquad (3\text{-c})$$

and

$$\varepsilon_{zz} = 1 - \frac{4\pi e^2}{m_0 \varepsilon_0 \omega} \left\{ \left(\frac{2\alpha_L + \alpha_T}{3}\right) \frac{N_A}{\omega + i\Gamma} \frac{(\omega + i\Gamma)^2 - \omega_5^2}{(\omega + i\Gamma)^2 - \omega_A^2} + \frac{\alpha_T N_B}{\omega + i\Gamma} \right\}$$
$$- \frac{4\pi e^2 N_l}{m_l \varepsilon_0 \omega} \frac{1}{\omega + i\Gamma} - \frac{4\pi e^2 N_h}{m_h \varepsilon_0 \omega} \frac{1}{\omega + i\Gamma} , \qquad (3\text{-d})$$

where α_T is the coefficient of the square of the component of wave vector perpendicular to the prolate axis of the spheroidal energy surface of electrons in Ge, and α_L, that of the component parallel to it; $\alpha_T = 12.2$ and $\alpha_L = 0.633$ and the subscripts A and B mean the two equivalent valleys where the prolate axis of the electron energy surface is perpendicular to the static magnetic field and the other two equivalent valleys; and ω_A and ω_B are the

217

angular cyclotron frequencies given by

$$\omega_A = [\alpha_T(\alpha_L+2\alpha_T)/3]^{1/2}\omega_0, \quad \omega_B = (\alpha_T\alpha_L)^{1/2}\omega_0, \text{ and } \omega_5 = [3\alpha_T^2\alpha_L/(2\alpha_L+\alpha_T)]^{1/2}\omega_0,$$

(ω_0: the free electron angular frequency). (3-e)

In addition, the subscripts l and h represent the light and heavy hole bands, therefore, for instance, N_h means the carrier density in the heavy hole band. Thus, in these calculations, the classical picture of the valence band was adopted; in other words, the independent heavy and light hole masses were used.

The normal dielectric tensor elements ε_m in H// [110] are given by

$$\varepsilon_\pm = (1/2)[(\varepsilon_{xx}+\varepsilon_{yy}) \pm i\{4\varepsilon_{xy}^2-(\varepsilon_{xx}-\varepsilon_{yy})^2\}^{1/2}], \text{ and } \varepsilon_{//} = \varepsilon_{zz}. \quad (3\text{-f})$$

The formula for the absorption intensities of EDR and MDR of EHD were given by Ford et al.[6] as follows:

$$p^E = (1/2)N\omega\varepsilon_0|\vec{E}|^2 \text{ Im } \alpha^E, \quad (4) \qquad p^M = (1/2\mu_0)N\omega|\vec{B}|^2 \text{ Im } \alpha^M, \quad (5)$$

where N is the number of EHD, $\vec{E}$ and $\vec{B}$ the electric and magnetic fields of the incident radiation, and α^E and α^M are represented by the followings:

$$\alpha_m^E = 4\pi a^3(\varepsilon_m-1)/(\varepsilon_m+2), \quad (6) \quad \alpha_m^M = (2\pi a^3/15)(2\pi a/\lambda)^2 \varepsilon_0 \varepsilon_m^{eff}, \quad (7)$$

with

$$\varepsilon_\pm^{eff} = 2\varepsilon_\pm \varepsilon_{//}/(\varepsilon_\pm + \varepsilon_{//}), \text{ and } \varepsilon_{//}^{eff} = 2\varepsilon_+\varepsilon_-/(\varepsilon_+ + \varepsilon_-), \quad (8)$$

where $\varepsilon_\pm$ and $\varepsilon_{//}$ are obtained from the above mentioned calculations of the complex dielectric functions of EHD, a the radius of EHD and λ the wavelength of the radiation.

By using these formula we calculated the theoretical transverse and longitudinal modes of the absorption. Examples of the calculations are shown in Fig.1 (a) and the second frame of Fig.2 for the cases of H// [110] and H// [100], respectively. In the calculations we used the bulk masses and the relaxation times obtained in this experiment.

3. Experimental Results and Discussions

3.1 The assignments of the spectral peaks

In the assignments of the experimental spectral peaks, first the order of the appearance of the peaks with increasing magnetic field is compared with the theoretical result. An example of the assignments are shown in Fig.1 (b) which should be compared with the calculation shown in Fig.1 (a) for the case of H// [110]. The main longitudinal EDR peak appears in the middle point between the main two transverse EDR peaks. This order is different from the cases of H// [100] and H// [111]. The assignments of the experimental peaks are thus confirmed by the comparisons with the calculations.

3.2 Mass renormalizations

As seen in the comparisons of the experimental results with the calculations (Fig.1 (b) with Fig.1 (a) at 570.6μm, or the first frame of Fig.2 with the second frame), the experimental resonance magnetic fields deviate appreciably from the theoretical results to higher magnetic fields. This fact suggests us the necessity of the effective mass renormalizations. The renormalizations were done by using mainly the data of H// [100] and [110] from the reason mentioned later. The best fits

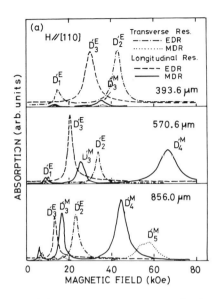 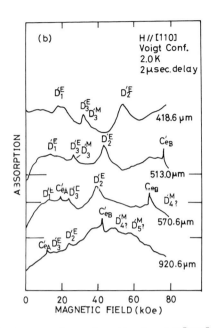

Fig.1 (a) Absorption spectra calculated at several wavelengths for H// [110]. The conditions are $n_d=2\times10^{17}$cm^{-3} and a=2μm. (b) Typical experimental spectra in the Voigt configuration of H// [110]. $D_n^{'E}$ and $D_n^{'M}$ represent the EDR and MDR magneto-plasma resonances of EHD, respectively, and $C_{eA}^{'}$ and $C_{eB}^{'}$ are the cyclotron resonance lines of free electrons in the A and B valleys

of most of the calculated resonant fields to the experimental ones can give more reliable results compared with the values so far obtained. The comparisons of the present results with the published experimental and theoretical results are tabulated in the followings:

Table 1. Renormalized Mass Parameters of carriers in EHD

		Calculation by Rice [7]	Interband lumi. by Martin et al. [8]	Magneto-plasma- by Gavrilenko et al. [2]	Present results
Electron	Transv. $\frac{m_{et}^*}{m_{et}}$	1.10	~1.10	1.15 ± 0.02	1.15
	Longit. $\frac{m_{el}^*}{m_{el}}$	0.99		1.0 ± 0.2	0.96
Light hole: m_{lh}^*/m_{lh}		1.10		~1.15	1.20
Heavy hole: m_{hh}^*/m_{hh}		1.14			1.15

The renormalization factors are believed to be independent of the magnetic field intensity, because most of the peak energies except for D_x show linear dependences upon the magnetic field [4].

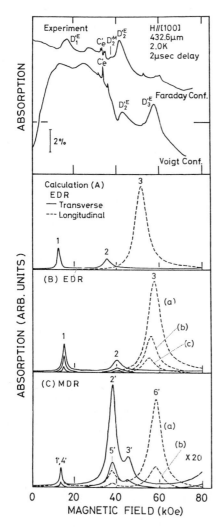
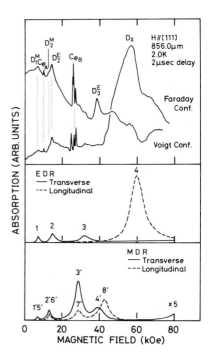

Fig.3 Comparison of the experimental results with the calculations. i) The upper part: The experimental absorption spectra in the Faraday and Voigt configurations of H// [111] for 856.0μm. ii) The middle part: The calculated absorption spectra of EDR using the renormalized masses of the carriers. iii) The lower part: The calculated absorption spectra of MDR using the renormalized masses of the carriers

Fig.2 Comparison of the experimental results with the calculations.
i) Experiment: The absorption spectra in the Faraday and Voigt configurations of H// [100] for 432.6μm. ii) Calculation (A): Calculated absorption spectra of EDR using the bulk masses of the carriers and for the carrier density of EHD, $n_d = 2 \times 10^{17} cm^{-3}$. iii) Calculation (B): Calculated spectra of EDR using the renormalized masses of the carriers, for the carrier densities in EHD; (a) $2 \times 10^{17} cm^{-3}$, (b) $3 \times 10^{17} cm^{-3}$, and (c) $4.5 \times 10^{17} cm^{-3}$. The resonant points shift to lower fields with increasing the density. iv) Calculation (C): Calculated spectra of MDR using the renormalized masses of the carriers, for two radii of EHD; (a) 2μm and (b) 1μm. The scale is enlarged 20 times compared with the cases of (A) and (B).

By using the results of mass-renormalizations and the relaxation times obtained in this study and adjusting the carrier density in plasma spheres and the size of EHD, all the spectra obtained in the present experiment are considered to be theoretically reproduced. An example for H// [100] is shown in Fig.2. In the third frame of the figure, the calculated spectra of EDR using the renormalized carrier masses are shown changing the carrier density, and in the fourth frame the theoretical MDR spectra are shown in a variety of the EHD size. In fact, the schedule of the theoretical reproduction has been fairly successful for the cases of H// [100] and H// [110], but for the case of H// [111], the several peaks could not be explained.

As shown in Fig.3, the giant absorption peak D_x does not appear in the calculation. The calculated spectra for EDR and MDR are shown in the middle and lower parts of the figure using the renormalized masses. The giant peak D_x can be observed only in the [111]-magnetic field direction at about 6T, and the resonant magnetic field of the peak D_x has little photon energy dependence. The origin of the giant peak D_x has not been made clear. However, in our recent experiment, the peak showed a shift to higher magnetic fields with increasing [111]-uniaxial stress. This result suggests us that the exhaustion of the electron population from the [111]-valley may cause the appearance of the giant peak, though the relation between the exhaustion and the appearance of the peak has remained unknown.

Another distinct peak, D_3^E, in the experimental data of the Voigt configuration in Fig.3 deviates considerably from the corresponding theoretical peak No.4, shown in the middle part of Fig.3. We have no reliable explanation for the large peak shift at present.

3.3 Increase of carrier density with magnetic field: In a previous paper [4], we reported an increase of the carrier density in the plasma sphere with increasing magnetic field estimated by a different method from those so far published. The increment was appreciably larger than that obtained by Störmer et al. [9] from the linewidth of the interband luminescence of EHD in a strong magnetic field. Our method is a comparison of the experimental absorption intensity with the theoretical results as shown in Fig.4 (a). The origin of the discrepancy was first supposed to be ascribed to our treatment of the renormalized masses in the classical picture. Störmer et al.[9] took into account the quantum effects upon the valence band structure in their calculation. However, we considered that there probably exists a problem in the estimation of the absorption intensity of the peak in the spectra. We changed the estimation method from a simple integration of the line shape of the absorption peak to the estimation by assuming the Lorentzian line shapes of the absorption peaks, so that we have been able to neglect ambiguities due to the differences of the absorptions at the skirts of the peaks. The revised absorption intensities are plotted against the magnetic field as shown in Fig.4 (a). The technique of the time resolved scheme enables us to hold the same condition of EHD. On the other hand, the theoretical absorption intensity decay is obtained by using (4) and (6) and calculating the total number of electron-hole pairs, N_{e-h}, according to

$$N_{e-h} = N n_d (4/3)\pi a^3 . \tag{9}$$

The experimental results cannot be explained without assuming the carrier density increase in the plasma sphere with increasing the magnetic field. The carrier density change obtained from this method is shown in Fig.4 (b). The slope of the increase in the present result is rather close to the data reported by Störmer et al.[9]. Although in the present method there is some

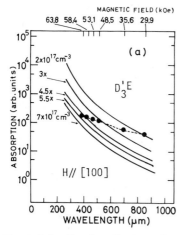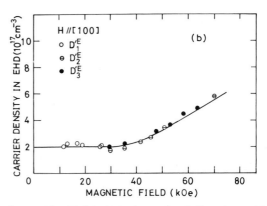

Fig.4 (a) Wavelength dependence (magnetic field dependence) of the absorption intensity of EDR calculated for several carrier densities in EHD (solid curves). The experimental results are shown by the closed circles. (b) Magnetic field dependence of the carrier density in EHD obtained from (a) and others

ambiguity in the estimation of the background absorption, the time resolved experiment enables us to make a discussion about the properties of the plasma sphere from a new angle.

The increase of the carrier density is believed to be ascribed to the cyclotron radius becoming smaller than the Bohr radius, which means that the carrier have to move closer to each other to achieve the wavefunction-overlap necessary for the EHD formations.

3.4 Decrease of radius of EHD with magnetic field: By using (5), (7) and (9), we obtain the absorption ratio of the incident radiation due to MDR:

$$P^M/(|B|^2/2\mu_0) = (N_{e-h}/10)(2\pi a/\lambda)^2 \, \varepsilon_0 \omega \, (\text{Im} \, \varepsilon_m^{eff}/n_d) \qquad (10)$$

and $\text{Im}\,\varepsilon_m^{eff}$ is almost perfectly proportional to n_d. Thus, under the same condition which is realized by the measurements at 2μsec delay after the short pulse excitation in the time resolved scheme, the absorption intensity due to MDR depends on the radius of EHD. On the other hand, using (4), (6) and (9) the absorption ratio of the incident radiation due to EDR is written by

$$P^E/(\varepsilon_0|E|^2/2) = 3\, N_{e-h} \omega \, (1/n_d) \, \text{Im}\,[(\varepsilon_m-1)/(\varepsilon_m+2)]. \qquad (11)$$

Therefore, we can estimate the radius of EHD by comparing the intensity ratio between MDR and EDR. The results are shown in Fig.5, where the decrease of the radius of the sphere with increasing magnetic field is seen.

In addition in our study of the time resolved spectroscopy, the absorptions associated with MDR seem to decrease more slowly than those with EDR with time. The fact is seen in the comparisons of D_1^M peak with D_1^E peak in Fig.6. We interpret the result as the increase of the radius of EHD by cohesions of drops with increasing the delay time after the pulse excitation, and, as a whole, the number of electron-hole pairs, N_{e-h}, decreases rapidly with time.

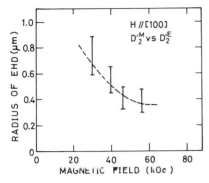

Fig.5 Magnetic field dependence of the radius of EHD. The error bars are obtained from the comparisons of the intensities of $D_2^{\prime M}$ with those of $D_2^{\prime E}$ in the spectra of H//[100]. The radius of EHD is seen to decrease with increasing magnetic field

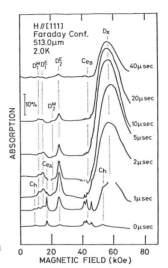

Fig.6 Difference of the decay speed between the magnetic dipole resonance absorption (MDR) and the electric dipole resonance absorption (EDR). A typical example is seen in the comparison of D_1^M with D_1^E in the spectra observed at 2.0K corresponding to several delay times after a pulse excitation in the Faraday configuration of H//[111]. You can also see that the intensity maximum of D_x appears at ∼10μsec delay after the excitation, which is different from other peaks

In conclusion, the far-infrared magneto-absorption spectroscopy has been applied to the study of the properties of electron-hole drops in Ge and the experimental results have shown fairly good correspondence with the theory.

References

1. K.Muro, Y.Nisida: J. Phys. Soc. Japan 40, 1069 (1976)
2. V.I.Gavrilenko, V.L.Konoenko, T.S.Mandel'shtam, V.N.Murzin: Sov. Phys. JETP Lett. 23, 645 (1976), and with S.A.Sauin: ibid. 26, 95 (1977)
3. H.Nakata, D.Fujii, E.Otsuka: J. Phys. Soc. Japan 45, 537 (1978)
4. M.Yamanaka, K.Muro, S.Narita: J. Phys. Soc. Japan 47, 1168 (1979)
5. P.R.Wallace: "Physics of Solids in Intense Magnetic Field" (edited by E.D.Haidemenakis, Plenum Press, New York, 1969) p.61
6. G.W.Ford, J.K.Furdyna, S.A.Werner: Phys. Rev. B12, 1452 (1975)
7. T.M.Rice: Nuovo Cimento 23B, 226 (1974)
8. R.W.Martin, H.L.Störmer, W.Rühle, D.Bimberg: J. Lumi. 12/13, 645 (1974)
9. H.L.Störmer, R.W.Martin: Phys. Rev. B20, 4213 (1979)

Electron Interactions in Bismuth

H.D. Drew, and S. Baldwin

Department of Physics and Astronomy, University of Maryland
College Park, MD 20742, USA

1. Introduction

The optical and magneto-optical properties of solids have generally been interpreted in terms of single particle excitations. With the exception of excitons at band edges in insulators and semiconductors it has been thought that many body effects are difficult to observe. In the case of metals considerable efforts were made to study many body effects associated with cyclotron resonance with very little success [1]. On the other hand measurements on conduction electron spin resonance in the alkali metals have shown clear evidence for electron-electron interaction effects that could be elegantly interpreted in terms of Landau's Fermi-Liquid theory [1,2]. More recently a number of many body effects have been observed in optical studies leading to the suspicion that electron-electron interaction effects may be far more prevalent in optical properties than has been previously supposed. HANKE and SHAM [3] have shown that the optical absorption spectrum of Si can be more successfully interpreted in terms of excitonic effects than in the single particle picture. In our laboratory we have found striking evidence for electron-electron interaction effects including excitonic modes in the magnetic subband transitions in bismuth and possibly in PbTe [4-7]. The Si MOSFET also has large many body corrections to the subband transitions in inversion layers; however the case for subband excitons is not so clear in this system [8,9].

Semimetallic bismuth has proven to be an ideal material for exploring electron-electron interaction effects in the optical response of a degenerate Fermion system. Bismuth has long carrier lifetimes and convenient Fermi energies and carrier effective masses for reaching extreme quantum conditions with reasonable magnetic fields. Also, the bismuth energy bands near the Fermi level are highly anisotropic and nonparabolic and this relaxes the optical selection rules permitting a wide variety of transitions under different polarization conditions which aid in the observation and identification of excitonic effects. Several years ago we reported on excitonic effects observed on magnetic subband transitions in the hole pocket and intepreted them successfully in terms of the final state Coulomb interaction between particles [4-6]. These measurements were mostly at $\sim$4 meV (337μ and 311μ). In recent measurements near the LO phonon frequency (12 meV) we have observed remarkable phonon mediated electron-electron interaction effects on the hole pocket cyclotron resonance [10]. Since the phonons are not polar in bismuth the electron-phonon coupling is relatively weak and the system is not complicated by the mixed phonon-plasma modes in other degenerate semiconductors. Consequently the experiments and the theory are relatively clear and lead to striking illustration of several many body processes.

In this paper we first review the Coulomb interaction effects in bismuth followed by a presentation of our results on the phonon mediated electron-electron interaction effects.

2. Subband Excitons

We are concerned with electron-electron interaction effects in subband transitions under degenerate Fermi conditions ($kT \ll \varepsilon_F$). Consequently the optical transitions involve the creation of electron-hole pairs and the possibility of final state interactions. Since some of the relevant states are near the Fermi level the sharpness of the Fermi surface plays a role in the excitonic modes as in the Cooper pair problem in metals. The subband exciton can be thought of as arising from final state interactions of the electron hole pairs produced in the optical transitions (see Fig. 1). A single pair state corresponding to the noninteracting electron gas is written as

$$\psi^+_{k,\vec{q}} = \Psi_{o,k_y,k_z} \Psi^+_{1,k_y+q_y,k_z+q_z}$$

where $\vec{q}$ is the momentum of the pair state which is the momentum of the incoming photon. The excitation energy of the pair is $\varepsilon(k,q) = \varepsilon_1(k+q) - \varepsilon_0(k)$ where $\varepsilon_n(k)$ is the subband energy given by $\varepsilon_n(k) = \hbar\omega_c(n + 1/2) + \hbar^2 k_z^2/2m$ for the example of parabolic bands. In the presence of the electron-electron interaction this pair state can scatter to other pair states as shown in Fig. 1 and therefore these states are no longer stationary states of the system.

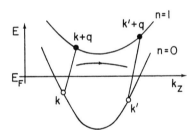

Fig.1 Electron hole pair states of momentum q_z generated by optical transitions.

The new stationary states may be written as

$$\Phi_{\vec{q}} = \sum_k{}' a_k \psi_{k,\vec{q}}$$

where the summation is restricted over the occupied states in the lower band. The Hamiltonian of the system is

$$H = \sum_{n,k} \varepsilon_{n,k} c^+_{n,k} c_{n,k} + \frac{1}{2} \sum_{\alpha,\beta,\gamma,\delta} V_{\alpha\beta\gamma\delta} c^+_\beta c^+_\delta c_\gamma c_\alpha \qquad (1)$$

where $V_{\alpha\beta\gamma\delta}$ includes both the Coulomb interaction and the phonon mediated electron-electron interaction. This is the form of the Hamiltonian after a canonical transformation to remove the electron-phonon interaction term to first order leaving an effective electron-electron interaction as in the theory of superconductivity [11].

The eigen equation for Φ_q can be obtained from $i\hbar(\partial\Phi/\partial t) = [H,\Phi]$ by making use of the random phase approximation with the result [6]

$$[\varepsilon_1(k+q) - \varepsilon_o(k) + \Delta_{10}(k) - \varepsilon] a_k + \sum_{k'}{}' V(k,k')a(k') = 0 \qquad (2)$$

where $V(k,k')$ is the matrix element for the scattering event shown in Fig. 1. The sum is restricted to k' corresponding to the occupied states, and $\Delta_{10}(k) = \Sigma_1(k+q) - \Sigma_o(k)$ where Σ is the exchange self energy.

Eq. 2 can be thought of as the momentum space representation of the Schrodinger equation for the electron-hole pair and a_k as the corresponding wave function in k_z space. An analogous equation can be derived for ordinary interband excitons which can be Fourier transformed into a real space Schrodinger equation for Φ which is of the hydrogenic form — leading to Mott-Wannier excitons. Such a transformation is not useful in the present case because the restricted sum $\sum'$ leads to a non-local effective potential in real space. Because of the large magnetic field and the extreme quantization of the electronic levels (2) is a one dimensional integral equation and its solution is straight forward but requires numerical solution for general $V(k',k)$. For the case of the phonon-mediated electron-electron interaction and the case where $V(k',k)$ can be considered constant (corresponding to a short range interaction in z space). Eq. 2 can be solved by elementary methods.

First consider the case $V(k,k') \simeq -I$ where I is a constant. In this case the sum over k' is independent of k leading to a self-consistency condition

$$1 = I \int_{-k_F}^{k_F} \frac{dk}{2\pi} \frac{1}{\xi_k - \varepsilon} \qquad (3)$$

where $\xi_k = \varepsilon_1(k+q) - \varepsilon_o(k) + \Delta_{10}(k)$ is the single particle-like excitation energy of a pair state at k. This equation can be put into a convenient form by transforming to an integral over ξ

$$I^{-1} = \int_0^\infty \frac{J(\xi)d\xi}{\xi - \varepsilon} \equiv \Lambda(\varepsilon) \qquad (4)$$

where $J(\varepsilon) = L/\pi \, dk/\partial\xi$ is the joint density of non-interacting pair states (the optical density of states) so defined that the Fermi level cutoff in (2) is included in $J(\varepsilon)$. $J(\varepsilon)$ and $\Lambda(\varepsilon)$ form a pair of Kramers-Kronig related functions. The excitonic mode frequencies are then given by $I^{-1} = \Lambda(\varepsilon)$. For a delta function joint density of states for example (parabolic bands and constant Δ_{10}) $\Lambda(\varepsilon) = k_F/\pi(\omega_c - \varepsilon)$ and the exciton energy is just $\varepsilon = \omega_c - k_F I/\pi$. These results show the plausability of forming excitonic states in subband transitions when there is an interparticle interaction. A thorough study of these electron-electron interaction effects was made on the hole pocket magnetic subband transitions, including cyclotron resonance, spin resonance and combined resonances, at frequencies between 1.5 and 7.3 meV [5,6]. The data were consistent with line shape calculations based on (2) taking a screened Coulomb interaction between the carriers with reasonable screening parameters.

3.1 The Phonon Mediated Electron-Electron Interaction

The conditions for which exciton-phonon interactions can be observed are generally such that the energy gap and therefore the exciton energy is large in comparison with phonon frequencies. We are concerned here with conditions in which the total exciton energy is comparable to the LO phonon frequency. Such a situation can arise either for a very narrow energy gap for ordinary excitons or for the case of inter-subband transitions in Fermi degenerate conducting systems. In bismuth, for H near the trigonal axis, both the hole pocket and the electron pocket cyclotron frequencies can be swept through the LO phonon frequency for fields below 100 kG. We have studied the subband exciton associated with the hole cyclotron resonance as a function of magnetic field and crystal orientation.

The measurements were performed on disk shaped single crystal bismuth samples in which the binary axis was coincident with the disk axis. The experimental configuration was the Voigt geometry – radiation $\vec{q}$ parallel to the binary axis and the applied static magnetic field in the binary plane. The angle θ between the trigonal axis of the sample and the magnetic field was continuously variable. Far infrared radiation from molecular gas laser sources absorbed by the sample was measured calorimetrically at 4.2° K by means of a Ge thermometer mounted on the back of the sample. Additional experimental details have been published elsewhere [5].

Experimental data for the case of 10.5 meV radiation are shown in Fig. 2. The major spectral features were identified by comparison with calculations based on a local quantum magneto conductivity tensor and the known band structure of bismuth [12]. The hole resonance is a tilted orbit resonance and it does not appear in the theory if the nonparabolicity of the bands is ignored. The sharp features on the high field side of the hole cyclotron resonance are identified as excitonic modes.

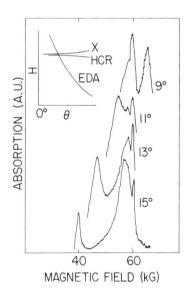

Fig.2 Absorption spectra for several angles, θ – the angle between the magnetic field and the trigonal axis. The inset diagram identifies the spectral features: dielectric anomaly following the electron cyclotron resonance (EDA), hole cyclotron resonance and dielectric anomaly (HCR), and exciton (x).

3.2 Intravalley Processes

For the phonon mediated interaction there are two types of scattering processes to be considered within a single valley. First, the exchange process in which the electron at k+q scatters to k'+q emitting a virtual LO phonon which then scatters the hole from k to k' (Fig. 1). In the direct process the electron at k+q falls into the hole at k emitting a phonon which then generates an electron-hole pair at k'. The matrix element for this process is of the form

$$V(k,k') = \frac{M^2}{\xi_k \pm \omega_{LO}} \tag{5}$$

where M is proportional to the electron-phonon matrix element. For nearly parallel bands $\varepsilon_1(k+q)-\varepsilon_0(k)+\Delta_{10}(k) = \xi_k$ is nearly constant $\xi_k \sim \omega_c$. Therefore for $\omega_c \sim \omega_{LO}$ we have a very large scattering amplitude over the whole Fermi surface. In the exchange process the energy denominator of the matrix element involves $\varepsilon_1(k+q) - \varepsilon_1(k'+q)$ which can vary between $\pm \omega$ and therefore the matrix element is a rapidly varying function that can be large at no more than two values of k corresponding to $\varepsilon_1(k)-\varepsilon_1(k') = \omega_{LO}$. Consequently this process will generally be smaller than the direct process and we will ignore it in this discussion. We will also ignore the + term in (5) because it too is small and keep only the resonant direct scattering process.

When the resonant phonon mediated process is included in the eigen equation the sum over k' is independent of k as before and the self-consistency condition can be readily found

$$1 = I\Lambda(\varepsilon) + \frac{M^2}{\omega_{LO}-\varepsilon}\left[\Lambda(\omega_{LO})-\Lambda(\varepsilon)\right]. \tag{6}$$

In the limit of parabolic bands $\xi_k = \omega_c$ and $\Lambda(\varepsilon) = 2k_f/\omega_c - \omega$. In this case the energy of the collective mode becomes

$$\varepsilon = \omega_c - 2k_F\left[1+M^2/(\tilde{\omega}_c-\omega_{LO})\right]. \tag{7}$$

Therefore the phonon process reduces the binding energy for $\tilde{\omega}_c < \omega_{LO}$ and it increases for $\tilde{\omega}_c > \omega_{LO}$. This is consistent with the idea that the phonon induced electron-electron interaction is attractive at low energies.

There is evidence for this effect on the exciton binding energy in the experimental data for the exciton binding energy vs magnetic field (or laser frequency). In particular the exciton binding energy appears suppressed at 10.5meV

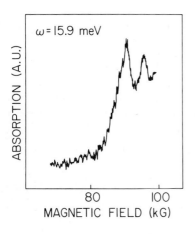

Fig.3 Far infrared magneto-absorption spectrum of bismuth at 15.9 meV. $\theta = 20°$.

in comparison with the 7.25 meV data and then it is enhanced at 14.8 meV (the enhancement can be seen by comparing Fig. 2 and Fig. 3). We also find that the exciton is broad at all angles at frequencies well above ω_{LO} (see Fig. 3). This is a consequence of the decay of the excitonic modes by the emission of real LO phonons and scattering of the electron-hole pairs into the T-pocket continuum.

3.3 Intervalley Processes

The absorption spectra at 10.5 meV show an orientation dependent broadening of the excitonic mode that is absent at lower frequencies (see Fig. 2). As θ is decreased from above the exciton is observed to broaden suddenly near $\theta = 12°$ and remain broad to the lowest observable angles. The dependence of the exciton linewidth on orientation for frequencies of 7.25 and 10.5 meV is shown in Fig. 4. In the range of angles where the exciton is broad it turns

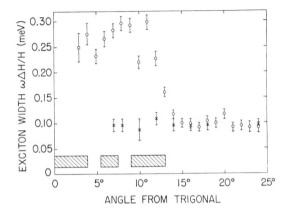

Fig. 4 Exciton linewidth as a function of the angle θ for two photon frequencies below ω_{LO}. The open circles are the 10.5 meV data and the crosses are the data for 7.25 meV. The cross hatched areas in the figure gives the expected range of large exciton damping as discussed in the text.

out that the hole subband exciton is degenerate with the cyclotron resonance continuum in one of the three electron pockets. This degeneracy occurs over a wide range of angles because the electron pockets are highly nonparabolic - so much so that the electron cyclotron resonance width is ∿25% of its mean at 10.5 meV. Using the two band model for the electron pocket and accepted band parameters [1,12] the calculated continua of the fundamental and second harmonic of cyclotron resonance are shown in Fig. 5. The angular range where degeneracy is predicted is indicated in Fig. 4 by the cross hatched bars. The critical angle is accounted for to within experimemtal error and the uncertainty in the band parameters. The data show dips in the linewidth at 5° and 10° that may be evidence for the small gaps in the range of degeneracy near 4° and 8°. Finally, the ∿2° width of the onset edge at 13° can be explained in terms of the temperature smearing of the continuum due to the Fermi occupation factors.

These data suggest a process by which excitons in the hole pocket decay into electron-hole pairs in the electron pocket cyclotron resonance continuum. The observed frequency dependence of the broadening effect suggests that the

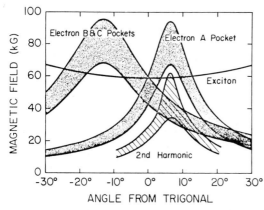

Fig. 5
The angular dependence of the electron pocket cyclotron resonance continua absorption spectrum at 10.5 meV.

process involves the zone center LO phonon at 12.5 meV. The appropriate decay process is the one in which the bound electron-hole pair in the T pocket, corresponding to the hole subband-exciton, annihilates emitting a virtual LO phonon which then generates an electron-hole pair in the cyclotron resonance continuum of the L pocket. This process can conserve energy and crystal momentum only when the T-pocket exciton and the electron pocket (L) cyclotron resonance continuum are energetically degenerate.

The exciton decay rate by this process is given by the golden rule

$$1/\tau = \frac{2\pi}{\hbar}|\tilde{M}^2|J_L(\omega) \quad . \tag{8}$$

where ω is the exciton energy and J_L the density of pair states in the electron pocket. $\tilde{M}$ is the matrix element for the phonon mediated exciton scattering which is of the form

$$\tilde{M} = \sum_{q_\perp} M^T_{n'n,\vec{q}} M^L_{n'm,-\vec{q}} / (\omega - \omega_{LO}) \tag{9}$$

where $M^T_{n/n\vec{q}}$ is the electron phonon matrix element for the electron-hole annihilation in the T pocket between subbands n' and n with a phonon of wavevector q and $M^L_{m'm}$ is the similar matrix element for the L pocket. $q_\perp$ is the phonon momentum perpendicular to the applied magnetic field and the $\perp$ photon momentum.

There is another phonon mediated process which could scatter excitons from T to L involving a zone edge phonon. An electron in the T pocket could emit a <u>zone edge</u> phonon and scatter into the L pocket and the virtual phonon could then scatter an electron at L into T. We can discriminate against this process in our experiment because of the resonant behavior of the data at $\omega_c = \omega_{LO}$.

Very near ω_{LO} the spectra become very dramatic and complex especially near the critical angle as illustrated in Fig. 6 for $\omega = 12.9$ meV. Dispersion in the exciton binding energy is seen at 12° and 17° and there are generally several distinct peaks associated with the cyclotron resonance absorption. Eq. 6 predicts multiple excitonic modes at this frequency but not the dispersion effects. This we believe is due to mode mixing effects between T-pocket and L-pocket subband excitons. Valley coupled excitonic modes are predicted when the eigen equations for the two pockets are considered with coupling via the

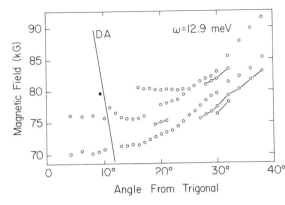

Fig.6 The angular dependence of the absorption spectral features are identified as excitonic modes. The points joined by solid lines are thought to be additional excitonic features arising because of the resonance phonon mediated electron-electron interaction for $\omega \approx \omega_{LO}$. The solid line labeled DA is the locus of the electron pocket dielctric anomaly

matrix element (9). The dispersion at 12° is consistent with the valley coupled cyclotron resonance excitons. The larger effect at 17° is not understood yet. It may be due to valley coupled excitons in some other subband transition in the electron pockets.

In conclusion, we have studied the subband transitions in the T-pocket in bismuth and found evidence for both the Coulomb interaction and the phonon-mediated electron-electron interaction. The experimental results near the LO phonon frequency provide striking evidence for several phonon mediated many body processes including: 1) subband-exciton decay by the phonon mediated inter-valley scattering of electron-hole pairs, 2) subband-exciton decay by phonon emission, 3) self energy corrections to cyclotron resonance, 4) phonon mediated electron-hole interaction, and finally 5) valley coupled subband-excitons. These results should stimulate similar studies on space charge layers in covalent semiconductors and on very narrow gap semiconductors ($E_g < \omega_{LO}$). They may also lead to an understanding of the more difficult problem of magneto-optical effects in degenerate semiconductors with polar phonons.

We wish to thank Ali Gungor for assistance with the optically pumped laser. This work was supported in part by the National Science Foundation under Grant No. 77-28399-A01. The computer time was supported in full through the facilities of the Computer Science Center of the University of Maryland.

References

[1] "Waves and Interactions in Solid State Plasmas", P. M. Platzman and P. A. Wolff, Academic Press (1973) New York.
[2] S. Schultz and G. Dunifer, Phys. Rev. Lett. 18, 283 (1967).
[3] W. Hanke and L. J. Sham, Phys. Rev. Lett. 43, 387 (1979).
[4] H. R. Verdun and H. D. Drew, Phys. Rev. Lett. 33, 1608 (1974).
[5] H. R. Verdun and H. D. Drew, Phys. Rev. B14, 1370 (1976).
[6] H. R. Verdun and H. D. Drew, Phys. Rev. B15, 5636 (1977).
[7] S. W. McKnight and H. D. Drew, Phys. Rev. B21, 3447 (1980).
[8] V. Vinter, Phys. Rev. Lett. 35, 598 (1975).
[9] T. Ando, Z. Physik B26, 263 (1977).
[10] S. Baldwin and H. D. Drew, submitted for publication.
[11] C. Kittel, Quantum Theory of Solids, Chpt. 8 (Wiley, New York, 1963).
[12] V. S. Edelman, Advances in Physics 25, 555 (1976).
[13] J. W. McClure and K. H. Chai, Solid State Comm. 21, 1015 (1977).

Part VIII

Narrow Gap Semiconductors

Free and Bound Magneto-Polarons in Narrow Gap Semiconductors

W. Zawadzki

Institute of Physics, Polish Academy of Sciences, 02-668 Warsaw, Poland

L. Swierkowski[1]

Department of Physics, University of Warwick, Coventry, U.K.

J. Wlasak

Institute of Physics, Wroclaw Technical University, 50-370 Wroclaw, Poland

1. Introduction

Since the discovery of magneto-optical effects due to resonant interaction of electrons with optic phonons in InSb by JOHNSON and LARSEN [1], resonant polarons have become subject of substantial interest. The resonant interaction with free and bound electrons was clearly observed in intraband transitions by KAPLAN and WALLIS [2] and by MCCOMBE and KAPLAN [3]. A striking broadening of linewidth as the cyclotron frequency ω_c passes the longitudinal optic-mode frequency ω_L was investigated by SUMMERS et al. [4] and explained by HARPER [5]. Later NAKAYAMA [6] used Green's function techniques to provide a more complete description of resonant behavior. VIGNERON et al. [7] demonstrated recently that, in order to obtain fully physical theoretical results, one should introduce an additional weak scattering, always existing in the material due to presence of impurities and imperfections.

Shortly after the discovery of the resonant polaron behavior a claim was made of an "offset" effect [8], by which the authors understood an energy shift of $\alpha\hbar\omega_L$ between lower and upper polaron branches, as ω_c passes ω_L. This was used to determine experimentally the polar coupling constant α in InSb [8], InAs [9] and HgCdTe [10].

All the work mentioned above is concerned with one-mode polarons, resulting from the interaction of electrons with single optic phonon branch. Mixed $Hg_{1-x}Cd_xTe$ crystals are known to possess two longitudinal optic phonon branches with their energies depending on the chemical composition x. MCCOMBE and WAGNER [11] reported an observation of a two-mode resonant behavior in $Hg_{0.797}Cd_{0.203}Te$ in combined resonance transitions.

In this review we consider polarons in weakly polar InSb-type semiconductors in the presence of a magnetic field, consistently taking into account the real band structure of these materials. First, free magneto-polarons are described in the resonant region of $\omega_c \approx \omega_L$. The Green function formalism is used to describe correctly both lower and upper polaron branches. The results are compared with experimental data for InAs. Second, free two-mode magneto-polarons are described in

---

[1] On leave from the Institute of Physics, Polish Academy of Sciences, 02-668 Warsaw, Poland.

the resonant region of $\omega_c \approx \omega_{L1}$ and $\omega_c \approx \omega_{L2}$. A multi-mode dielectric theory of TOYOZAWA [12] is used here to determine different constants of electron-phonon coupling for the two optic modes in question. The results are compared with experimental data for $Hg_{0.72}Cd_{0.28}Te$. Further, a nonresonant contribution to polaron energy is calculated as a function of magnetic field. It is shown that the "offset" effect does not exist. Finally, polarons bound to shallow donors at high magnetic fields are described. A short characterization of shallow donors in narrow gap materials at high fields is given. The effects of resonant electron-phonon interaction are considered and the results compared with experimental data on the combined cyclotron-spin resonance in InSb.

2. One-mode Resonant Polarons

First, we consider a weakly polar narrow gap semiconductor of InSb-type with one longitudinal optic phonon branch in the presence of a magnetic field. We shall be interested in the resonant region of magnetic fields, for which ω_c passes ω_L. The initial Hamiltonian for our problem reads

$$H = P^2/2m_o + V_o(\vec{r}) + H_{so} + H_{Fr} + H_{phn} \qquad (1)$$

where m_o is the free electron mass, $\vec{P} = \vec{p}+(e/c)\vec{A}$ is the kinetic momentum, $\vec{A}$ is the vector potential of an external magnetic field $\vec{H}$, V_o is the periodic potential of the lattice, H_{so} is the spin-orbit interaction, H_{Fr} denotes the FRÖHLICH polar electron-phonon interaction, and H_{phn} stands for the free phonon field.

The first three terms describe band structure of the material in the presence of a magnetic field. This electronic part is solved for the vicinity of the Γ point in the BRILLOUIN zone using a three-level model. The latter takes into account exactly the k·p interaction between Γ_6, Γ_8 and Γ_7 levels, including explicitly the small energy gap ε_g and the large spin-orbit energy Δ (cf. BOWERS and YAFET [13], ZAWADZKI [14]). All other distant bands are neglected in this approximation. The resulting electron energies are (assuming for simplicity $\Delta \gg \varepsilon_g$),

$$\varepsilon(n, k_z, \pm) = -\varepsilon_g/2 + [(\varepsilon_g/2)^2 + \varepsilon_g D(n, k_z, \pm)]^{1/2} \qquad (2)$$

where

$$D(n, k_z, \pm) = \hbar\omega_c(n+1/2) + \hbar^2 k_z^2/2m_o^* \pm \mu_B |g_o^*| H/2 \qquad (3)$$

with $\omega_c = eH/m_o^* c$. Here m_o^* and g_o^* denote the effective mass and the spin g-factor at the bottom of the conduction band, as resulting from the three-level model. Plus and minus signs refer to the two effective-spin states (in the presence of spin-orbit interaction spin is not a good quantum number). The electron wave functions obtained from the same model are

$$\Psi^{\pm}_{nk_z}(\vec{r}) = \Sigma\, c_l(n,k_z,\pm)\, f_l^n(\vec{r})\, u_l(\vec{r}) \qquad (4)$$

where the summation l is over the three Γ levels in question (8 states including spin). f_l are slowly varying envelope functions of the harmonic-oscillator type and u_l denote the LUTTINGER-KOHN periodic amplitudes. The wave functions (4) are mixtures of s-like and p-like amplitudes, as well as spin-up and spin-down states (explicit expressions are given by ZAWADZKI [14,15]).

In the standard procedure the electronic part of the Hamiltonian (1) is transformed into the one-band effective mass approximation. In this case one should in principle transform also the FRÖHLICH Hamiltonian into the same scheme. This is possible under two conditions: 1) The interaction Hamiltonian must be slowly varying in space, i.e. $\lambda \gg a$ (λ is phonon wavelength and a interatomic distance). 2) The optic phonon energy must be much smaller than the gap: $\hbar\omega_L \ll \varepsilon_g$. In our approach, however, the FRÖHLICH interaction Hamiltonian in (1) can be used directly as a perturbation, because the electron wave functions (4) have been obtained without the effective mass transformation.

We consider a resonant situation, in which $\varepsilon_1^+ - \varepsilon_o^+ \approx \hbar\omega_L$. The upper electron state b(n=1, $k_y=k_z=0$, s=+) can decay by a virtual emission of an optic phonon $\vec{q}$ to the ground state a(0, k_y, k_z, +). The resonant part of the electron selfenergy Σ of the state b is (in the lowest order of perturbation theory),

$$\Sigma(E) = \sum_{k_y,k_z} \sum_q \frac{|(q,a|H_{Fr}|0,b)|^2}{E+i\Gamma_o-\varepsilon_o^+-\hbar\omega_L} - i\Gamma_o \qquad (5)$$

We have introduced here an additional weak scattering (independent of energy) by a phenomenological constant parameter $\Gamma_o > 0$. This eliminates a divergence of the selfenergy at $E = \varepsilon_o^+ + \hbar\omega_L$, leading to fully physical results at all magnetic fields. The selfenergy is now calculated, using the electron energies (2), (3) and the wave functions (4), as a function of a magnetic field in the resonant region. The s-p mixing in the wave functions results in an effective weakening of the electron-phonon interaction for a given coupling constant α. (Similar result has been obtained in the analysis of electron mobility limited by optic phonon scattering [16]).

Next, we calculate the spectral function A(E) for the state b(1,0,0,+), defined as

$$A(E) = \frac{1}{\pi} \frac{\Gamma}{(E-\varepsilon_1^+-\Delta)^2+\Gamma^2} \qquad (6)$$

Where Δ and Γ are real and imaginary parts of self-energy, respectively, $\Sigma(E) = \Delta(E)-i\Gamma(E)$. In pure crystals intraband magneto-optical transitions occur at low k_z values. In interband transitions low k_z values are strongly favored by the joined density of states. In both cases maxima of the spectral function at $k_z=0$ correspond to maxima of the density of states, i.e. to the observable polaron energies. Omitting the detailed calculations for the lack of space we note that below the resonant energy $E_c = \varepsilon_o^+ + \hbar\omega_L$, the real part of self-energy has a strong maximum (it behaves as $\Delta \sim -(E_c-E)^{-1/2}$ if $\Gamma_o \equiv 0$) and the imaginary part $\Gamma \equiv 0$ (if $\Gamma_o \equiv 0$). Above the resonant energy the roles reverse: Δ has a finite value and Γ has a strong maximum ($\Gamma \sim (E-E_c)^{-1/2}$ if $\Gamma_o \equiv 0$). The alternating singularities of real and imaginary parts push the polaron energies away from the resonant value E_c, causing the "pinning" effect both below and above the resonance (cf. Fig. 2).

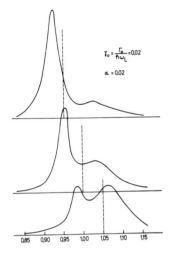

Fig.1 Spectral function (in arbitrary units) of the $|n=1, k_y=0, k_z=0, s=+\rangle$ polaron state as a function of energy (in $\hbar\omega_L$ units) for three values of magnetic field, near the resonance with optic phonons. Upper curve is for $x=0.95$, middle curve for $x=1.00$, and lower curve for $x=1.05$, where $x=(\varepsilon_1^+-\varepsilon_0^+)/\hbar\omega_L$. The dashed lines indicate respective positions of the Landau level $|1,0,0,+\rangle$ unperturbed by the electron-phonon interaction.

Fig. 1 shows the spectral function of $b(1,0,0,+)$ state at $k_z=0$ for three values of the magnetic field: below, at, and above the resonance. It can be seen that two maxima of the spectral function (polaron branches) appear, the upper one being broader. For $\Gamma_o \equiv 0$ the lower maximum is a singularity of δ-type (cf. [6]). Far below and far above the resonant magnetic field only one branch is observed in practice. This result strongly resembles that of VIGNERON et al. [7] for the optical absorption, confirming our reasoning on the close relation between maxima of the spectral function at $k_z=0$ and observable transitions.

In Fig. 2 calculated positions of the maxima are compared with experimental data of HARPER et al. [9] for InAs. Unfortunately, this experiment, investigating directly the cyclotron resonance transition between the $|0,+\rangle$ and $|1,+\rangle$ levels in question, is obscured in the pinning region of interest by the Rechtstrahl absorption, since at the resonance there is $\hbar\omega \approx \hbar\omega_c \approx \hbar\omega_L$.

MCCOMBE and KAPLAN [3] investigated combined cyclotron-spin resonance transition in InSb: $\hbar\omega \approx \varepsilon_1^- - \varepsilon_0^+$, while the resonant polaron occurs for $\hbar\omega_L \approx \varepsilon_1^- - \varepsilon_0^-$, perturbing the $|1,-\rangle$ Landau state. This allowed the authors to avoid the Rechtstrahl absorption and to investigate the pinning region. The theory of NAKAYAMA [6] for a parabolic band describes quite well the experimental data [17].

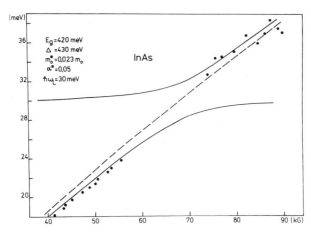

Fig.2 Polaron energies in InAs as functions of a magnetic field in the resonant region. Experimental points are after Harper et al. [9], solid lines are theoretical. Dashed line indicates the energy of the Landau level $|1,0,0,+)$ in absence of electron-phonon interaction.

3. Two-mode Resonant Polarons

Now we consider two-mode resonant polaron behavior in mixed HgCdTe solid solutions. Behavior of two-mode crystals poses new interesting questions, e.g. are the strengths of electron-phonon polar interaction for the two longitudinal modes the same ar different?

We shall be specifically concerned with the $Hg_{0.72}Cd_{0.28}Te$ crystal (cf. SWIERKOWSKI et al. [18]). Since the polaron states have been observed by means of two separate measurements involving different magneto-optical transitions to the same upper level, the band structure description must be quite precise, including bands nonparabolicity and the LUTTINGER [19] effects in the valence bands. Thus, in the calculations of unperturbed Landau energies the corrected PIDGEON and BROWN [20] model has been used, taking exactly into account the $k \cdot p$ interaction between Γ_6, Γ_8, Γ_7 levels and including all other bands to k^2 approximation. The overall fit to unperturbed energy levels has been obtained with the following values of band parameters: $\varepsilon_g = 208$ meV, $E_p = 19$ eV, $\gamma_1 = 5$, $\bar{\gamma} = 1.5$, $\kappa = -0.5$, $\Delta = 1$ eV, $F = 0$.

As mentioned above, mixed $Hg_{1-x}Cd_xTe$ crystals have two longitudinal optic phonon branches. BAARS and SORGER [21] investigated this system using reflectivity technique and determined separately longitudinal and transverse phonon energies in a wide composition range. For x = 0.28 in question their data give: $\hbar\omega_{L1} = 17.0$ meV, $\hbar\omega_{T1} = 15.1$ meV, and $\hbar\omega_{L2} = 19.5$ meV, $\hbar\omega_{T2} = 18.5$ meV. In order to calculate the polaron properties one needs to know the strengths of electron-phonon polar interaction for the two modes. To this purpose a multi-mode dielectric function approach proposed by TOYOZAWA [12] is used. The two-mode frequency-dependent dielectric funtion is taken in the simplest form

$$\kappa(\omega) = \frac{\omega_{L1}^2 - \omega^2}{\omega_{T1}^2 - \omega^2} \frac{\omega_{L2}^2 - \omega^2}{\omega_{T2}^2 - \omega^2} \kappa(\infty) \qquad (7)$$

where $\kappa(\infty)$ is the high-frequency dielectric constant. Using TOYOZAWA's generalized definitions of the electron-phonon interaction one derives

$$\alpha_1 = \frac{e^2}{\hbar \kappa(\infty)} \left(\frac{m_o^*}{2\hbar\omega_{L1}}\right)^{1/2} \left(1 - \frac{\omega_{T1}^2}{\omega_{L1}^2}\right) \left(\frac{\omega_{T2}^2 - \omega_{L1}^2}{\omega_{L2}^2 - \omega_{L1}^2}\right) \tag{8}$$

$$\alpha_2 = \frac{e^2}{\hbar \kappa(\infty)} \left(\frac{m_o^*}{2\hbar\omega_{L2}}\right)^{1/2} \left(1 - \frac{\omega_{T2}^2}{\omega_{L2}^2}\right) \left(\frac{\omega_{L2}^2 - \omega_{T1}^2}{\omega_{L2}^2 - \omega_{L1}^2}\right), \tag{9}$$

where m_o^* is the electron effective mass at the bottom of the conduction band [2]. Taking $m_o^* = 0.015\, m_o$ and $\kappa(\infty) = 11.9$ [22] one can calculate the coupling constants to get $\alpha_1 = 0.037$ and $\alpha_2 = 0.046$. Thus, the strengths of electron-phonon polar interactions for the two longitudinal modes are different and they depend on the chemical composition x.

To describe a resonant polaron behavior in the vicinity of $\omega_c \approx \omega_{L1}$ and $\omega_c \approx \omega_{L2}$ resonances the spectral function (6) is again used. We note that, for $\Gamma_o \equiv 0$, below the first resonant energy $E_{c1} = \varepsilon_o + \hbar\omega_{L1}$ the real part has a singularity $\Delta \sim -(E_{c1}-E)^{-1/2}$ and the imaginary part $\Gamma \equiv 0$. Above the resonant energy the roles reverse: Δ has a finite value and $\Gamma \sim (E-E_{c1})^{-1/2}$. Approaching from below the second resonant energy $E_{c2} = \varepsilon_o + \hbar\omega_{L2}$, we have again $\Delta \sim -(E_{c2}-E)^{-1/2}$ and above there is $\Gamma \sim (E-E_{c2})^{-1/2}$. The alternating singularities of real and imaginary parts of self-energy push away the polaron energies from the resonant values E_{c1} and E_{c2}, causing the "pinning" effects both below and above the resonances. In actual calculations a phenomenological contribution to the imaginary part $\Gamma_o = 0.5$ meV was included, accounting for other scattering modes in the crystal.

Fig.3 presents experimental and theoretical results for the two-mode resonant polaron in $Hg_{0.72}Cd_{0.28}Te$. The resonant behavior occurs due to virtual electron transitions between $a_c(1)$ and $a_c(0)$ states, accompanied by an emission of optic phonons belonging to L1 and L2 modes. Behavior of the $a_c(1)$ state is observed experimentally in two independent interband transitions from the valence states $a_{1h}(0)$ (in σ_- configuration) and $b_{hh}(2)$ (in π configuration). Thus, the different magnetic field dependences of the above hole levels is subtracted, in order to determine the net comportment of the $a_c(1)$ level. Three polaron branches are observed for magnetic fields between 27 and 32 kG, clearly demonstrating the two-mode character of the polaron. It can be seen that, with the use of material parameters determined from independent experiments, the theory describes qualitatively and quantitatively the experimental observations. This confirms in particular the validity of TOYOZAWA's generalized approach to multi-mode electron-phonon interaction in polar crystals.

---

[2] In agreement with the usual notation we define coupling constants including the electron mass, although the complete FRÖHLICH interaction Hamiltonian is mass independent.

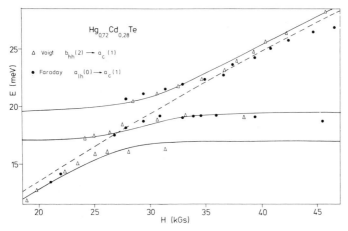

Fig.3. Two-mode resonant polaron energies in the $Hg_{0.72}Cd_{0.28}Te$ mixed crystal as functions of an external magnetic field. Dashed line indicates unperturbed energy difference between $a_c(1)$ and $a_c(0)$ Landau levels, solid lines are theoretical including the electron-phonon interaction.

4. Nonresonant Part of Polaron Energy

The nonresonant contribution to polaron energy is calculated by taking into account the electron-phonon coupling of a given Landau state with all other states by means of eq. (5). For n=1 state the summation excludes contribution of n=0 state, which has been calculated above. The sums can be converted into integrals (here we assume a parabolic energy band) [15]. The results are shown in Table 1. It can be seen that the nonresonant contribution is a smooth function of a magnetic field, not undergoing any dramatic change as ω_c passes ω_L.

l	s_o	s_1
0.1	1.008	0.911
0.3	1.024	0.886
0.5	1.040	0.880
0.7	1.055	0.881
0.9	1.070	0.884
1.0	1.077	0.886
1.1	1.084	0.888
1.3	1.098	0.894
1.5	1.111	0.900
1.7	1.124	0.906
1.9	1.136	0.915
2.0	1.142	0.916

Table 1 Nonresonant contribution to polaron energy for n=0 and n=1 Landau states as functions of magnetic field ($l = \hbar\omega_c/\hbar\omega_L$). The energy shifts are

$$\Sigma_o = -\alpha\hbar\omega_L s_o \qquad \Sigma_1^{nr} = \alpha\hbar\omega_L s_1$$

Thus we conclude that the "offset" effect does not exist. The claims of having it observed are, in our opinion, due to the following reasons:

1. Incorrect calculation of the upper polaron branch, not predicting the pinning effect from above. This results from the use of the WIGNER-BRILLOUIN perturbation theory, assuming implicity that the imaginary part of selfenergy vanishes, while actually it has a singularity at the resonance.

2. Improper extrapolation of experimental results (of the lower polaron branch into higher magnetic fields and of the upper branch into lower fields) in the region obscured by the Rechtstrahl absorption.

5. Bound Resonant Polarons

Above we considered free polarons in a magnetic field. In narrow gap semiconductors at low fields the electrons are always free because the binding energy of shallow donors (equal to $Ry^* = m^*e^4/2\hbar^2\kappa_o^2$) is very small. However, as first shown in a variational calculation by YAFET et al. [23], the binding energy quickly increases with increasing magnetic field and electrons become bound to impurity atoms (magnetic freeze-out). The characteristic parameter for this problem: $\gamma = \hbar\omega_c^*/2Ry^*$, is very small for the free hydrogen atom at achievable magnetic fields (the Zeeman effect regime) but can reach large values in narrow gap materials. WALLIS and BOWLDEN [24] demonstrated that, at large γ values, one deals with series of bound magneto-donor states "attached" on the energy scale to each Landau subband. LARSEN [25] generalized the theory into the case of InSb-type band structure using the three-level model. Recently the theory was put into analytical form, extended to several donor states for each Landau subband, and used in a successful description of the existing experimental data for InSb (ZAWADZKI and WLASAK [26,27]).

This theory provides a good starting point for treating the problem of bound resonant polarons, in which the energy difference of two bound magneto-donor states becomes equal to optic phonon energy $\varepsilon_2^d - \varepsilon_1^d \sim \hbar\omega_L$. The initial Hamiltonian is

$$H = P^2/2m_o + V_o(\vec{r}) + H_{so} + U(r) + H_{Fr} + H_{phn} \qquad (10)$$

where $U(r)$ represents the Coulomb potential of the donor. The first four terms describe magneto-donor with full account of the real band structure. This problem is formulated within the three-level model and solved variationally. The resulting wave functions are given by an expression similar to (4). They are classified (in the symmetric gauge for the vector potential) by the quantum numbers $(NM\lambda\pm)$, where $N=0,1,2,\ldots$; $M=\ldots-1,0,+1,\ldots$; $\lambda=0,1,2,\ldots$ Numbers N,M quantize the motion transverse to $\vec{H}$ and λ that parallel to $\vec{H}$. A given donor state belongs to the $n = N + (M+|M|)/2$ Landau subband.

In case of free magneto-polarons one deals with one-dimensional magnetic subbands (k_z is quasi continuous), so that the upper polaron branch is degenerate with the continuum of magnetic states, resulting in the nonvanishing imaginary part of selfenergy and the necessity to treat the problem by Green's function formalism. The case of bound polarons is theoretically considerably simpler, because one deals with discrete energy levels and the continuum problem does not arise. In this

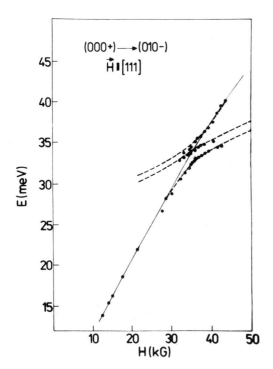

Fig.4 Resonant bound magneto-polaron in InSb. Experimental points (after MCCOMBE and KAPLAN [3]) show magneto-optical transition between (000+) and (010-) donor states in the region, where the (010-) state resonates via electron-phonon polar interaction with (OMλ-) states (M$\leqslant$0). The solid line is theoretical for magneto-donor energies, the dashed lines include electron-phonon interaction.

situation the Wigner-Brillouin perturbation theory can be used for both lower and upper polaron branches.

We are interested in the experiment of MCCOMBE and KAPLAN [3] on InSb, where a combined resonance transition was observed between the ground magneto-donor state (000+) and the excited state (010-), at magnetic fields such that $\varepsilon(010-) - \varepsilon(OM\lambda-) \sim \hbar\omega_L$, with $M \leqslant 0$. It turns out that dominating matrix elements of the electron-phonon interaction are $(000-|H_{Fr}|010-)$ and $(0\bar{1}0-|H_{Fr}|010-)$, so that we deal with a quasi-degenerate system of three levels: $E_1 = \varepsilon(010-)$; $E_2 = \varepsilon(000-) + \hbar\omega_L$; $E_3 = \varepsilon(0\bar{1}0-) + \hbar\omega_L$. The perturbation theory for quasi-degenerate levels leads to the following eigenvalue equation for the energy E,

$$(E_1 - E) - \sum_q \frac{|(\Psi_2|H_{Fr}|\Psi_1)|^2}{E_2 - E} + \sum_q \frac{|(\Psi_3|H_{Fr}|\Psi_1)|^2}{E_3 - E} = 0 \quad (11)$$

where $\Psi_1 = (010-)$, $\Psi_2 = (000-)$, $\Psi_3 = (0\bar{1}0-)$. Assuming dispersionless optic phonons one obtains cubic equation for the energy as function of a magnetic field. Following material parameters were used in the calculations: $\hbar\omega_L = 24.45$ meV, $R_y^* = 0.605$ meV, $\kappa(\infty) = 15.68$, $\kappa(o) = 17.88$ (resulting $\alpha = 0.022$). The results are shown in Fig. 4. It can be seen that the theory describes very well the experimental data. This confirms the validity of our approach to band structure, magneto-donors and electron-phonon interaction in narrow-gap semiconductors.

Problem of bound polarons in InSb in the two-phonon region was treated recently by DEVREESE et al. [28].

References

1. E.J.Johnson and D.M.Larsen: Phys.Rev.Lett. 16 (1966) 655.
2. R.Kaplan and R.F.Wallis: Phys.Rev.Lett. 20 (1968) 1499.
3. B.D.McCombe and R.Kaplan: Phys.Rev.Lett. 21 (1968) 756.
4. C.J.Summers, P.G.Harper, and S.D.Smith; Solid State Commun. 5 (1967) 615.
5. P.G.Harper: Proc.Phys.Soc. London 92 (1967) 793.
6. M.Nakayama: J.Phys.Soc. Japan 27 (1969) 636.
7. J.P.Vigneron, R.Evrard, and E.Kartheuser: Phys.Rev. B 18 (1978) 6930.
8. D.H.Dickey, E.J.Johnson, and D.M.Larsen: Phys.Rev.Lett. 18 (1967) 599.
9. P.G.Harper, S.D.Smith, M.A.Arimondo, R.B.Dennis and B.S. Wherrett: Proc.Int.Conf.Phys.Semicond., Boston 1970,p.166
10. M.A.Kinch and D.D.Buss: Proc.Conf.Phys.Semicond. and Semimet. Pergamon (1971) p.461.
11. B.D.McCombe and R.J.Wagner: Proc.Int.Conf.Phys.Semicond. Warsaw 1972, p.321.
12. Y.Toyozawa: in "Polarons in Ionic Crystals and Polar Semiconductors" (Ed. J.T.Devreese), North Holland 1972,p.1.
13. R.Bowers and Y.Yafet: Phys.Rev. 115 (1959) 1165.
14. W.Zawadzki: in "New Developments in Semiconductors" (Ed.P.R. Wallace), Noordhoff 1973, p.441.
15. W. Zawadzki:"Intraband Magnetooptics in Narrow Gap Semiconductors", in *Narrow Gap Semiconductors, Physics and Applications*, Proceedings, Nimes, 1979, ed by W. Zawadzki, Lecture Notes in Physics, Vol. 133 (Springer, Berlin, Heidelberg, New York 1980) p. 85
16. W.Zawadzki and W.Szymańska: Phys.Stat.Solidi (b) 45 (1971) 415.
17. R.Kaplan and K.L.Ngai: Comments Sol.St.Phys. 5 (1973) 157.
18. L.Swierkowski, W.Zawadzki, Y.Guldner and C.Rigaux: Solid State Commun. 27 (1978) 1245.
19. J.M.Luttinger: Phys.Rev. 102 (1956) 1030.
20. C.R. Pidgeon and R.N.Brown: Phys.Rev. 146 (1966) 575.
21. J.Baars and F.Sorger: Solid State Commun. 10 (1972) 875.
22. D.M.Larsen: Phys.Rev. 142 (1966) 428.
23. Y.Yafet, R.W.Keyes and E.N.Adams: J.Phys.Chem.Solids 1 (1956) 137.
24. R.F.Wallis and H.J.Bowlden: J.Phys.Chem.Solids 7 (1958) 78.
25. D.M.Larsen: J.Phys.Chem.Solids 29 (1968) 271.
26. W.Zawadzki and J.Wlasak: Proc.Int.Conf.Phys.Semicond. Edinburgh 1978, p.413.
27. W.Zawadzki and J.Wlasak: Proc.Adv.Study Inst. on Magnetooptics, Antwerp 1979, Plenum Press (in print).
28. J.T.Devreese, J.De Sitter, E.J.Johnson and K.L.Ngai: Phys. Rev. B17 (1978) 3207.

Investigation of Strain Effects in Epitaxial Semiconductor Films by Interband Magnetooptical Transitions: IV-VI Compounds

H. Pascher

Physikalisches Institut der Universität, 8700 Würzburg, Fed. Rep. of Germany

E.J. Fantner and G. Bauer

Institut für Physik, Montnuniversität, 8700 Leoben, Austria

A. Lopez-Otero

Institut für Physik, Universität Linz, 4045 Linz, Austria

Introduction

Since the early magnetooptical investigations of IV-VI compounds by Palik et al. [1] it has been known, that epitaxially grown single crystalline films on various alkali halide substrates exhibit strain effects. These are caused by the difference in the temperature dependence of the thermal expansion coefficients of semiconductor film and substrate. This mismatch leads to a tetragonal (on (100) oriented substrates: NaCl,KCl) or rhombohedral (on (111) oriented substrates: BaF_2, CaF_2) distortion [2]. e.g. in Ref. [1] it was shown that the energy gap of PbTe/NaCl is different from bulk PbTe at low temperatures. Despite these shortcomings heteroepitaxial semiconductor films are of considerable interest for devices like infrared detectors.

Apart from the magnetooptical data, quantum magnetotransport experiments (SdH and Magnetophon-experiments [3]) were interpreted as being determined by a decrease of stress-induced damping parameters with increasing film thickness in the 2...30μm range, where the main effect was attributed to strain relaxation with increasing sample thickness. The existence of shear strain was proven experimentally by a redistribution of carriers between valleys parallel and oblique to the surface normal [4]. A very elegant determination of the distortions away from the cubic structure was obtained by low-field magnetoresistance measurements [2]. However, no information on the dependence of the strain on crystal growth conditions and on strain profile could be obtained from the experiments quoted above.

We report about a study of the influence of growth conditions of the $PbTe/BaF_2$ system on strain by interband-magnetooptical transitions. In addition, a study of the strain and its profile has been carried out by a finite element calculation technique [5].

Experimental

Epitaxially grown $PbTe/BaF_2$ films were produced by the hot wall technique [6] on cleaved {111} surfaces. The growth parameters used are listed in table I together with the sample identification. With as-grown films, magnetooptical interband transitions were observed in Faraday configuration ($\vec{B}||\vec{k}||$ [111], σ_+ and σ_-) using polarized CO-laser radiation (190-240 meV). Magnetotrans-

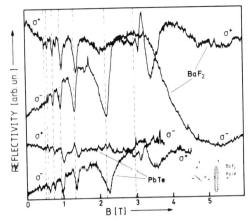

Fig.1 Magnetoreflectivity vs. magnetic field of sample 176.2 (set 1, d =10.33μm) at $\tilde{\nu}$ = 1644.3 cm^{-1} for σ^+ and σ^-; broken lines: averaged position of the transmission minima.

mission and magnetoreflectivity were measured simultanously in magnetic fields up to 6.5T. The laser radiation was impinging onto the samples both (i) from the film side and by rotating the sample by 180° (ii) from the substrate side. (See insert Fig.1).

The experimental results can be summarized as follows:

a) The resonance positions of the reflectivity spectra occur at different magnetic fields for the configurations (i) and (ii) as well as for σ^+ and σ^- (Fig. 1).

b) An extrapolation (B → 0) of the transition energies obtained from the reflectivity data (config. (i) and (ii)) indicates, that the extrapolated ε_g from the BaF$_2$/PbTe interface is typically lower by about 2 meV than from the PbTe-surface. The difference in the energy gaps extrapolated from the transition energies in the two Landau ladder systems ($\vec{B}$ || [111]) is about 1 meV both in config. (i) and (ii). This is much lower than the energy separation between valleys parallel and oblique to the [111] -surface normal proposed by Burke and Carver [4] from SdH-data.

c) Although magnetotransmission experiments should yield data on the strain averaged over sample thickness, different positions of transmission minima were observed for both configurations of beam incidence. Fig. 2 shows typical transmission minima due to interband transitions within the <$\bar{1}$11> valleys for |1,+> → |1,->, |1,-> → |1,+> Landau states, for three samples. The difference in position between σ^+ and σ^- minima is more pronounced for config. (i) compared to config. (ii), increasing film thickness and decreasing deposition rate. It should be noticed here, that the transitions between [111] - valleys show the same systematic behaviour.

d) A smaller deposition rate - which means in our case a lower substrate temperature - increases the differences between the magnetoreflectivity data for config. (i) and (ii). Especially the modulation of the magnetoreflectivity spectra, but also the splitting for σ^+ and σ^- is much more pronounced for reflection at the BaF_2/PbTe-interface (config.(ii)). It seems to be noteworthy, that samples, which show a pronounced splitting for σ^+ and σ^- in the transmission spectra, also show the analogous behaviour in the reflectivity data.

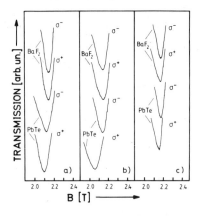

Fig.2: Transmission minima at $\tilde{\nu}$ =1644.3cm due to interbandtransitions of the <111> - valleys for $|1, +> \to | 1,->$, $|1,-> \to | 1,+>$ Landau states
a) 188.2:set 2 (d=6.53 μm)
b) 188.3:set 2 (d=10.29 μm)
c) 176.3:set 1 (d=5.23 μm)

Discussion

The experimental findings listed above confirm the magnetotransport experiments in the fact, that the PbTe-films grown on alkali halide substrates are subjected to a considerable strain. As it was not possible to fit the considerable splitting of the interband transitions (σ^+, σ^-) by a reasonable set of cubic PbTe-bandparameters, we believe, that there is a considerable rhombohedral distortion as anticipated by Allgaier et al. from low field magnetoresistance measurements [2].

As mentioned above, the interpretation of previous quantum magneto transport experiments suggests a strain profile, which reflects the relaxation of the elastic expansion with increasing film thickness. The magnetoreflectivity data obtained from the reflection at the PbTe-film surface and the BaF_2/PbTe-interface, which for the first time have allowed an experimental investigation of this suggested strain profile, show that the difference of the strain-induced splitting for σ^+ and σ^- in config. (i) and (ii) is relatively small compared to its absolute value. This means that our reflectivity data are in contrast to previous interpretations of quantum magneto transport experiments. If the film thickness is much smaller than the substrate thickness - in our samples 3-20 μm compared to 1.5mm, the decrease of strain with increasing film thickness seems to be always much smaller than its absolute value.

To proof these controversial experimental findings we have calculated the strain profile in the BaF_2 - PbTe - system, using a cylinder symmetrical model with real sample dimensions (see insert of Fig.3) by the method of finite elements [5]. For the elastic moduli of PbTe values given in Ref. [7] and for BaF_2 in Ref. [8] were used. The results are shown in Fig. 3. The diagram shows the absolute values of the elastic displacements of the nodes - this means certain lattice points - whose positions within our model are given in the insert of Fig. 3

The profile of the elastic displacements, which also reflects the profile of the strain, shows, that the difference of the strain for the BaF_2 - PbTe - interface and the PbTe-film surface is of the order of 1% of their absolute values. Therefore the result of our calculation is that the strain within the PbTe-film induced by the mismatch of the coefficients of thermal expansion of substrate and film at low temperatures is nearly independent of the sample thickness (in our case: film thickness: 3-20μm, substrate thickness: 1,5mm) if the crystal growth conditions are equal. This is confirmed by our magnetoreflectivity data from the BaF_2-PbTe-interface and the PbTe-film surface

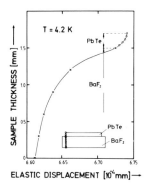

Fig.3 Absolute values of the elastic displacements of characteristic nodes (xxx BaF_2, ooo PbTe) of the finite element network for the BaF_2 - PbTe sample with film thickness of 20μm, substrate thickness 1,5mm and 10x10 mm^2 film area. The PbTe-film thickness is extended by a factor of 10 for a better illustration. Thermal expansion coefficients for PbTe and BaF_2 according to [9]

Table I
Growth conditions and transport parameters of PbTe/BaF_2 films

	Source-temp.	Substrate-temp.	Deposition-rate	n(300K) [cm^{-3}]	μ(300K) [cm^2/Vs]
set 1	530 C	470 C	9 μm/h	1.7×10^{17}	1100-1400
set 2	530 C	380 C	3.5 μm/h	$1.3-1.9 \times 10^{17}$	1100-1200

Acknowledgements

We express our appreciation to Prof.F. Sturm and Dr.N.Seifter for performing the finite element numerical calculations. Work supported by the Fonds zur Förderung der wissenschaftlichen Forschung, Austria and by the Deutsche Forschungsgemeinschaft.

References

1 E.D.Palik, D.L.Mitchell and J.N.Zemel, Phys.Rev.A,763,135 (1964).
2 R.S.Allgaier, J.B.Restorff, B.Houston, J.D.Jensen and A.Lopez - Otero, Phys.Rev., to be published.
3 G.P.Carver, B.B.Houston, J.R.Burke, H.Heinrich and A.Lopez-Otero, Solid State Commun. 30, 461 (1979).

4 J.R.Burke and G.P.Carver, Phys.Rev.B 17, 2719 (1978).
5 F.Sturm and R.Harreither, Arch.Eisenhüttenwesen 47, 357 (1976).
6 A.Lopez-Otero, J.Appl.Phys. 48, 446 (1977).
7 Yu.Ravich, B.A.Efimova and I.A.Smirnov, Semiconducting Lead Chalcogenides (Plenum Press, New York, 1970).
8 C.Kittel, Introduction to Solid State Physics (3rd edition, J. Wiley, N.Y. 1966).
9 R.F.Bis, NOL Report 1972.

Study of Electronic and Lattice Properties in (Pb, Sn, Ge) Te by Intra-band Magneto-Optics

T. Ichiguchi, and K. Murase

Department of Physics, Osaka University, 1-1 Machikane-yama
Toyonaka, 560 Japan

1. Introduction

In $Pb_{1-x}Sn_xTe$, the energy gap decreases and the NaCl-type lattice structure becomes unstable with increasing alloy composition x. The lattice instability or the softening of the TO-phonon is related to strong electron-phonon interactions. To understand this mechanism in detail, the simultaneous investigation of the band-edge structure and the lattice property is necessary. The band-edge structure itself also arouses another interest concerning the narrow energy gap [1]. In spite of the importance of the magnetic energy levels, there have been only a few reliable works on the g-tensor, and the value has not been well established.

We report the determination of the electronic and the lattice parameters in PbTe and $Pb_{0.7}Sn_{0.3}Te$ by magnetoplasma-phonon excitations. The g-value obtained from the combined resonance in PbTe is presented. It was confirmed that the TO-phonon energy ($\hbar\omega_{TO}$) is lower and the transverse mass (m_t) is lighter in $Pb_{0.7}Sn_{0.3}Te$ than in PbTe. Experimental results will be discussed with the six-band model.

2. Electronic and Lattice Parameters

The experiment was made at 4.2K with a strip-line using 513μm-, 433μm-, 394μm-, 337μm- and 311μm-laser lines, whose energies are close to the TO-phonons in $Pb_{1-x}Sn_xTe$ [2]. Every sample has a (100)-surface, on which the submillimeter waves propagate being partially absorbed into the sample. The transmitted power through the strip-line was recorded as a function of magnetic field.

The spectra in PbTe with $\vec{H}$ along the <011>- and <100>-directions are shown in Fig.1(a) and (b), respectively. The waves penetrate into the sample almost perpendicularly to the surface because of the high dielectric constant, so that Fig.1(a) is for the Voigt configuration ($\vec{q}$// <100> and $\vec{H}$// <011>), and Fig. 1(b) for the Faraday configuration ($\vec{q}$// $\vec{H}$// <100>). The dispersion relation of strip-line modes and the shape of the spectrum were numerically calculated with the classical magnetoplasma theory. The calculated line shapes agree well with the experiment with suitable damping factors of the magnetoplasma and the phonon (ν=1.0cm^{-1} and Γ=3.5cm^{-1}, respectively). It was confirmed that the main dips correspond to the dielectric anomalies. Usually, we have three dips in the Voigt configuration, two dips in the Faraday configuration and four dips in the intermediate angle configuration. No dips, however, were observed with 513μm (=19.5cm^{-1}) line in any configuration. The reason is that the photon energy is very close to the $\hbar\omega_{TO}$. The additional small dips pointed by arrows in Fig.1(b) are due to the combined resonance, which will be

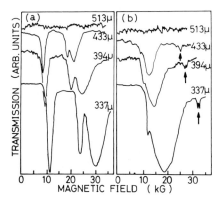

Fig.1 Typical strip-line transmission spectra in PbTe at 4.2K

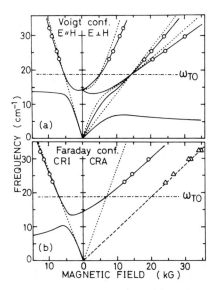

Fig.2 Dip position (o ; main dip, △ ; combined resonance), dielectric anomaly (solid line) and pole (dotted line) in H-ω plane in PbTe

mentioned in the following section.

The dip positions in PbTe are marked against the magnetic field and the photon energy in Fig.2. The lattice parameters (ω_{TO} and ε_∞) were obtained accurately with best fitting between the dip positions and the calculated dielectric anomaly. Hereupon, $\hbar\omega_{LO}$=112 cm^{-1} is assumed in PbTe, but it does not seriously affect the band parameters or the $\hbar\omega_{TO}$ to be determined. We can estimate the $\hbar\omega_{TO}$ more easily from the frequency where two modes are degenerate in the H-ω plane. The angular dependence of the spectra was taken in (011)- and/or (001)-plane with each laser line. The transverse mass m_t, mass anisotropy ratio K and dielectric constant at the laser frequency $\varepsilon(\omega)$ were obtained on the classical magnetoplasma model with four ellipsoidal mass tensors. The parameters m_t and K do not depend on the laser frequency we used. The calculations of the dip positions in the H-ω plane at the intermediate angle configurations are also in good agreement with the experiment.

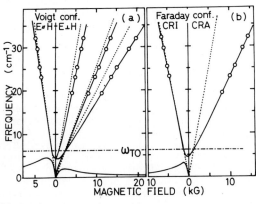

Fig.3 Dip position in Pb$_{0.7}$Sn$_{0.3}$Te at 4.2K

The dip position in Pb$_{0.7}$Sn$_{0.3}$Te is shown in Fig.3. It exhibits remarkable contrasts to PbTe at following two points: lower ω_{TO} and lighter cyclotron mass. The obtained band and lattice parameters are listed in Table 1. The static

250

Table 1 Experimentally obtained lattice and band parameters, and calculated ones (in parentheses) with six-band model

	PbTe	$Pb_{0.7}Sn_{0.3}Te$
$\omega_{TO}[cm^{-1}]$	18.8	6.2
$\omega_{LO}[cm^{-1}]$	112	116
ε_∞	34	39
ε_s	1210	13700
carrier dens.	$n=2.0\times10^{17}$	$p=1.3\times10^{17}[cm^{-3}]$
m_t/m_0	$\{\begin{array}{c}0.0218\\(0.0219)\end{array}$	$\begin{array}{c}0.0115\\(0.0115)\end{array}$
K	$\{\begin{array}{c}11.0\\(10.5)\end{array}$	$\begin{array}{c}9.9\\(10.8)\end{array}$
m_t/m_0 at edge	c-(0.0202)	v-(0.0040)
K at edge	(10.6)	(10.8)
E_g	187[meV]	24.3[meV]

dielectric constant ε_s is determined from the LST-relation, and it is in good agreement with the microwave helicon measurement [3]: 1500 for PbTe and ~10000 for $Pb_{0.7}Sn_{0.3}Te$.

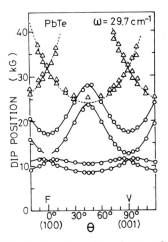

Fig.4 Angular dependence of dip position in PbTe. Main dip (o), combined resonance (△)

3. g-factor from Combined Resonance and Six-band Model

We observed new fine structures at high magnetic field (pointed by arrows in Fig.1(b)). Judging from the dip positions and the laser frequency dependence (open triangles in Fig.2), the spectrum can not be identified as an impurity cyclotron resonance. We conclude it to be due to the combined resonance ($0^+ \to 1^-$) from the position of the Fermi level at the resonance field. The angular dependence of the dip position is shown in Fig.4. We obtained $g_\parallel=60.0$ and $g_\perp=17.5$ using the angle dependent g-value as follows,

$$g_\theta = (g_\parallel^2 \cos^2\theta + g_\perp^2 \sin^2\theta)^{1/2}, \qquad (1)$$

where θ is the angle from the <111>-axis. Our g-value is compared with others in Table 2. The anisotropy of g-value is 3.4 and is in good agreement with that of the cyclotron mass (i.e. $\sqrt{K}=3.3$). We found the following relation irrespective of the orientation of the magnetic field;

Table 2 Comparison of the g-value in PbTe

	$g_\parallel$	$g_\perp$	g(100)	carrier dens.$[cm^{-3}]$	experiment
this work	60.0±1.0	17.5±1.0	37.5±0.5	$n=2\times10^{17}$	FIR-ESR
SCHABER & DOEZEMA[4]	$\{\begin{array}{c}59.5\pm0.7\\58.3\pm1.0\end{array}$	$\begin{array}{c}15.3\pm1.5\\18.8\pm1.5\end{array}$		$\{\begin{array}{c}n=3\times10^{16}\\p=4\times10^{16}\end{array}$	FIR-ESR
PATEL & SLUSHER[5]	57.5±2	15 ±1	35.3	$n=1\times10^{17}$	spin-flip Raman
BURKE et al.[6]	32 ±2	7 ±2	23 ±5	$p=3\times10^{18}$	SdH
APPOLD et al.[7] (cited from[4])	$\{\begin{array}{c}59.3\pm2.7\\61.3\pm2.3\end{array}$	$\begin{array}{c}18.2\pm0.9\\19.1\pm0.7\end{array}$		$\{\begin{array}{c}p\\n\end{array}$	six-band model

$$m_c \cdot g/2m_0 = 0.65 \pm 0.02. \tag{2}$$

That is, the ratio of the spin splitting to the Landau level separation is almost constant. The result may directly justify the DIMMOCK's convenient assumption [1].

From about a dozen of samples ($x \leq 0.4$), we obtained six-band parameters as follows [8],

$$2P_\perp^2/m_0 = 5.80 \text{ eV}, \quad 2P_{//}^2/m_0 = 0.540 \text{ eV}, \quad m_0/m_t^- = 18.5,$$

$$m_0/m_\ell^- = 1.8, \quad m_0/m_t^+ = 11.0, \quad m_0/m_\ell^+ = 1.0 \; . \tag{3}$$

The masses calculated using these parameters are included in Table 1. The two-band contributions to the $g_{//}$ and $g_\perp$ are given by $(4P_\perp^2/m_0)/E_g = 62.0$ and $(4P_\perp P_{//}/m_0)/E_g = 18.9$, respectively, in PbTe. It should be noted that an assumption $P_\perp^2/P_{//}^2 = m_\ell^-/m_t^- = m_\ell^+/m_t^+$ is well realized in (3), which safely guarantees an ellipsoidal shape of the Fermi surface. This fact is consistent with an agreement in anisotropies about the g-value and the cyclotron mass, when the remote band contributions to the g-value (i.e. $g_{//}^r$ and $g_\perp^r$) are small enough. The decrease of $m_c \cdot g/2m$ from unity is caused by remote band contributions to the mass rather than those to the g-value.

We have reported that the transmission spectrum through the strip-line is drastically changed below the phase transition temperature T_c in $Pb_{1-x}Ge_xTe$ [9]. The band edge structure should be remarkably changed below T_c by two types of lattice distortions: the strain (ε_{ij}) and the relative sublattice displacement ($\vec{u}$) [10]. The typical fine structures may be explained by quantum transitions with a six-band model by taking the lattice distortions into consideration.

Acknowledgements

The authors would like to acknowledge the joint effort of Dr.S.Nishikawa. They wish to thank Prof.H.Kawamura for encouragement through this work, and Dr.S.Sugai for experimental suggestions, and Dr.S.Nishi and T.Fukunaga for sample supplying. One of authors (K.M.) was supported in part by the Kurata Foundation and by the Grant-in-aid for Scientific Research from the Ministry of Education.

References

1. J.O.Dimmock, Proc.Conf.Phys.Semimetals and Narrow Gap Semiconductors, p.319, Dallas, Texas (1970).
2. T.Ichiguchi and K.Murase, Solid State Commun. **34**, 309 (1980).
3. S.Nishi, H.Kawamura and K.Murase, phys.stat.sol.(b)**97**, 581 (1980).
4. H.Schaber and R.E.Doezema, Solid State Commun. **31**, 197 (1979).
5. C.K.N.Patel and R.E.Slusher, Phys.Rev. **177**, 1200 (1969).
6. J.R.Burke, B.Houston and H.T.Savage, Phys.Rev.**B2**, 1977 (1970).
7. G.Appold, R.Grisar, G.Bauer, H.Burkhard, R.Ebert, H.Pascher and H.G.Häfele, Proc. 14th Int.Conf.Phys.Semiconductors, Edinburgh (1978), p.1101; or H.Burkhard, G.Bauer and W.Zawadzki, Phys.Rev.**B19**, 5149 (1979).
8. S.Nishikawa, K.Murase, T.Ichiguchi and H.Kawamura, submitted in Phys.Rev.
9. H.Kawamura, S.Nishikawa and K.Murase, Proc.Int.Conf.on Appl.High Mag.Field in Semiconductor Phys., Oxford (1978), p.170.
10. K.Murase, Proc.15th Int.Conf.Phys.Semiconductors, Kyoto (1980).

Magnetic Freeze-Out and High Pressure Induced Metal-Nonmetal Transition in Low Concentration n-Type InSb[*]

R.L. Aulombard, A. Raymond, L. Konczewicz[**], J.L. Robert

C.E.E.S. associé au CNRS (LA 21)
Université des Sciences et Techniques du Languedoc
34060 Montpellier Cedex, France, and

S. Porowski

High Pressure Research Center, Polsih Academy of Sciences, Unipress
Warsaw, Poland

It is well known that the conductivity of n-type InSb is metallic in zero magnetic field when the carrier density is larger than 10^{14} cm^{-3}. Some authors [1] have suggested that such a behaviour is governed by an impurity band separated from the conduction band. Others [2] have estimated that impurity states have been merged with the conduction band.
In paper [3] we have reported some results obtained on samples with an excess donor density larger than 10^{14} cm^{-3}. We have shown that the temperature dependence of the resistivity-with and without pressure- for a sample with an excess donor density of 4×10^{14} cm^{-3} could be explained by the variation of the mobility only. As a result, we have concluded that the impurity states must be considered to be merged with the conduction band.
On the other hand, a metal-non metal transition can be observed when a magnetic field or an hydrostatic pressure is applied : it allows the investigation of the low temperature conduction processes ($\varepsilon_1, \varepsilon_2, \varepsilon_3$). As we have shown, it was difficult to conclude the existence of the ε_2 process because no common intercept of the curves logσ_{xx} versus 1/T at 1/T = 0 has been observed, which would be characteristic of this ε_2 regime.
Actually we have performed similar experiments for a sample with a lower carrier density (InSb F.d n≃7.5×10^{13} cm^{-3}). This value is very close to the critical one, deduced from the Mott criterion ($n_c \simeq 6\times10^{13}$ cm^{-3} for n-type InSb).
To clarify the existence of an activation energy in zero magnetic field, experiments under hydrostatic pressure have been performed. To investigate the low temperature conduction processes in the high magnetic field condition, experiments have been performed both in the transverse and in the longitudinal configuration - while they had been previously made only in one configuration. All the experiments have been made in magnetic fields up to 200 kG, under hydrostatic pressure up to 535 MPa between 1.8 K and 8 K.

a) Analysis of the results in zero magnetic field

We can noticed that the determination of the free carrier density from Hall measurements in the low temperature range is practically impossible. First, in consequence of the possibility of freeze-out even in very low magnetic field and then, because the Hall factor r is unknown since the sample is at the boundary of degeneracy. To avoid these effects the excess donor density of the sample F.d have been determined from Hall measurements at 77 K. The

---

[*] The experiments have been performed at the S.N.C.I.-CNRS -GRENOBLE-FRANCE
[**] On leave from High Pressure Research Center, Polish Academy of Sciences
UNIPRESS - WARSAW - POLAND.
Acknowledgments : we would like to thank Dr Picoche and Rub of S.N.C.I. for their helpful assistance.

Hall coefficient R_H has been measured as a function of the magnetic field and the carrier concentration $n = \frac{1}{R_H e} = 7.5 \times 10^{13}$ cm^{-3} has been deduced from the R_H value in the saturation range.
Between 1.8 K and 8 K the resistivity at p = 0 MPa decreases by a factor nearly equal to 4 (Fig. 1). To explain this variation the temperature dependence of the conduction band mobility (ionized impurity scattering [4]) has been considered. Between 3 K and 8 K a good agreement is obtained if it is assumed that the carrier density is equal to the excess donor density(7.5×10^{13} cm^{-3}) and

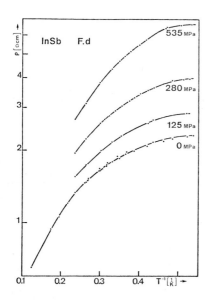

Fig.1 Resistivity ρ versus 1/T without pressure

remains constant. Thus, it gives no evidence of an ionization energy. When the temperature is lowered, to describe the temperature dependence of ρ, it would be necessary to assume a decrease of 20% of the carrier concentration. This effect could be explained by the existence of a small activation energy. However because the impurity density is larger than the critical value of the Mott criterion, the carriers are probably trapped in potential wells of about 0.1 meV deep and connected to long range fluctuations. In any case, the decrease of the resistivity is essentially due to the variation of the mobility. If an activation energy does exist, its value is less than 0.1 meV and certainly is not detectable from these experiments.
On the contrary, the conclusions are quite different when a pressure is applied. Under pressure, the changes in the resistivity with the temperature are much stronger and cannot be explained by the variations of the mobility only.
It is worthwhile to notice that in the case of a sample with larger carrier density for which no activation energy has been observed [3], the curves log ρ(T) at different pressures remain parallel.
These results indicate that for the investigated sample, an activation energy appears when the pressure is high enough to reduce the overlap of the wave functions of the impurity states.

b) Analysis of the results in presence of magnetic field

ε_2 region : one of the criteria used to etablish the existence of the ε_2 process is the common intercept of the curves log σ versus 1/T at 1/T = 0 [5,6] . On Fig.(2) and (3) the variations of the longitudinal magnetoresistance and those of the tranverse component of the magnetoconductivity tensor versus 1/T are presented. No common intercept is observed at 1/T = 0 and as, in our previous work [3] , the existence of the so-called ε_2 process cannot be proved.

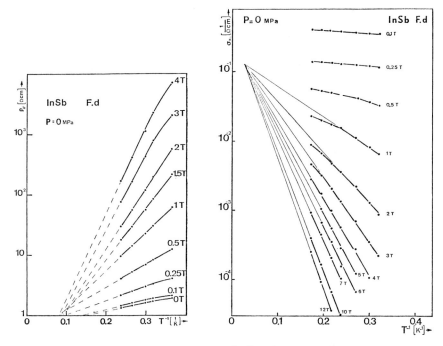

Fig.2 The longitudinal magneto-resistance ρ versus 1/T without pressure

Fig.3 The transverse component σ_{xx} versus 1/T without pressure

Similar results have been observed under pressure (Fig.4). As expected due to the increase of the effective mass, the slopes of the curves increase when the pressure is applied.

ε_1 region : assuming that the conduction takes place in the conduction band only, the magnetic field dependence of the ionization energy ε_1 is deduced from the plot log $nT^{-1/2}$ vs 1/T (Fig.5) as well as the order of magnitude of the compensation ratio (70%). On Fig.(6), the magnetic field dependence of ε_1 at p = 0 MPa is compared to the Y.K.A. theory [7] . In the high magnetic field range the experimental value is larger than the one predicted by this theory and this effect is probably connected with the nonparabolicity of the band. Fig.(7) presents the pressure dependence of ε_1 vs magnetic field. It can be seen that the activation energy increases with the pressure.

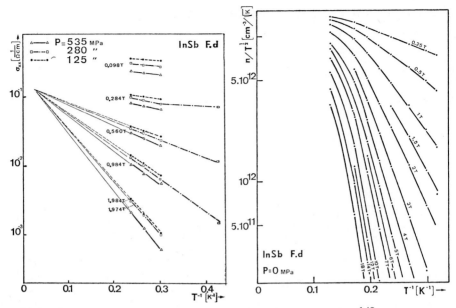

Fig.4 the transverse component σ_{xx} vs 1/T at different pressures

Fig.5 The log $nT^{-1/2}$ vs 1/T without pressure

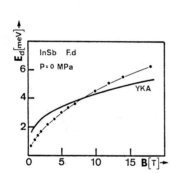

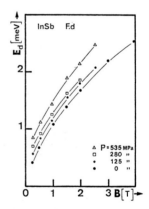

Fig.6 The energy ε_1 vs magnetic field p=0MPa. The solid line-The YKA theory

Fig.7 The energy ε_1 vs magnetic field at different pressures

1. J.R. Sandercock Solid State Com. 7(1969) 721
2. H.Miyazawa and H. Hikoma J. Phys. Soc. Japan 23 (1967) 290
3. JL.Robert,A.Raymond,RL.Aulombard,C.Bousquet Int. Conf. on the Impurity Bands- Wurzburg 1979 To be published in Phil. Magazine
4. A.Raymond - Thesis Univ. of Sc. and Tech. of Languedoc-Montpellier(1977)
5. M.Pepper,J. Non Crystalline Solids 32 (1979) 161
6. D.Ferré,H.Dubois,G.Biskubski Phys. Stat. Sol. B 70 (1975) 81
7. Y.Yafet,RW.Keyes, EN.Adams J. Phys. Chem. Solids 1 (1956) 137

A Sensitive HgCdTe Bolometer for the Detection of Millimeter Wave Radiation

B. Schlicht, and G. Nimtz

II. Physikalisches Institut der Universität zu Köln, Zülpicherstr. 77
5000 Köln 41, Fed. Rep. of Germany

The detection of low-power millimeter and submillimeter wave radiation is often carried out with n-InSb hot carrier bolometer mixers. In pure n-InSb at liquid-helium temperatures the temperature T_e of the free electrons can be raised by the absorption of RF power. Since the electron mobility under these conditions is determined by ionized impurity scattering the dependence of the resistivity on the electron temperature is described by $\rho \propto T_e^{-3/2}$.

Recently a magnetotransport anomaly was discovered in the narrow-gap semiconductors PbTe, $Pb_{0.8}Sn_{0.2}Te$, and $Hg_{0.8}Cd_{0.2}Te$ [1,2], which in the latter material exhibits an extraordinarily strong bolometer effect. In each semiconductor the anomaly is characterized by a critical temperature and a critical mag-

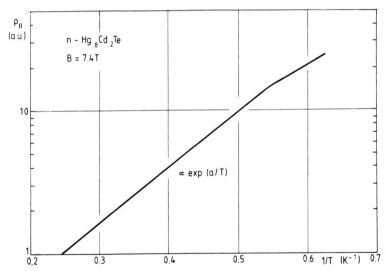

Fig. 1 Longitudinal magnetoresistance versus reciprocal temperature for an n-type $Hg_{0.8}Cd_{0.2}Te$ crystal. The extrinsic carrier density at 4.2 K is $n = 6.4 \cdot 10^{20}$ electrons / m^3

netic field, at which a phase transition was observed to take place. Presumably this phase transition is due to a condensation of the free electrons into an ordered state, e.g. a three-dimensional CDW or a Wigner lattice or a Wigner glass. The anomalous behaviour is characterized by several effects : for instance by a pronounced deviation from the magnetoresistance power law, which is expected from the single-electron model, by the loss of the strong anisotropy of longitudinal and transverse magnetoresistance, and by a strong increase of the longitudinal and transverse magnetoresistance within a small temperature interval. The latter effect turned out to be most pronounced in n-type $Hg_{0.8}Cd_{0.2}Te$ at carrier densities below $10^{21}m^{-3}$. In such crystals the longitudinal magnetoresistivity increased by a factor of 20 to 40 when cooling down the sample from 4.2 K to 1.5 K in a high magnetic field. A typical result is shown in Fig. 1. The temperature dependence of the longitudinal magnetoresistivity can be described by $\rho_\| \propto \exp(a/T)$ with $a_\|$ 9K in the temperature range between 4 K and 1.8 K. Accordingly , the resistivity shows a much stronger temperature dependence in HgCdTe than in InSb in the liquid-helium temperature regime. This favours the application of HgCdTe as an efficient microwave mixer device [3].

The strong temperature dependence of the longitudinal resistivity at high magnetic fields proved to be a pure free carrier

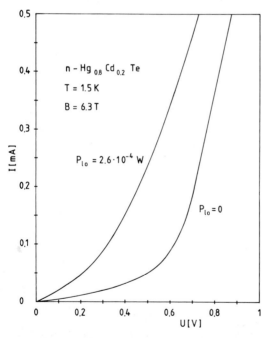

Fig. 2 Current-voltage characteristic of an $n-Hg_{0.8}Cd_{0.2}Te$ bolometer with and without local oscillator bias power

effect : Keeping the lattice temperature low at 1.5 K the high resistance state is switched to a low resistance state by heating the electrons in electric fields (dc or RF) of approx. 100 to 400 V/m. Fig. 2 shows the current-voltage characteristic for the sample of Fig. 1 measured with and without local oscillator power. Without absorption of RF power the ratio of high resistance to low resistance is about 36, the low resistance state corresponds to a temperature near 4.2 K.

The current-voltage characteristic displayed in Fig. 2 does not show any region of negative differential resistivity as has been observed with InSb under magnetic fields due to a freeze out of carriers [4]. In HgCdTe and in the lead salts a magnetically induced freeze out of carriers does not take place [2].

The transition from the high to the low resistance state requires some microseconds. A relaxation time τ of 10^{-7}s was determined from the experimental data which equals the energy relaxation time observed in other semiconductors at low temperatures. This value permits IF response up to about 2 MHz when utilizing the material as a bolometer mixer.

In order to measure the sensitivity of the new bolometer a homodyne scheme similar to that described in [5] and operated at 37 GHz was used. The microwave signal was derived from the local oscillator and amplitude modulated at 10 kHz or 20 kHz. It was mixed in the bolometer with the unmodulated local oscillator power of 5×10^{-5}W. The sample was mounted in a rectangular waveguide (7.11 x 3.56 mm² cross section) as an inductive

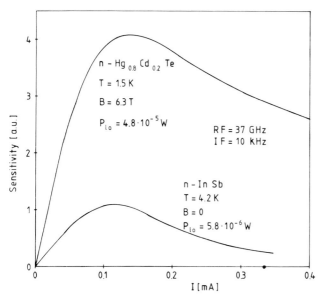

Fig. 3 Mixer sensitivity versus bias current for a $Hg_{0.8}Cd_{0.2}Te$ bolometer and an InSb bolometer obtained with equal signal powers

post at a distance of $\lambda/4$ in front of a short circuit. The bolometer was placed in the center of a superconducting coil and immersed in a liquid helium bath.

Fig. 3 shows the signal voltage as a function of the bias current for the $Hg_{0.8}Cd_{0.2}Te$ bolometer of Fig. 2. The maximum signal voltage has been optimized with respect to the local oscillator power. For comparison the sensitivity characteristic of a high quality InSb bolometer measured in the same waveguide at 4.2 K and again optimized with respect to the local oscillator power is plotted. The investigated InSb material is normally used for the low-power microwave radiation detection in astrophysical equipment[6]. The measured responsivities were $3 \cdot 10^4$ V/W for the HgCdTe bolometer and $8 \cdot 10^3$ V/W for the InSb bolometer. For a $Hg_{0.8}Cd_{0.2}Te$ bolometer of smaller cross section an even better responsivity of $1.3 \cdot 10^5$ V/W was measured[3].

Utilizing the hot load/cold load technique described in [5] preliminary noise measurements were carried out with both the HgCdTe and the InSb bolometer. They revealed that the noise of the investigated HgCdTe sample was at most twice the value obtained for the InSb sample, mainly due to the higher resistance of the HgCdTe sample.

From these results one can conclude that the detectivity of the new HgCdTe bolometer is improved by at least a factor of two if compared with the commonly used InSb bolometers. Considering the fact that due to the long integration time it takes about 10 hs to record an astrophysical microwave spectrum the improvement provided by the new bolometer more than outweighs the need of the magnetic field and of the low temperature.

We are grateful to A.R.Gillespie, who performed the noise measurements.

References

1. G.Nimtz, B.Schlicht, E.Tyssen, R.Dornhaus, L.D.Haas: Solid State Commun. 32, 669 (1979)
2. G.Nimtz, B.Schlicht: *Experiments Concerning the Magnetic Field Induced Wigner Condensation in Semiconductors* in: Festkörperprobleme (Advances in Solid State Physics), Vol. XX, 369 (1980)
3. G.Nimtz, B.Schlicht, H.Lehmann, E.Tyssen: Appl. Phys. Letters 35, 640 (1979)
4. F.Arams, C.Allen, B.Peyton, E.Sard: Proc. IEEE 54, 612 (1966)
5. T.G.Phillips, K.B.Jefferts: Rev. Sci. Instrum. 44, 1009 (1973)
6. A.R.Gillespie: private communication

Part IX

Space Charge Layer and Superlattice

Electric Subbands in a Magnetic Field

F. Koch

Physik-Department E 16, Technische Universität München
8046 Garching, Fed. Rep. of Germany

> We survey the theoretical ideas related to surface binding in combined electric and magnetic fields. Considering the magnetic effects as a weak perturbation on the electric binding we quote specific results and examine the possibilities of resonant excitation. The resonances observed in experiments are compared and discussed with reference to these expectations.

1. Introduction

The states of mobile carriers confined in a semiconductor surface charge layer are referred to as electric subbands. Their spectroscopic observation in tunneling, as a bolometric response in the surface conductivity, in far-infrared absorption and emission experiments, most recently also in inelastic light scattering, is an established part of semiconductor physics.

The subbands are electrically quantized states associated with the Coulomb potential and surface electric field. When we examine here the influence of a magnetic field, it is most often in the role of a perturbation. The magnetic field is an auxiliary variable, an additional handle on the problem that provides further information. The present review looks at the many and varied ways that magnetic fields have been employed in subband spectroscopy and the information derived from them. We conclude by mentioning some potential applications in future investigations.

2. Electric Subbands vs. Magnetic Surface Levels

Before there was subband spectroscopy there were magnetic surface levels, and a comparative perspective is instructive.

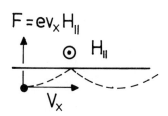 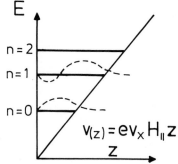

Fig. 1: Surface Landau levels. Classical trajectory and potential well.

Consider in Fig. 1 the electron moving parallel to the surface
with velocity v_x. In the field $H_\parallel$ directed along the surface it
experiences a force $F = ev_x H_\parallel$ binding it to the surface. The
classical trajectory is a shallow skipping orbit. In the limit
of near straightline motion, the binding potential is the triangular potential $V(z) = ev_x H_\parallel z$ with quantized states E_n described by Airy functions and known as surface Landau levels.
In the folklore of metal Fermi surface physics these states
have proved to be a detailed probe of the conduction electrons.
They are observed in microwave resonance experiments, an example of which appears in Fig. 2. The sequence of peaks at 23, 9,

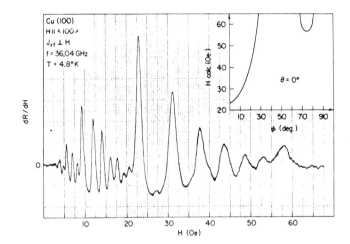

Fig. 2:
Surface Lan-
dau level
resonances
observed in
microwave
absorption
on Cu (100)
(ref. [1]).

6, 4 ... Oe represents transitions from n = 0 to 1, 2, 3 ...
etc. A similar series starting at 31 Oe results from transitions from occupied n = 1 states to 2, 3, 4 ... etc. The entire
spectrum of peaks is essentially from a group of electrons
with a unique parallel velocity v_x.

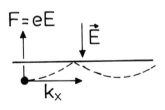

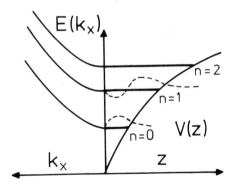

Fig. 3: Electric subbands.
Classical trajectory, surface potential well, and
dispersion $E_n(k_x)$.

The semiconductor subband is formed from quantized states of $\hat{z}$-directed motion in the presence of a surface electric field $\vec{E}$ (Fig. 3). The potential $V(z) \sim eEz$ is a poor approximation, because the presence of charge in the surface layer requires curvature in accord with Poisson's equation. The problem has to be solved self-consistently, taking account of the occupancies of the levels. Up-to-date calculations of the subband energies [2] include an image potential in the neighboring dielectric. They take account of exchange and correlation contributions to the potential as well as the long-range depletion charge potential. Because electric surface binding is independent of the parallel velocity v_x, a given state n is associated with the two-dimensional band

$$E_n(k_\|) = E_n + \frac{\hbar^2}{2m^*}(k_x^2 + k_y^2). \qquad (1)$$

At T = 0 and reasonable values of the surface charge density N_s (a few times 10^{12}cm^{-2}), only a few subbands are partially filled. In Si (100) the n = 0 band alone is occupied. The sequence of transition peaks in Fig. 4 are from the ground state

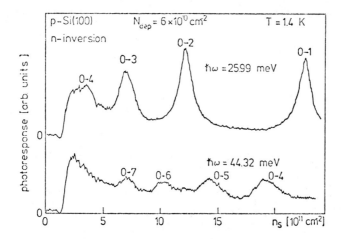

Fig. 4: Subband resonance transitions observed as a bolometric response in the surface channel conductivity on Si (100) (ref. [3]).

to successively higher levels. Because they are recorded as bolometric response of the surface conductivity, higher order transitions are enhanced by the increasing $d\sigma/dT$ at low N_s.

3. The Magnetic Perturbation of Electric Subbands

Having examined separately magnetic surface levels and electric subbands it is a simple matter to compound the electric and magnetic forces that produce the bound state. We consider the limit of an electric subband perturbed weakly by $H_\|$. The subband represents carriers moving with all possible velocities v_x parallel to the surface. In a given state n they are all confined to the surface layer within a length $\langle z \rangle_n$, which is a quantum-mechanical expectation value for the nth subband.

The perturbation $H_{//}$ adds to each state k_x a different paramagnetic energy $ev_x H_{//} \langle z \rangle_n$, where $v_x = \hbar k_x/m^*$. The result is the broken line parabola in Fig. 5. However, this is not all. There is an additional diamagnetic energy term that we had ignored for the surface Landau level problem with its large $v_x \sim v_F$. As we now examine all carriers, including stationary $k_x = 0$ electrons, the diamagnetic energy comes to be of some importance. The $k_x = 0$ particle experiences a force in the field $H_{//}$ by virtue of its $\hat{z}$-directed motion and is deflected parallel to the surface. This induced parallel motion gives rise to the diamagnetic energy $e^2 H_{//}^2 \langle z^2 \rangle_n /2m^*$, which is the same for all states k_x. We write altogether for the perturbed subband

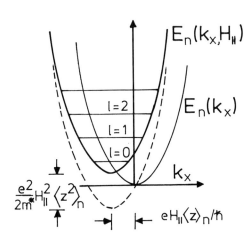

Fig. 5: The magnetically perturbed electric subband.

$$E_n(k_x, H_{//}) = E_n + \frac{\hbar^2}{2m^*}\left(k_x + \frac{eH_{//}}{\hbar}\langle z \rangle_n\right)^2 + \frac{e^2}{2m^*} H_{//}^2 \left(\langle z^2 \rangle_n - \langle z \rangle_n^2\right). \quad (2)$$

We recognize in this expression the hybrid electric-magnetic subband as a parabola shifted to $-eH_{//}\langle z \rangle_n/\hbar$ and raised by the small, positive definite quantity which is the net diamagnetic energy. The latter contains a factor known as the spread of the bound state $\Delta z^2 = \langle z^2 \rangle_n - \langle z \rangle_n^2$ and depends sensitively on the exact form of the wave-function.

Within the bounds of perturbation theory the other two possibilities that can be realized experimentally, namely perpendicular field $H_\perp$ and the tilted field geometry $(H_{//}, H_\perp)$, are readily described. The first of these gives rise to the usual Landau quantization, such that

$$E_{n,\ell} = E_n + (\ell + \tfrac{1}{2})\hbar\omega_c. \quad (3)$$

The carriers are wholly quantized and are described by numbers n and ℓ. The wave-functions are the usual harmonic oscillator states. Each subband becomes a ladder of Landau levels as sketched in Fig. 5. The tilted field configuration just compounds the features of the two previous cases. $H_\parallel$ shifts the subband parabola and raises it by the diamagnetic term. $H_\perp$ leads to Landau quantization with a ladder of discrete states of oscillator wave-functions with shifted k_x value.

Perturbation or not, is a question that needs still to be examined in dealing with magnetic fields. The relevant magnetic energy $\hbar\omega_c$ needs to be compared with subband energies E_{mn} to decide on that question. For Si, in typical fields of a few Tesla, $\hbar\omega_c/E_{01} \sim 0.1$. At least for the ground state a perturbation approach is justified. For most other materials such as InSb, or for PbTe, or for higher order subbands in Si, the magnetic effects can no longer be regarded as weak. There are theories [4,5] that have given a more general description. Nevertheless, perturbation theory remains an instructive guide to thinking about the problem.

Examining potentials and the bound states in Figs. 1 and 3 we realize that binding lengths $\langle z \rangle_n$ are linked to binding energies. The perturbation description has assumed that the $\hat{z}$-directed motion is unchanged. For this to be a reasonable limit we require that the relevant magnetic length, the cyclotron radius R_c, be much greater than the electric binding length. For (100) Si in the magnetic quantum limit $R_c \sim 80$ Å in the large, but reasonable, 10 T field. The electric length is generally smaller, but not by very much. Once again we realize that a quantitatively correct description will be found only beyond perturbation.

4. Resonances and Modes

There are many distinct resonant excitation modes of carriers in the combined electric-magnetic subbands. The three different configurations $H_\perp$, $H_\parallel$, and $(H_\parallel, H_\perp)$ coupled with the possible rf modes of polarization, namely $E_\perp$ and the two parallel modes $E_\parallel^x$, $E_\parallel^y$, make for a number of different responses. In principle one could also consider each of the three limits $\hbar\omega_c \gtreqless E_{mn}$ and for good measure include "nonparabolic" and "tilted ellipsoid" cases, to have sheer infinite variety.

To gain some insight we confine ourselves here to the perturbation limit and the parabolic subband. It is the case of electrons on (100) Si in a modest H-field that we are thinking about in constructing the table below. We ignore at present the complications caused by depolarization and exciton-shifts [6] and adhere to the spirit of the theory in considering the $\hat{z}$-directed motion to be unaffected by $H_\parallel$. Nevertheless, while the states are not altered in z, we should phenomenologically include the fact that with $E_\parallel^x$ and $H_\parallel$ an rf Hall-field $E_\perp(H_\parallel)$ will be induced in the surface layer.

A particular asymmetry is realized in this table. Combined resonance, although classically a coupled parallel-perpendicular motion of the surface carriers, appears directly in $E_\perp$

Table 1: Excitation modes

rf Polarization	$E_\perp$	$E_\parallel^x$	$E_\parallel^y$
$H_\perp$	subband resonance E_{mn}	cyclotron resonance $\hbar\omega_c$	
$H_\parallel$	magnetically broadened and shifted subband resonance $E_{mn}(H_\parallel)$	subband resonance $E_{mn}(H_\parallel)$ induced by the Hall-field $E_\perp(H_\parallel)$	- 0 -
$(H_\parallel, H_\perp)$	subband resonance $E'_{mn} = E_{mn} + \frac{e^2 H_\parallel^2}{2m^*}(\Delta z_n^2 - \Delta z_m^2)$ combined subband-Landau level resonances $E'_{mn} \pm \Delta \ell \hbar \omega_c$ $\Delta \ell = 0, \pm 1, \pm 2,$ etc.	cyclotron resonance $\hbar\omega_c$ $E'_{mn} \pm \Delta \ell \hbar \omega_c$ induced by Hall-field $E_\perp(H_\parallel)$	- 0

polarization only. The eigenstates of distinct subbands are in the limit of perturbation theory orthogonal. They are not coupled in $E_\parallel^{x,y}$ excitation. A weak signal will appear in $E_\parallel^x$, if a Hall-field is induced.

The combined resonance in the $E_\perp$ mode is an allowed transition in that the oscillator wave-functions of an adjacent pair of subbands are centered at different k_x. Because of the relation between k_x and y, this is equivalent to different centers of the oscillators along the y-direction. Matrix elements of functions with different ℓ do not vanish. The allowed resonance transitions with $E_\perp$ and $E_\parallel$ are sketched in Fig. 6.

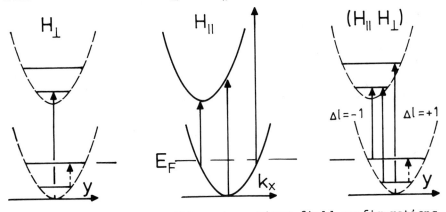

Fig. 6: Resonance transitions in various field configurations (solid line $E_\perp$; broken line $E_\parallel$).

5. The Experimental Evidence

The resonant excitation of subbands in infrared spectroscopy is known to involve a shift away from the pure energy splitting E_{mn}. In [6] the two contributions to the shift, a local-field, depolarization correction and the exciton-like shift, have been linked and described by a single factor γ. The two are inherently opposite in sign and depend on N_S. One writes the excitation $\tilde{E}_{01}$ of the $0 \to 1$ transition as

$$\tilde{E}_{01} = (1 + \gamma)^{1/2} E_{01}. \tag{4}$$

γ is of order 0.1 for electrons on (100) Si at typical N_S values.

We come to examine here the experiments, primarily those on (100) Si, and the conclusions drawn from them.

a) Perpendicular Field $H_\perp$

This is the simplest of the configurations that can be realized. As in Fig. 6, the subband resonance in $E_\perp$ consists of Landau-level conserving ($\Delta \ell = 0$) vertical transitions. The energy splitting remains E_{01}. We have studied [7] the $\tilde{E}_{01}$ resonance in accumulation and inversion samples at 10.5 and 15.8 meV excitation energy up to $H_\perp = 7$ T. The result is a straightforward "no effect". Neither line-position nor amplitude and width are found to depend on $H_\perp$.

The "null" result is not as trivial as it appears at first sight. It had been speculated [8] that a periodic broadening and shift would occur because the intersubband relaxation should depend on Landau quantization. Questions have been raised about the possible $H_\perp$-dependence of the many-body contributions to the resonance.

b) Parallel Field $H_{\parallel}$

For (100) Si this case has been explored and discussed in some detail in theory and experiment [4,5,9]. The experiments show a definitive shift. At fixed $\hbar\omega$ of 10.5 meV the $\tilde{E}_{01}$ peak moves down to nearly half the N_S value in a 10 T field. Fig. 7 is typical of the results. A primary fact is the broadening of the resonance that results from a dependence of E_{01} on k_x (compare Fig. 6). It has been recognized that the perturbation description is not entirely sufficient to describe the observed $H_{\parallel}$-shift. ANDO [5] has calculated a model case, one that simulates the conditions of the experiment, fully including the $H_{\parallel}$ in a calculation with exchange-correlation and γ-shift contributions to the resonance. For comparison the theoretical result appears in Fig. 8. One has to translate between the different N_S and $\hbar\omega$ sweeps of experiment and theory. A more exacting comparison requires experiments done with a frequency tunable source. We note that the $H_{\parallel}$ shifts provide a sensitive test of a theoretical description of the subbands.

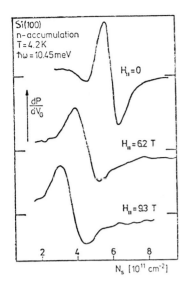
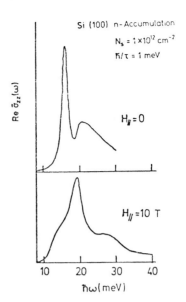

Fig. 7: $H_{\parallel}$-induced shift of the subband resonance [9].

Fig. 8: The effect of $H_{\parallel}$ as calculated in [5].

They are quite large, easily measured and depend sensitively on $\langle z \rangle_n$.

There is an alternative, nonresonant way to observe the influence of a parallel field on subband energy. The onset of occupancy of a new subband is often linked with distinct structure in the surface conductivity. Such features as a local minimum or maximum in $d\sigma/dV_G$ will be found to shift sensitively on the gate voltage scale when a parallel field is applied. The effect, as observed in Fig. 9 for InSb, has been taken to signify a shift in subband energy [10]. A quantitative evaluation of the effect for the nonparabolic, multi-subband situation of InSb is still lacking. Rather similar observations have recently been made for InAs [11]. The shift of the tunneling structures for InAs with $H_{\parallel}$ [12] are a direct measure of changes in subband energy with $H_{\parallel}$.

The case of parallel rf excitation ($E_{\parallel}^x$ or $E_{\parallel}^y$) in the presence of a field $H_{\parallel}$ has not yet been realized. We may expect with $E_{\parallel}^x$ to induce a dynamic Hall-field $E_{\perp}(H_{\parallel})$ and thus observe a resonance similar to the direct $E_{\perp}$ signal.

c) Tilted Field ($H_\parallel$, $H_\perp$)

This is the most interesting configuration and has been actively discussed in the literature [7,13-17]. The excitement stems from the fact that it allows a precise determination of the γ-parameter. Both the sign of γ and its magnitude can be derived from the measurement of the combined resonances.

ANDO [13] made the remarkable discovery that if $\gamma \neq 0$, then there will be a striking asymmetry of the $\Delta \ell = \pm 1$ combined resonances in $E_\perp$ excitation. The essential point is contained in the Fig. 10 below. For $\gamma = 0$, except for incidental differences in matrix elements and occupation factors, the $\Delta \ell = \pm 1$ sidebands have similar amplitudes and are symmetrically positioned relative to the main, $\Delta \ell = 0$ resonance peak at 31.5 meV in the figure. For positive γ, the $\Delta \ell = 0$ peak shifts in accordance with eq. 4 to higher energy. The $\Delta \ell = +1$ sideband is enhanced in amplitude at the expense of the $\Delta \ell = -1$ peak. For

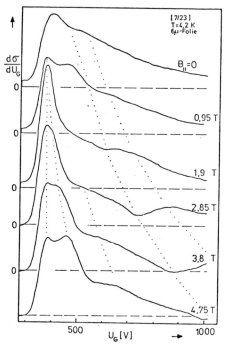

Fig. 9: Field-effect mobility structures observed for (111) InSb as a function of V_G applied across a 6 μ thick mylar insulator [10].

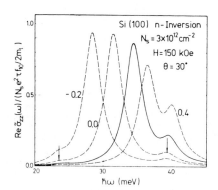

Fig. 10: Subband absorption spectrum in a tilted field for various values of the shift-parameter γ. The solid line is for $\gamma = +0.2$, a value near that realized in (100) Si [13].

negative γ, the reverse situation is encountered. In both cases the sidebands occur practically at the same energy as for $\gamma = 0$. Thus the measurement of the combined resonance sensitively and precisely determines γ.

The prediction is confirmed in experimental work. The comparison is made somewhat difficult by the need to move from the $\hbar\omega$ variation of the theory to the N_S-domain spectroscopy of the experiments. There is no doubt, however, in examining the set of data curves in Fig. 11 about the essential correctness of the calculation. The $\Delta \ell = +1$ sideband is much enhanced, it is positioned measurably closer to the main resonance $\Delta \ell = 0$. The experiments give a positive γ of order 0.15.

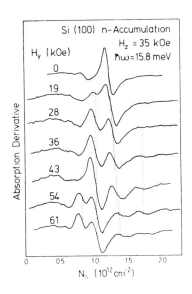

Fig. 11: Tilted-field subband resonance [7,11,12].

ANDO [14] has since improved the calculation in an attempt to fit the data more precisely, choosing parameters that more nearly match the experiments. The overall agreement is good, but in some relevant detail there are potentially interesting differences. For example, the $\Delta \ell = 0$ resonance is tuned out experimentally with increasing $H_{\parallel}$ more quickly than theory would suggest. Moreover, it moves to increased N_S as it disappears, opposite to what has been calculated. There are two possible reasons for the shifting. One is the diamagnetic energy term in E_{01}. The second is a turning-off of the shift parameter γ with tuned-out amplitude. The net shift is such as to suggest that the second effect dominates. This provides for an alternative measurement of γ.
ANDO [14] speculates that the remaining disagreements reflect inaccuracies in wave-funtions and binding lengths.

The discussion so far has concerned itself with $E_{\perp}$ excitation. We have recently carried out a similar set of experiments at fixed tilt angle of 45° and using $E_{\parallel}^{x}$ [13]. Genuine parallel excitation is possible only, if there is a breakdown of the perturbation description. Only if $\langle z \rangle_n$ is $H_{\parallel}$-dependent and the $\hat{z}$-motion of the carriers is modified by the field, can one expect to observe the resonances. The experiments [13] for $\hbar\omega = 10.5$ meV and fields to 9 Tesla have been carried out with various n- and p-type samples. The conclusion is that the $\Delta \ell = 0, \pm 1$ resonances can be identified for H > 4-5 Tesla. They increase in amplitude up to the highest fields. Except for the fact that the $\Delta \ell = -1$ peak has a larger relative strength for the n-type sample, the overall spectrum resembles that for $E_\perp$ excitation. The positions of all peaks observed are identical with those for $E_\perp$. This leaves room for the speculation that the coupling occurs in reality via the Hall component $E_\perp(H_\parallel)$ which should accompany the $E_\parallel^x$ applied field. To decide this question additional measurements with $E_\parallel^y$ need to be done.

6. Of Things to Come

Scanning the past few years work, the impression is one of "overall" agreement and by and large understanding of the role a magnetic field plays in electric subband spectroscopy. It is clear, however, that we have confined the discussion to the simplest case - parabolic, high symmetry subbands and a magnetic perturbation of electrically bound levels. There are alternative possibilities, and there are some loopholes in our understanding. Here are some things considered for future work.

- $H_\perp$ and electron condensation into the "highly correlated state". The application of a field, it has been suggested, would lead to various more highly ordered states of the electron gas. Valley condensation, charge-density waves and Wigner lattice would most likely affect the resonance E_{01}.

- $H_\perp$ and intersubband relaxation. It was a surprise not to find any effect at fields up to 7 T in (100) Si. Subband resonance linewidths are not adequately understood, and $H_\perp$ can help.

- $H_\perp$ and breakdown of the $k_\parallel$-selection rule. The field has the effect of replacing the subband parabolas in $k_\parallel$ by a ladder of discrete states. Phonon scattering or other transitions between the subbands become independent of $k_\parallel$. The effect has been utilized in the recent observation of the electro-phonon effect [18].

- $H_\perp$ and magnetically tunable subband resonance. If subbands are nonparabolic, then ω_c is dependent on subband index. It follows that $E_{01}(H_\perp) = E_{01} + 1/2\,\hbar(\omega_c^1 - \omega_c^0)$, for a transition between the lowest Landau levels of the n = 0 and 2 subbands. The related case of electrically tunable cyclotron resonance has been studied in InSb and InAs.

- $H_\perp$ and tilted ellipsoid geometries. Electric quantization is linked to carriers with $v_z = 0$. Magnetic quantization requires k_z = constant. The net effect depends on what is stronger.

- $H_\parallel$ and the electric perturbation limit. If $\hbar\omega_c > E_{01}$, electrically perturbed cyclotron resonance and magnetic surface level resonance will be observed. An indication of this behavior has been found for PbTe.

- $H_\parallel$ and the field effect mobility features. Additional work is necessary to understand the rather sizable shifts.

- ($H_{||}$, $H_{\perp}$) and the γ-shift. The remaining differences, although small, need to be resolved. The effect must be examined in other parabolic subband structures. For example Ge or InP.

Evidently, and this is a good note to end the story, magnetic fields will be a useful aid in subband spectroscopy for some time to come.

References

1. R.E. Doezema and F. Koch, Phys. Rev. B 5, 3866 (1972)
2. T. Ando, Phys. Rev. B 13, 1468 (1976)
3. F. Neppl, Ph.D. Thesis, Tech. Univ. München (1979)
4. T. Ando, J. Phys. Soc. Japan 39, 411 (1975)
5. T. Ando, J. Phys. Soc. Japan 44, 475 (1978)
6. T. Ando, Sol. State Commun. 21, 133 (1977); Z. Physik B 26, 263 (1977)
7. W. Beinvogl, Ph.D. Thesis, Tech. Univ. München (1977)
8. T. Ando, Z. Physik B 24, 33 (1976)
9. W. Beinvogl, Avid Kamgar, and F. Koch, Phys. Rev. B 14, 4274 (1976)
10. H.P. Grassl, Diplom Thesis, Tech. Univ. München (1975)
11. R.E. Doezema, M. Nealson, and S. Whitmore, Sol. State Commun., submitted for publication (1980)
12. D.C. Tsui, Sol. State Commun. 9, 1789 (1971)
13. T. Ando, Sol. State Commun. 21, 801 (1977)
14. T. Ando, Phys. Rev. B 19, 2106 (1979)
15. W. Beinvogl and F. Koch, Phys. Rev. Lett. 40, 1736 (1978)
16. B. Tausendfreund, Diplom Thesis, Tech. Univ. München (1979)
17. F. Stern and W.E. Howard, Phys. Rev. 163, 816 (1967)
18. S. Komiyama, H. Eyferth, and J.P. Kotthaus, Proc. of the 15th Int. Conf. on the Phys. of Semiconductors (Kyoto, 1980)

Electron Transport in Silicon Inversion Layers at High Magnetic Fields

Th. Englert

Max-Planck-Institut für Festkörperforschung
Hochfeldmagnetlabor Grenoble 166 X
38042 Grenoble-Cedex, France

1. Introduction

In the surface space charge layer of a MOSFET (Metal-Oxide-Semiconductor-Field-Effect-Transistor) a quasi-two-dimensional (2d) electron gas of variable carrier density is present. Because of its versatility this system has been studied both experimentally and theoretically in great detail during the last years. In the presence of a strong magnetic field perpendicular to the surface the system is fully quantized due to Landau and surface quantization. Oscillations of the Shubnikov-de Haas type (SdH) occur at low temperatures when either the magnetic field or the carrier density in the surface channel is varied. From the analysis of SdH oscillations, which were first observed by FOWLER et al. [1], the electronic properties of space charge layers on silicon of either p- or n-type and of different surface orientation were determined. Magnetic fields of the order 10 T, which are available in many laboratories, have proved to be very useful and often sufficient to study transport properties in 2d systems with a high carrier mobility. Higher magnetic fields, up to about 20 T, are necessary for the investigation of energy splittings not resolved in lower fields. In a so-called tilt field experiment, which was first employed by FANG et al. [2] for the determination of the effective spin splitting, such high fields allow measurements at both high tilt angles and sufficiently strong components of the magnetic field perpendicular to the surface. This type of experiment is one of the most important techniques to determine the magnitude of energy splittings in a 2d system from magnetotransport measurements.

The purpose of the present paper is to review some recent experimental work on silicon MOSFETs performed at the Hochfeld-Magnetlabor Grenoble. It will be restricted to transport measurements on n-type inversion layers.

2. (100) Surface Layers in a Tilted Magnetic Field: The Effective Spin Splitting

In a 2d system the period of SdH oscillations is only related to the degeneracy of a Landau level and not to the energy separation between levels. Rotating the sample in a magnetic field is an experimental method to determine energy splittings. Due to the 2d behavior of the carrier motion the Landau orbits are restricted to the plane parallel to the surface. Therefore, when the sample is tilted, only the component perpendicular to the surface, $B_\perp$, determines the Landau splitting $\hbar\omega_c$ ($\omega_c = eB/m_c$, where m_c is the cyclotron mass). The spin Zeeman splitting $g\mu_B \cdot B$ (μ_B: Bohr magneton) depends on the total magnetic field and can thus be varied independently). The tilted field method can be used to measure the effective spin splitting by comparing it to the Landau splitting, which is known either from cyclotron resonance experiments [3] or from the temperature dependence of the SdH amplitudes [4]. It was found previously in such experiments [2,5,6] that the effective g-factor is considerably enhanced over its bulk value, which has been explained theoretically in terms of exchange and correlation effects [7].

The first coincidence, $\hbar\omega_c \cdot \cos\phi = g\mu_B B$, ($\phi$ is the angle between the surface normal and the magnetic field), is determined when the Fermi level lies in the middle of a superimposed, broadened Landau level (N↑, N+1↓). The condition $gm_c/(2m_0 \cdot \cos\phi_0) = 1$ is assumed to be fulfilled when the conductivity peak σ_{xx}Max belonging to this level has a relative maximum. Previous experiments [2,5,6] were performed at constant total magnetic field. We have reinvestigated the spin splitting under the condition $B_\perp$ = const, which can be done by increasing the total magnetic field by $1/\cos\phi$ for each angle ϕ. This technique has the advantage that the oscillations do not shift on the gate voltage scale. Effects due to the dependence of the linewidth on the carrier density and the variation of energy splittings do not play a role. It is sufficient to make the following assumptions, which can be justified in the case of electrons at a silicon surface: a) the carrier motion is strictly 2-dimensional, b) the coupling between the orbital motion and the spin is negligible, c) the valley splitting (see section 4) and the linewidth Γ do not depend on the parallel component of the magnetic field.

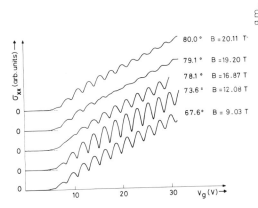

Fig.1 SdH oscillations for a circular device at different tilt angles. The perpendicular component of the magnetic field is $B_\perp \cong 3.5$ T for all curves.

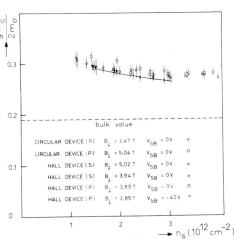

Fig.2. $gm_c/2m_0$ values as a function of the surface carrier concentration

Figure 1 shows several recordings of the conductivity σ_{xx} obtained on a Corbino type MOSFET at different tilt angles. A perpendicular field of about $B_\perp = 3.5$ T was chosen in order to allow tilt angles as high as 80° for a total field of 20 T. The exact tilt angle was calculated from the period of the oscillations. At $\phi = 78.1°$ the peak conductivity of the oscillations is smaller than at $\phi = 73.6°$. At small quantum numbers ($\phi=78.1°$) a splitting is resolved which indicates that the spin splitting is larger than the Landau splitting, $1 < gm_c/(2m_0 \cos\phi) < 3/2$. The data obtained at $\phi = 79.1°$ demonstrate the variation of the spin splitting with carrier density. At low quantum numbers spin levels differing by 2 in the Landau quantum number have already merged. At gate voltages around $V_g=20V$ the condition $gm_c/(2m_0 \cos\phi)$

= 3/2 is fulfilled whereas at higher gate voltages the phase of the oscillations is that of the first coincidence. At $\phi = 80.0°$ the condition for the second coincidence is almost fulfilled. When σ_{xx} is plotted as a function of the tilt angle a maximum occurs at angles around 73° depending on the Landau quantum number [8]. It should be noted that the valley splitting was not resolved in this experiment and considered as an additional broadening of the levels. For the Landau levels investigated in detail no change in phase as a function of carrier density was observed. Such a phase change is thought to be due to the valley splitting [9].

Similar experiments were performed with samples of the Hall type. In this case ρ_{xx}Max instead of σ_{xx}Max is measured at various tilt angles. Since the Hall component ρ_{xy} is constant for $B_\perp$ = const (apart from an oscillatory part which can be neglected), both ρ_{xx}Max and σ_{xx}Max have a maximum at the angle of coincidence ϕ_0. Data obtained from the first coincidence with samples of different source, different geometry and at different substrate bias voltages are summarized in Fig.2. If a cyclotron mass of $m_c = 0.2\ m_0$ is taken [3] the effective g-values are around g = 2.75 with a slight tendency to increase at low carrier concentrations in agreement with previous results [2,6]. It should be noted that a measurement of the second coincidence, $(gm_c/(2m_0 \cdot \cos\phi) = 2)$, yields the same g-values.

3. Electrons on a (111) Surface

Silicon has six conduction band minima and a valley degeneracy factor of $\gamma = 6$ is expected for electrons in a surface space charge layer on a (111) surface. In fact, such a 6-fold degeneracy has been observed by TSUI et al. [10] on MOSFETs prepared with high temperature preoxidation annealing in Ar ambient. Prior to this work only $\gamma = 2$ was observed in n-channel MOSFETs fabricated on the (111) surface [11-14]. A 2-fold degeneracy was also observed on (110) surfaces and on (100) samples under conditions where $\gamma = 4$ was expected [11,14-17]. Three different explanations have been offered for this apparent discrepancy. The first (T.K.) [18] invokes the existence of strain domains at the Si-SiO$_2$ interface. A large amount of built-in strain is required in this model for a (111) surface to lower two of the six valleys sufficiently in energy, so that the other valleys cannot be occupied. The second model (K.F.) [19] invokes electron-electron intervalley interaction which leads to a symmetry breaking charge-density-wave (CDW) ground state with $\gamma = 2$. The third model (V.O.) [20] is based on a reinterpretation of SdH results under the assumption of a narrow intrinsic linewidth. A relatively small amount of strain randomly distributed in the sample is required to make the six valleys inequivalent and give rise to a mimic degeneracy of 2 in SdH oscillations.

We have studied the electronic properties of both 6-fold and 2-fold samples in some detail and compared the results. At B = 0 the low temperatures there is a large difference in the effective electron mobility between the two types of samples. The 6-fold samples have a maximum mobility typically of the order 1000 cm^2/Vs which is about a factor 3 less than that of good 2-fold samples. It has been suggested [10] that this results from additional intervalley scattering. At low temperatures intervalley scattering is only possible through electron-electron interaction or through a lateral perturbation of the translational symmetry which can be caused by surface roughness. It is not clear at present whether the difference in mobility is significant for the valley degeneracy problem.

The valley degeneracy factor is determined from SdH oscillations by making use of two properties of a 2d system. The carrier density can be determined from the gate voltage if the capacitance of the MOS device is known. It can independently be checked from the Hall effect. The orbital degeneracy of a Landau level is given by $e/h \cdot B$. When the conductance oscillations are observed at constant magnetic field the total degeneracy of a particular peak is obtained by simply measuring the amount of gate voltage, ΔV, necessary to fill this level. However, this total degeneracy may include a factor 2 for the spin. In order to get information about the valley degeneracy it is necessary to distinguish spin splitting from Landau splitting. This can be done by rotating the sample in the magnetic field.

Figure 3 shows experimental results for a circular sample in a constant magnetic field of $B = 21$ T perpendicular to the surface. Plotted are the conductivity and the derivative $d\sigma/dV_g$ as a function of the gate voltage. The large period of the oscillations has a 6-fold (total) degeneracy. The additional splitting of the first peak will be discussed in section 5.

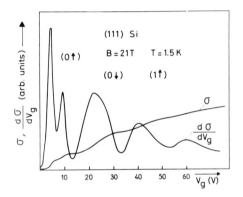

Fig.3 Conductivity and its derivative $d\sigma/dV_g$ for a (111) Si-MOSFET with a six-fold valley degeneracy. The oxide capacitance per unit area per unit charge is $1.65 \times 10^{11}/V cm^2$.

In Fig.4 σ and $d\sigma/dV_g$ are shown at two different tilt angles. In both cases the perpendicular component of the magnetic field is $B_\perp = 11.8$ T, which implies that the Landau splitting is always the same. In the upper curve the spin splitting was increased by a factor $1/\cos 55°$. As a result the period of the oscillations has doubled. The total degeneracy of each peak is 12 except for the lowest one, which is in agreement with a Landau level scheme in case of $gm_c/(2m_0 \cos\phi) = 1$: (0↑); (0↓ 1↑); (1↓ 2↑); etc. Due to the low mobility neither the accurate angle of coincidence nor the cyclotron mass could be determined. However, an angle of $\phi \sim 55°$ corresponds to $gm_c/2m_0 \sim 0.57$ which is compatible with the effective g-factors on (100) surfaces and a cyclotron mass of $m_c = 0.4\ m_0$ [11].

For comparison the same type of experiment was performed with 2-fold samples. Fig.5 shows some experimental results for ρ_{xx} obtained on a sample of the Hall type. The magnetic field component perpendicular to the surface was kept constant ($B_\perp = 10$ T), when the sample was rotated from 0° to 49° and 59°. The splitting of the main oscillation peaks which is clearly resolved at $\phi = 0°$ disappears when the total magnetic field is increased ($\phi = 49°$) and appears again at higher tilt angles ($\phi = 59°$). Since in this type of experiment the spin splitting increases and the Landau splitting remains constant, the experimental results strongly suggest that the split peaks observed at

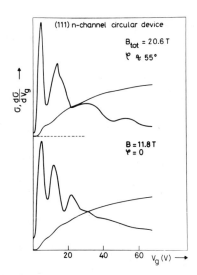
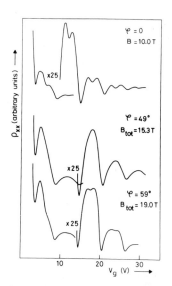

Fig.4 Conductivity versus gate voltage for a (111) MOSFET with a 6-fold valley degeneracy. The maximum mobility is 1300 cm^2/Vs.

Fig.5 Resistivity versus gate voltage for a (111) MOSFET with a 2-fold valley degeneracy. The maximum mobility is 3000 cm^2/Vs. The thickness of the oxide is 1200 Å.

$\phi = 0°$ belong to spin split Landau levels of different Landau quantum number. The coincidence occurs in the same range of angles as for the 6-fold samples in agreement with former results [11]. The large period of oscillations in Fig.5 corresponds to a total degeneracy of 4 including a factor 2 for the spin. Since the ratio of effective spin splitting to Landau splitting is close to one in case of $\phi \sim 50°$, the first peak (0↑) should be 2-fold degenerate whereas the subsequent levels (0↓; 1↑); etc. should have a total degeneracy of 4. This is experimentally observed.

In the model by VINTER et al. [20] a small built-in strain splits the six conduction band ellipsoids into 3 pairs of valleys (a,b,c). In a magnetic field there are three different sets of Landau levels. When the intrinsic level broadening is sufficiently small it is not possible to distinguish between the N-th level of valley system a and the N'-th level of system b or c, because in a 2d system only the degeneracy of levels and not their energetic separation is measured. In such a case SdH oscillations exhibit a mimic degeneracy of 2. Principally, the model is attractive because only a small stress induced splitting is necessary which cannot be excluded in real devices and it provides a simple explanation for a variety of phenomena not restricted to the (111) surface. There is, however, apart from the required narrow linewidth, at least one other argument against it. When the ratio of spin to Landau splitting is taken into account a series of superimposed levels with degeneracies of 2,4,4,... is expected for each valley pair a,b or c. Therefore more than one 2-fold peak should occur when the conductance oscillations are observed at constant magnetic field as a function of V_g. However, only one such peak is observed experimentally, which implies that the stress induced splitting must be so large that valleys b and c are not occupied. In this case the V.O. model is equivalent to the explanation proposed by TSUI et al. [18]. Moreover, the simulations shown in [20] demonstrate that, although the oscillatory structure is dominated by a period reflecting a 2-fold degeneracy, a larger period corresponding to $\gamma = 6$ should also be present under realistic assumptions. This is not observed.

As mentioned above three models have been proposed for an explanation of the long-standing valley degeneracy problem. The third model (V.O.) is equivalent to the first (T.K.) in the limit of high strain. In the second model (K.F.) the strong intervalley phonon exchange interaction is problematic. The fact that samples with a 6-fold degeneracy exist also seems to indicate that there is no need for such a sophisticated model. On the other hand, if additional intervalley scattering caused by a lateral perturbation of the effective potential is responsible for the mobility in 6-fold samples, this might destroy possible electron-electron interaction effects.

4. An Investigation of the Valley Splitting in (100) Inversion Layers

For (100) inversion layers the lowest subband arises from the two conduction band valleys lying along the [100] direction in k-space. The lifting of the degeneracy of these two valleys was first observed by FOWLER et al. [1]. Each level was found to result in four conductivity peaks due to the lifting of spin and valley degeneracies. Theoretical descriptions of the mechanism of valley splitting have been proposed by OHKAWA et al. [21,22] and SHAM et al. [23,24] and by KÜMMEL [25]. OHKAWA et al. proposed the electric breakthrough model where the splitting arises from the overlap of wave functions in k-space at the Γ-point. Using a different physical picture SHAM et al. developed the so-called surface scattering model which later turned out to be essentially equivalent to the OHKAWA model [26]. KÜMMEL has proposed a different mechanism in which the valley splitting arises from spin-orbit interaction. Recent calculations by CAMPO [27] show that this effect is only important if an additional perturbation of the crystal is assumed in a finite transition region from Si to SiO_2. In all theories the valley splitting ΔE_v depends to first order linearly on the charge density at the surface:

$$\Delta E_v = \alpha \cdot (N_{inv} + \gamma^{-1} N_{depl}) \tag{1}$$

where the prefactor α has values between 0.15 and 0.33 meV/10^{12}cm^{-2} [21,24] and γ^{-1} is 2 or 3 depending on the trial wave function. There will in addition to this be a strong enhancement of the valley splitting due to exchange and screening [21] in a similar way to that found for the spin splitting.

The magnitude of the valley splitting was estimated from a cusp-like structure in experimental data by KAWAJI [21] which arises from an overlap of valley split levels of the same quantum number. KOHLER et al. [9] deduced the valley splitting from a change in phase of SdH oscillations in a tilted magnetic field. In both cases the valley splitting is not directly observable and such measurements have the disadvantage that they depend critically on the lineshape of the density of states. In order to avoid these difficulties we have determined the valley splitting from the condition $\Delta E_v = \Delta E_s/2$ (2), where ΔE_s is the spin splitting, in magnetic fields such that the valley splitting was always resolved [29]. This condition was assumed to result in a lineshape consisting of four equally well resolved peaks. It can be shown that this condition does not critically depend on the lineshape assumed. The variation of the Fermi level is not significant because of the relatively small modulation of the peak values within the same level observed experimentally.

In order to find condition (2) it is necessary to vary either ΔE_v or ΔE_s without changing too many other parameters. This is possible at a fixed inversion layer concentration and magnetic field by altering ΔE_v through a variation of the depletion charge N_{depl}. The spin splitting ΔE_s can be changed by tilting the sample under the condition that $B_\perp$ remains constant. In this the assumption is made that the valley splitting does not depend on the component of the magnetic field parallel to the surface. Experimental evidence for this comes from observing the depth of a valley split minimum as a function of the total field with $B_\perp$ = const. No change was observed.

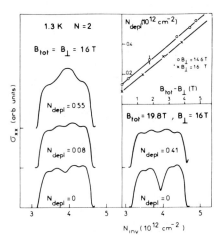

Fig.6 The conductivity (σ_{xx}) variation for the N = 2 Landau level as a function of inversion layer concentration (N_{inv}). Curves are shown for different values of depletion charge (N_{depl}) in units of $10^{12}cm^{-2}$, with the perpendicular component of magnetic field ($B_\perp$) constant and two different values of total magnetic field (B_{tot}) corresponding to tilt angles of 0° and 36°. The inset shows a plot of the depletion charge concentration necessary for equal resolution of the spin and valley splitting, as measured by ($B_{tot} - B_\perp$), for constant perpendicular fields of 14.6 and 16 T.

The influence of substrate bias on the Landau level N = 2 where the levels are clearly not far from equidistant is shown in Fig.6. At $N_{depl} = 0.08 \cdot 10^{12}$ cm^{-2} and B = 16 T condition (2) is obeyed and four equally well resolved peaks are observed. The magnetic field was then rotated while keeping $B_\perp$ constant and the increase in N_{depl} necessary to achieve condition (2) was measured. N_{depl} was determined from the shift in gate voltage of the oscillation from that observed when the substrate was forward biassed into quasi-accumulation. It is shown in Fig.6 that there is a linear dependence of the increase in spin splitting (proportional to $B_{total} - B_\perp$) on the depletion charge necessary for equal spacing of the levels. The result may be expressed as

$$\Delta E_v(N_{depl}) = \Delta E_v(o) + 0.33 \cdot g \cdot N_{depl} \cdot meV \qquad (3)$$

where N_{depl} is in units of $10^{12}cm^{-2}$, and g is the effective g-factor. From section 2 an average value of g = 2.75 can be used. It should be noted that many body effects may enhance the valley splitting and that this enhancement may vary with the position of the Fermi level. It is therefore concluded that an average valley splitting $\overline{\Delta E_v}$ is measured in our experiment which depends on N_{depl} with a factor given by

$$\Delta E_v(N_{depl}) = \Delta E_v(o) + (0.90 \pm 0.15) N_{depl} \, meV \qquad (4)$$

The prefactor is larger than predicted by the theories for the bare valley splitting. The origin of this descrepancy may lie in enhancement effects. It should be mentioned that the presence of a magnetic field is not included implicitly in the theories and since there seems to be a small dependence of the effective splitting on the perpendicular magnetic field this may well be the origin of this difference. Equation (4) may be used to make the small correction necessary to deduce the absolute magnitude of the valley splitting at $N_{dep1} = 0$. If we write $\Delta E_v = c \cdot N_{inv}$ then c is 0.29 meV at 14.6 T and 0.30 meV at 16 T, again larger than the theoretically calculated bare splittings.

5. Valley Splitting in (111) and (110) Surfaces

In sufficiently strong magnetic fields we observed an additional splitting of conductivity peaks in SdH oscillations on (111) silicon which we attribute to valley splitting analogous to the effect in (100) surfaces. The occurrence of valley splitting in (111) surfaces has some theoretical implications because the models currently accepted as an explanation in case of (100) samples cannot be applied. The reason for this is the 2-d Brillouin zone for a (111) surface. The valley configuration can be visualized as a projection of the six conduction band valleys onto a (111) surface. States belonging to different, energetically equivalent valleys have different 2-d k-vectors $k_\parallel$. In the model by OHKAWA et al. [21] e.g., the valley splitting in (100) surfaces arises from an overlap of wave functions with the same $k_\parallel$ belonging to different valleys at $+k_z$, $-k_z$.

We have studied (111) samples of the 6-fold type and 2-fold type and in both cases we have observed the additional splitting when the magnetic field was sufficiently strong [30]. In Fig.3 data obtained on a (111) circular MOSFETs are shown. Both Landau and spin splittings are resolved experimentally. The assignment of the conductivity peaks was determined in a tilted field experiment and is discussed in part 3. Each spin split level has a six-fold valley degeneracy. The first peak which corresponds to the $N = 0$ Landau level shows a distinct additional splitting. A rough estimate of the magnitude of this valley splitting based on the argument that it has to be larger than the linewidth yields $\Delta E_v \sim 3$ meV at $B = 21$ T if the SCBA (Self-Consistent Born Approximation) linewidth is taken.

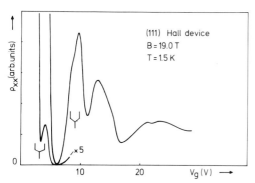

Fig.7 Resistivity versus gate voltage for a (111) MOSFET with a 2-fold valley degeneracy. The additional valley splitting is indicated.

A similar splitting was observed on 2-fold samples provided by different manufacturers. Fig.7 shows an example for a Hall type sample where ρ_{xx} is plotted as a function of V_g at $B = 19.0$ T. Splittings are resolved for the first (0↑) and second (0↓) peaks as indicated in the figure. A weak shoulder is present in the (1↑) peak. For these samples all possible degeneracies are lifted ($\gamma = 1$) as it is the case for a (100) surface. It should be mentioned that this valley splitting could only be observed in magnetic fields above about 10 T depending on the mobility. Therefore, the methods described in part 4 were not applicable for the determination of the magnitude of the splitting.

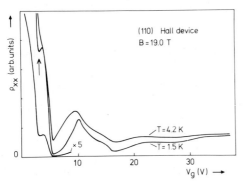

Fig.8 Resistivity versus gate voltage for a (110) MOSFET. The valley splitting is indicated by the arrow.

Recently we have observed valley splitting on (110) surfaces, too. Fig.8 shows experimental results in a magnetic field of 19.0 T at two different temperatures. The splitting is indicated by an arrow. As in the case of (111) surfaces this splitting was only resolved in rather strong magnetic fields ($B \geq 12$ T).

Summarizing the results obtained with n-type inversion layers on surfaces of the main symmetry orientations one may say that there is a tendency to lift all possible degeneracies in sufficiently strong magnetic fields. Except for the 6-fold (111) samples the resulting valley degeneracy factor is always $\gamma = 1$, at least in the range of relatively low carrier concentrations. It is not clear at present whether the higher apparent degeneracy ($\gamma = 3$) in 6-fold (111) samples is due to the low mobility or whether the valley splitting in this case originates from a different mechanism.

As already mentioned the currently accepted models for the valley splitting in (100) surfaces are not applicable for a (111) surface layer. Recently BLOSS et al. [31] have proposed that intravalley exchange and correlation lowers the valley degeneracy, a model which is principally not restricted to a particular surface. In their theory many body effects are the dominant mechanism and no "bare" splitting is required. This might be an adequate general explanation for the valley splitting observed experimentally on different surfaces. For particular surface orientations, however, like the (100) surface, other mechanisms may also be important.

Acknowledgements

The work presented here was largely based on a collaborative effort. I want to thank G.Dorda, K.von Klitzing, G.Landwehr, R.J.Nicholas, M.Pepper,

D. C. Tsui and Ch.Uihlein for valuable experimental contributions and stimulating discussions. The technical assistance of A.Klaschka during the experiments is acknowledged.

References

1. A. B. Fowler, F.F.Fang, W.E.Howard and P.J.Stiles, Phys. Rev.Lett. 16, 901 (1966).
2. F.F.Fang and P.J.Stiles, Phys.Rev. 174, 823 (1968).
3. G.Abstreiter, J.P.Kotthaus, J.F.Koch and G.Dorda, Phys.Rev. B14, 2480 (1976).
4. J.L.Smith and P.J.Stiles, Phys.Rev.Lett. 29, 102 (1972).
5. M.Kobayashi and K.F.Komatsubara, Jap.J.Appl.Phys.Suppl. 2, Pt 2, 343 (1974).
6. G.Landwehr, E.Bangert and K.von Klitzing, In: Physique Sous Champs Magnetiques Intenses, Grenoble 1974, Colloques Internationaux du C.N.R.S., No. 242, P.177 (1975).
7. T.Ando and Y.Uemura, J.Phys.Soc.Japan 37, 1044 (1974).
8. Th.Englert, K.von Klitzing, R.J.Nicholas, G.Landwehr, G.Dorda and M.Pepper, Phys. Stat. Sol. (b) 99, 237 (1980).
9. H.Köhler and M.Roos, Phys. Stat. Sol. (b) 91, 223 (1979).
10. D.C.Tsui and G.Kaminsky, Phys.Rev.Lett. 42, 595 (1979).
11. T.Neugebauer, K.von Klitzing, G.Landwehr and G.Dorda, Solid State Commun. 17, 295 (1975).
12. A.Lakhani and P.J.Stiles, Phys.Lett. 51A, 117 (1975).
13. G.Dorda, H.Gesch and I.Eisele, Solid State Commun. 20, 429 (1976).
14. G.Dorda, I.Eisele and H.Gesch, Phys.Rev. B17, 1785 (1978).
15. Th.Englert, G.Landwehr, K.von Klitzing, G.Dorda and H.Gesch, Phys.Rev. B18, 794 (1978).
16. K.von Klitzing, Th.Englert, G.Landwehr and G.Dorda, Solid State Commun. 24, 703 (1977).
17. J.Wakabayashi, K.Hatanaka and S.Kawaji, Physics of Semiconductors 1978, edited by B.L.H.Wilson, Conference Series 43, The Institute of Physics, Bristol and London, p. 1251.
18. D.C.Tsui and G.Kaminsky, Solid State Commun. 20, 93 (1976).
19. M.J.Kelly and L.M.Falicov, Phys.Rev.Lett. 37, 1021 (1976); Phys.Rev. B15, 1974 (1977).
20. B.Vinter and A.W.Overhauser, Phys.Rev.Lett. 44, 47 (1980).
21. F.J.Ohkawa and Y.Uemura, J.Phys.Soc.Japan 43, 907, 917, 925 (1977).
22. F.J.Ohkawa, Solid State Commun. 26, 69 (1978).
23. L.J.Sham and M.Nakayama, Surf.Sci. 73, 272 (1978).
24. L.J.Sham and M.Nakayama, Phys.Rev. B20, 734 (1979).
25. R.Kümmel, Z.Phys. B22, 223 (1975).
26. T. Ando, Surf.Sci., in press.
27. A. Campo, Thesis, Würzburg (1980).
28. H.Rauh and R.Kümmel, to be published in Solid State Commun.
29. R.J.Nicholas, K.von Klitzing and Th.Englert, Solid State Commun. 34, 51 (1980).
30. Th.Englert, D.C.Tsui and G.Landwehr, Solid State Commun. 33, 1167 (1980).
31. W.L.Bloss, L.Sham and B.Vinter, Phys.Rev.Lett. 43, 1529 (1979).

Temperature Dependence of Transverse and Hall Conductivities of Silicon MOS Inversion Layers under Strong Magnetic Fields

S. Kawaji, and J. Wakabayashi

Department of Physics, Gaskushuin University, Mejiro, Toshima-ku
Tokyo, Japan

An n-channel MOS inversion layer on Si (100) surface is a typical two-dimensional electron system. When a strong magnetic field is applied along a direction normal to the Si-SiO$_2$ interface, the continuum of electron energy levels in two-dimensional motion coalesces into a series of perfectly quantized Landau levels [1]. In such a system in an extreme-quantum-limit condition, there exist gap regions in the density of states of electrons between the boundaries of each Landau level.

In 1975 we observed finite gap regions of the gate voltage V_G where the magnetoconductivity σ_{xx} vanishes for weak source-drain fields E_{sd} in H = 140 kOe at T = 1.4 K [2]. The vanishing conductivity ($\sigma_{xx} < 10^{-9}$ mho) region suggests the possibility of the existence of immobile electrons at the edge of each Landau level in 2D systems. The behaviour of the lowest Landau level is particularly interesting because the σ_{xx} peak increases with increasing T or with increasing E_{sd} at low T while σ_{xx} peaks in other Landau levels decrease. As a possible explanation of the immobile electrons at low T and weak E_{sd}, we have proposed pinning of the Wigner solid or Wigner glass due to impurities [2,3,4].

There exist several experiments for such systems which suggest to have observed highly correlated states [5,---8] or the existence of a mobility edge [9,10] in strong magnetic fields. There also exist theoretical investigations which indicate that the existence of the crystalline state or the CDW state is enhanced by a strong magnetic field [11,---16]. The existence of a mobility edge [17] and the existence of an amorphous state or a Wigner glass state in the intermediate case [18] in a strong magnetic field have also been investigated.

However, experiments so far performed for the localization by DC conductivity measurements have been limited to σ_{xx} only until recently. Without Hall conductivity σ_{xy} data, effects of the electron concentration N_S and the width of the Landau level Γ or relaxation time τ cannot be discussed separately in transport properties [19,---23]. We have introduced the Hall current method and obtained first successful results for σ_{xy} as well as for σ_{xx} [24,25]. In the following, some results of our recently performed investigations in the localization regime in strong magnetic fields will be described.

Figure 1 shows gate-voltage dependences of σ_{xx} and σ_{xy} in the lowest four Landau levels, (0↑+), (0↑-), (0↓+) and (0↓-) level, in 150 kOe and at several temperatures. The sample, an n-channel Si-MOSFET fabricated on a

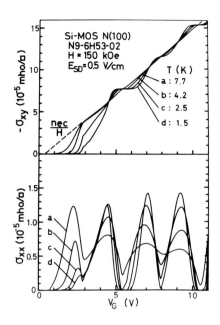

Fig. 1 Gate-voltage dependences of σ_{xx} and σ_{xy}

(100) surface of a p-type wafer, has a peak mobility of μ_{peak} = 14,000 cm^2/V.s at 1.5 K and the minimum density of electrons for metallic conduction at H = 0, N_{min} = 2.0 x 10^{11}/cm^2.

The flat part in σ_{xy} at 1.5 K in the spin gap between the (0↑-) level and the (0↓+) level shows clearly the existence of localized electrons in the upper edge of the (0↑-) level and in the lower edge of the (0↓+) level as predicted theoretically by Ando, Matsumoto and Uemura [19]. The theory has shown that localized states in the upper edge of a Landau level are produced by repulsive scatterers and localized states in the lower edge of a Landau level are produced by attractive scatterers, and that when the Fermi level lies in energy gap between the adjacent N-th and N+1-th Landau level, σ_{xy} becomes - $N_s$ec/H = -e^2(N + 1)/2π$\hbar$ if the localized states associated with these Landau levels are well separated from the edge of main Landau levels. This means that - σ_{xy} exceeds $N_s$ec/H when the Fermi level lies above the edge of the main Landau level until the upper localized states produced by repulsive scatterers are completely filled with electrons. When σ_{xy} becomes flat, σ_{xx} becomes zero because added electrons by the increase in the gate fields are immobile. Similar behaviour of σ_{xy} and σ_{xx} is observed in the gap regions between the (0↓+) level and the (0↓-) level as well as between the (0↓-) level and the (1↑+) level.

The behaviour of σ_{xx} and σ_{xy} in the lowest Landau level and in the gap region between the lowest two Landau levels is different from the one in higher Landau levels and in higher gap regions. Electrons in the lowest Landau level look like to be immobilized completely at T → 0, i.e. σ_{xx} = 0 as well as σ_{xy} = 0 at absolute zero.

We evaluate the density of conduction electrons N_C by the use of

$$\frac{N_c ec}{H} = -\frac{\sigma_{xx}^2 + \sigma_{xy}^2}{\sigma_{xy}} . \qquad (1)$$

Figure 2 shows temperature dependence of N_C. We note here that the saturation value of N_C agrees with N_s, the gate-voltage induced electron density determined by the plateaus in σ_{xy} at 1.5 K and sharp minima in σ_{xx} at higher temperatures, at V_G > 2V. The temperature dependence of N_C cannot be described by a simple activation type expression as shown in Fig. 3. Therefore, the present electron localization in strong magnetic fields

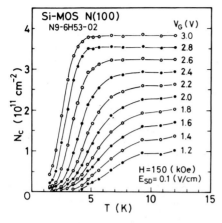
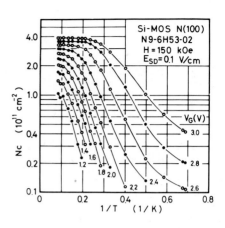

Fig. 2 Temperature dependence of N_C

Fig. 3 Temperature dependence of N_C

can not be explained by the mobility edge model.

When we define a characteristic temperature T_S as $N_C(T_S) = N_C$(saturation)/2, T_S changes as shown in Fig. 4.

A Hartree-Fock calculation has shown that a CDW instability occurs at temperatures well above the classical Wigner solid in two dimensions in a strong magnetic field [15]. A calculation of the quantum effects has also shown that the region of densities and temperatures is very limited for the existence of the Wigner solid even in strong magnetic fields [16]. Our results of the conductivities are in accord with a theoretically predicted characteristic feature of pinned CDW state, i.e. $\sigma_{xx} = \sigma_{xy} = 0$ at $T = 0$ [26]. Recent observations of an extra absorption peak in CR experiments [8] are also in accord with the theoretical prediction [26]. However, the CDW transition temperature calculated in the framework of the Hartree-Fock theory has a peak at N_S for a half-filled Landau level [14,16]. Our results shown in Fig. 4 do not agree with the theoretical prediction.

Further experimental and theoretical investigations including other possibilities will be necessary to make clear the present problem.

The authors thank Professor S. Tanuma and Mr. I. Oguro, ISSP, for their extending to us the use of high magnetic field facilities, and Mr. A. Yagi, Sony Corporation, for his providing us with samples. A part of this work is supported by the Grants-in-Aids for Fundamental Scientific Research from the Ministry of Education.

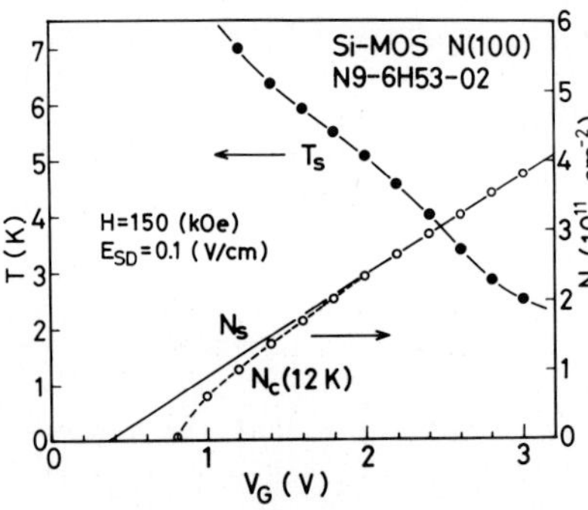

Fig. 4 Gate voltage dependences of T_S and N_C at 12 K.

Note Added

Quite recently v. Klitzing, Dorda and Pepper [27] have reported results of high-precision measurements of a plateau in the Hall resistance ($R_H = h/4e^2$) of a Si MOSFET in a strong magnetic field and its application for determinimg the fine structure constant α.

References

1. Y. Uemura, Proc. 2nd Int. Conf. on Solid Surfaces, Kyoto, 1974; Japan. J. Appl. Phys. Suppl. 2 (1974) 17.
2. S. Kawaji and J. Wakabayashi, Surface Sci. 58 (1976) 238.
3. S. Kawaji and J. Wakabayashi, Solid State Commun. 22 (1977) 87.
4. S. Kawaji, J. Wakabayashi, Y. Namiki, and K. Kusuda, Surface Sci. 73 (1978) 121.
5. D. C. Tsui, Solid State Commun. 21 (1977) 675.
6. T. A. Kennedey, R. J. Wagner, B. D. Mc Combe, and D. C, Tsui, Solid State Commun. 21 (1977) 459.
7. D. C. Tsui, H. L. Störmer, A. C. Gossard, and W. Wiegmann, Proc. Yamada Conf. II on Elecronic Properties of 2D Systems, Lake Yamanaka, 1979, Surface Sci. 98 (1980).
8. B. A. Wilson, S. J. Allen, and D. C. Tsui, Proc. Yamada Conf. II on Electronic Properties of 2D Systems. Lake Yamanaka, 1979, Surface Sci. 98 (1980), and Phys. Rev. Letters 44 (1980) 479.
9. R. J. Nicholas, R. A. Stradling, S. Askenazy, P. Perrier, and J. C. Portal, Surface Sci. 73 (1978) 106.
10. M. Pepper, Phil. Mag. B 37 (1978) 83.
11. Yu. E. Lozovik, and V. I. Yucson, Pisma ZH. Eksp. Theor. Fiz. 22, 26 (1975) [Soviet Physics-JETP Letters 22 (1975) 11].
12. H. Fukuyama, Solid State Commun. 19 (1976) 551.
13. M. Tsukada, J. Phys. Soc. Japan 41 (1976) 1466.
14. H. Fukuyama, P. M. Platzman, and P. W. Anderson, Phys. Rev. B 19 (1979) 5211.
15. D. Yoshioka, and H. Fukuyama, Proc. Yamada Conf. II on Electronic Properties of 2D Systems, Lake Yamanaka, 1979, Surface Sci. 98 (1980).
16. H. Fukuyama, D. Yoshioka, J. Phys. Soc. Japan 48 (1980) 1853.
17. H. Aoki and H. Kamimura, Solid State Commun. 21 (1977) 45.
18. H. Aoki, J. Phys. C 12 (1978) 633.
19. T. Ando, Y. Matsumoto, and Y. Uemura, J. Phys. Soc. Japan 39 (1975) 278.
20. T. Igarashi, J. Wakabayashi, and S. Kawaji, j. Phys. Soc. Japan 38 (1975) 1549.
21. S. Kawaji, T. Igarashi and J. Wakabayashi, Progr. Theoret. Phys. (Kyoto) Suppl. 57 (1975) 176.
22. S. Kawaji, Surface Sci. 73 (1978) 46.
23. J. Wakabayashi and S. Kawaji, J. Phys. Soc. Japan 44 (1978) 1839.
24. J. Wakabayashi and S. Kawaji, Proc. Yamada Conf. II on Electronic Properties of 2D Systems, Lake Yamanaka, 1979; Surface Sci. 98 (1980).
25. J. Wakabayashi and S. Kawaji, J. Phys. Soc. Japan 48 (1980) 333.
26. H. Fukuyama, and P. A. Lee, Phys. Rev. B 18 (1978) 6245.
27. K. v. Klitzing, G. Dorda and M. Pepper, Phys. Rev. Letters 45 (1980) 449.

Two-Dimensional Charge Density Wave State in a Strong Magnetic Field

D. Yoshioka, and H. Fukuyama

The Institute for Solid State Physics, The University of Tokyo
Roppongi, Minato-ku
Tokyo 106, Japan

1. Introduction

Two-dimensional electron systems are realized at the interface of a metal-oxide-semiconductor (MOS) structure and on the free surface of liquid helium. Due to the Coulomb repulsive force between electrons these systems have a possibility to form a Wigner lattice or charge density wave (CDW) state.
In the electron system on the liquid helium the formation of a Wigner lattice has actually been observed recently[1]. In this system the density of electrons is low ($\simeq 10^8 \text{cm}^{-2}$), and the observed transition temperature agrees closely with the theory of dislocation mediated melting of classical electrons[2]. On the other hand, the density of electrons is high ($\geq 10^{12}\text{cm}^{-2}$) in the case of MOS, and the system cannot be treated as classical. In such cases the formation of the Wigner lattice is difficult due to quantum fluctuations[3]. It is possible, however, to suppress these quantum fluctuations by the application of strong magnetic fields, and thus to encourage the localization of electrons[4-7]. Several experiments have already been reported[8-11] which prefer such interpretation based on strongly correlated electronic states. The localization under such circumstances can not be so strong to be viewed as a Wigner lattice. It is more appropriate to treat such a state as a CDW state. In this paper we investigate the phase diagram of the CDW state in a strong magnetic field by use of the Hartree-Fock approximation.
In (100)Si-MOS structure, which is of interest to us here, two degenerate valleys out of six have lower energy, and only these two valleys are occupied in ordinary situations. In the presence of the magnetic field the energy spectrum of each valley splits into discrete Landau levels with the energy of $(N+1/2)\hbar\omega_c$, where $N=0,1,2\cdots$, and $\omega_c (=eH/mc)$ is the cyclotron energy. Each of these Landau levels has a degeneracy of $S/2\pi\ell^2$ ($=eHS/2\pi c$), where S and ℓ are area of the system and the Larmor radius, respectively. Previous theories of the CDW state in a strong magnetic field neglected the presence of the two valleys and took into account only the lowest (N=0) Landau level (hence $\nu \equiv 2\pi\ell^2 n$ is less than one, n being the eletron density)[12-15]. In this paper we examine the effect of the valley degeneracy, higher Landau levels and impurity scattering. We compare the results with those of the lowest Landau level in a single valley.
We adopt in this paper the following approximations. Since it is impossible to take into account all Landau levels in the calculation, we consider only two Landau levels. When we investigate the effect of the valley degeneracy, we take account of two N=0 Landau levels from the two valleys, and when we consider the effect of the higher Landau levels, we take into account N=0 and N=1 Landau levels from one of the valleys. The transition temperature to the CDW state is determined by the mean field theory assuming a second

order phase transition. The effect of impurities is considered in the self-consistent Born approximation (SCBA). The inter-valley scattering is neglected, since the matrix element for the scattering is smaller than that of intravalley scattering due to orthogonality of the different valleys. The small valley splitting due to the interface scattering effect[16,17] is neglected, but the valley splitting due to the many-body effects[18,19] is taken into account by the Hartree-Fock approximation (HFA) which is consistent with the mean field theory for the CDW transition. We do not consider the spin degree of freedom, since the degeneracy is lifted in the magnetic field.

2. The Effect of Valley Degeneracy

In this section we consider only two N=0 Landau levels. First we must determine the distribution of electrons in the two valleys assuming that the system is not in the CDW state. If we denote the density of electrons in N=0 Landau levels of the σ-valley (σ=±) as $n_\sigma = \nu_\sigma/2\pi\ell^2$, the self-consistency equation for ν_σ is given in HFA and in SCBA as follows.

$$\nu_\sigma = T \sum_n 2 / [X_\sigma(i\omega_n) + \sqrt{X_\sigma(i\omega_n)^2 - \Gamma^2}] , \qquad (1)$$

$$X_\sigma(i\omega_n) = i\omega_n - \frac{1}{2}\omega_c + \sqrt{\frac{\pi}{2}} \frac{e^2}{\varepsilon\ell} \nu_\sigma + \mu , \qquad (2)$$

where $\omega_n = (2n+1)\pi T$, μ is the chemical potential to be determined to satisfy $\nu_+ + \nu_- = \nu$, and Γ gives the level width due to impurity scattering. Above some critical temperature T_V, $\nu_+ = \nu_- = \nu/2$ gives the only solution for (1) and (2). Below T_V, however, another solution of $\nu_+ \neq \nu_-$ appears and it has the lower free energy than that of $\nu_+ = \nu_-$. Below T_V the degeneracy of the two valleys is lifted. If Γ=0, T_V is given by $T_V = \sqrt{\pi/32}\nu(2-\nu)e^2/\varepsilon\ell$. On the other hand if $\Gamma > 0.8 e^2/\varepsilon\ell$, T_V becomes zero.

Now that the distribution of electrons is determined, we can calculate the CDW transition temperature T_C. It turns out that above T_V the CDW transition does not occur. In Fig.1 T_C for Γ=0 is shown as a function of ν. Since T_C is lower than T_V, there is a large energy splitting of the two levels, and T_C is very close to that for a single valley system which is given by $T_C = 0.557\nu(1-\nu)$ and shown by a broken curve in Fig.1. Due to the electron-hole symmetry in the present approximation, T_C at ν is the same as that at $\nu'=2-\nu$. In Fig.2 T_V and T_C is shown for $\Gamma/(e^2/\varepsilon\ell)$=0, 0.2, and 0.4. In this figure Tc is for the single valley system, since it gives a good approximation to the actual T_C. T_C vanishes if Γ is greater than $0.43 e^2/\varepsilon\ell$. At Tc the wave number of the CDW is about 1.5/ℓ.

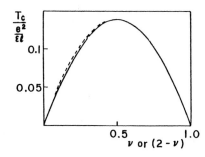

Fig.1 The transition temperature to the CDW state T_C vs. $\nu = 2\pi\ell^2 n$ for a pure system. The solid line is T_C for the two valley system and the broken one is that for a single valley system. In the calculation higher Landau levels are neglected. The temperature is scaled by a characteristic Coulomb energy $e^2/\varepsilon\ell$.

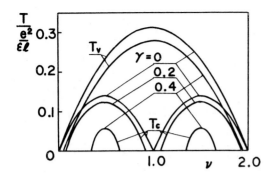

Fig.2 The critical temperature for valley splitting T_v and the CDW transition temperature T_c for two-valley system. Choices of the level width are $\gamma \equiv \Gamma/(e^2/\varepsilon\ell)=0.0$, 0.2, and 0.4. The higher Landau levels are not considered.

3. The Effect of Higher Landau Levels

In this section we consider only N=0 and N=1 Landau levels of one of the valleys. The difference between the present case and the case in 2. is that here we must consider inter-Landau-level scattering by the Coulomb interaction. In this section we consider only the pure case (Γ=0). The transition temperature depends on the cyclotron energy $\hbar\omega_c$. The result of the numerical calculation is shown in Fig.3 for $\hbar\omega_c/(e^2/\varepsilon\ell)$=0.0, 0.1 and 1.0. The result for $\hbar\omega_c=e^2/\varepsilon\ell$ is already close to the result of $\omega_c=\infty$ which is given by

$$T_c = 0.557\, \nu(1-\nu)\frac{e^2}{\varepsilon\ell} \quad (\nu \leq 1)$$
$$= 0.517\, (\nu-1)(2-\nu)\frac{e^2}{\varepsilon\ell} \quad (1<\nu\leq 2) \quad . \qquad (3)$$

The result for $\omega_c=0$ is also not so different from that of $\omega_c=\infty$, because even for $\omega_c=0$, there is an energy difference between the two Landau levels due to the self-energy shifts.

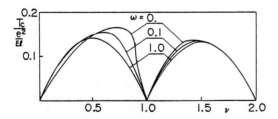

Fig.3 The CDW transition temperature T_c for a single valley system vs ν. Here the lowest and the second lowest Landau levels are taken into account and the results are given for $\omega \equiv \hbar\omega_c/(e^2/\varepsilon\ell)$=0.0, 0.1, and 1.0.

4. Conclusion

We have calculated the CDW transition temperature considering the effect of valley degeneracy or higher Landau levels in the framework of the mean field theory. In the calculation we took into account two Landau levels. We have shown that the valley degeneracy is lifted at temperature lower than T_v due to the exchange interaction. Such lifting of the valley degeneracy has been considered by BLOSS et al.[19] at H=0 and T=0. Here we considered the same phenomenon in a strong magnetic field and at finite temperatures. It is shown that the CDW transition occurs always at temperature lower than T_v. Hence at T_c there is a large energy difference between the lowest Landau levels in each valley, and T_c is close to that of a single valley system. The exchange interaction not only causes the valley splitting but also enhances the energy difference between the N=0 and N=1 Landau levels of the same valley. Hence at $\nu<1$ we saw that the effect of the N=1 Landau level on T_c is small even if ω_c is small.

The details of the present theory will be published elsewhere[20].

References

1. C.C. Grimes, G. Adams: Phys. Rev. Lett. $\underline{42}$, 795 (1979)
2. R.H. Morf: Phys. Rev. Lett. $\underline{43}$, 931 (1979)
3. H. Fukuyama: J. Phys. Soc. Japan $\underline{48}$, 1841 (1980)
4. Yu.E. Lozovik, V.I. Yudson: Pis'ma Zh. Eksp. and Teor. Fiz. $\underline{22}$, 26 (1975). translation: Sov. Phys. JETP Lett. $\underline{22}$, 11 (1975)
5. H. Fukuyama: Solid State Commun. $\underline{19}$, 551 (1976)
6. M. Tsukada: J. Phys. Soc. Japan $\underline{42}$, 391 (1977)
7. H. Fukuyama, D. Yoshioka: J. Phys. Soc. Japan $\underline{48}$, 1853 (1980)
8. S. Kawaji, J. Wakabayashi, M. Namiki, K. Kusuda: Surf. Sci. $\underline{73}$, 121 (1978)
9. D.C. Tsui: Solid State Commun. $\underline{21}$, 675 (1977)
10. R.J. Wagner, D.C. Tsui: J. Magn. Magn. Mat. $\underline{11}$, 26 (1979)
11. B.A. Wilson, S.J. Allen, Jr., D.C. Tsui: Phys. Rev. Lett. $\underline{44}$, 479 (1980)
12. H. Fukuyama, P.M. Platzman, P.W. Anderson: Phys. Rev. B$\underline{19}$, 5211 (1979)
13. Y. Kuramoto: J. Phys. Soc. Japan $\underline{45}$, 390 (1978)
14. A. Kawabata: Solid State Commun. $\underline{28}$, 547 (1978)
15. D. Yoshioka, H. Fukuyama: J. Phys. Soc. Japan $\underline{47}$ 394 (1979)
16. F.J. Ohkawa, Y. Uemura: J. Phys. Soc. Japan $\underline{43}$, 907 (1977), ibid $\underline{43}$, 917 (1977).
17. L.J. Sham, M. Nakayama: Phys. Rev. B$\underline{20}$, 734 (1979)
18. F.J. Ohkawa, Y. Uemura: J. Phys. Soc. Japan $\underline{43}$, 925 (1977)
19. W.L. Bloss, L.J. Sham, B. Vinter: Phys. Rev. Lett. $\underline{43}$, 1529 (1979)
20. D. Yoshioka, H. Fukuyama: to be submitted to J. Phys. Soc. Japan.

InAs-GaSb Superlattices in High Magnetic Fields

M. Voos[*], and L. Esaki

IBM Thomas J. Watson Research Center, P.O. Box 218
Yorktown Heights, NY 10598, USA

1. Introduction

Semiconductor superlattices, which have been first proposed by ESAKI and TSU in 1970 [1], consist of thin alternate layers of two different semiconductors which closely match in lattice constant. In these one-dimensional periodic structures, the layer thickness ranges roughly from 10Å to a few hundred Å, smaller than or comparable to the electron mean free path but larger than the interatomic spacing. In this case, "macroscopic quantum effects" are expected to occur and to change the electronic structure of the involved materials [2], leading to unusual transport and optical properties.

The recent advance of molecular-beam-epitaxy (MBE) in ultra-high vacuum has made possible the growth of superlattice structures of high-quality compound semiconductors in a controlled manner, providing in particular very sharp boundaries at each interface. This technique was first used to prepare $GaAs-Ga_xAl_{1-x}As$ superlattices, which have been extensively studied during the last ten years [2]. However, quite recently MBE has proven successful in growing a new type of structure [3]; namely, InAs-GaSb superlattices, which can be distinguished from the $GaAs-Ga_xAl_{1-x}As$ system by the relative position of the band-edge energies of the host semiconductors.

We review here some recent results obtained under high magnetic field in semiconducting and semimetallic InAs-GaSb superlattices. Indeed, as discussed later, these superlattices undergo a semiconductor to semimetal transition when the layer thickness reaches a critical value of the order of 100Å [4].

2. Theoretical Aspects

The different band gaps of the host semiconductors in a superlattice give rise to a one-dimensional periodic potential in the z direction perpendicular to the layers. Such a potential causes a folding of the Brillouin zone, resulting in a series of mini-bands or subbands in both the conduction and valence bands. In the InAs-GaSb system, electron affinity measurements [5] indicate that the InAs conduction bandedge E_{c1} lies at lower energy than the GaSb valence bandedge E_{v2}, as illustrated in Fig. 1. It follows that the InAs and GaSb layers serve as potential wells

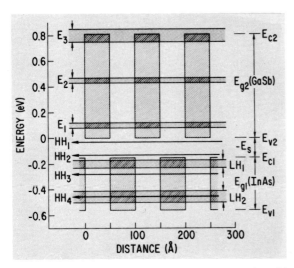

Fig. 1 Potential energy profile and subband energies for an InAs-GaSb superlattice with $d_1 + d_2 = 100$ Å

for electrons and holes, respectively. InGaAs-Ga_x-$Al_{1-x}As$ structures, E_{c1} and E_{v1} for GaAs are respectively at lower and higher energies than E_{c2} and E_{v2} for $Ga_xAl_{1-x}As$ so that GaAs layers are potential wells for both electrons and holes, as illustrated in Fig. 2. In the InAs-GaSb system, due to the relative position of E_{c1} and E_{v2}, there is a strong interaction between the InAs conduction band and the GaSb valence band, leading to a greater anomaly in the E-k relationship of the subbands than in GaAs-Ga_xAl_{1-x} superlattices.

Band calculations in InAs-GaSb superlattices were performed using the linear-combination-of-atomic-orbitals (LCAO) method [6]. The calculated subband structure was found to be strongly dependent on the period d which is equal to $d_1 + d_2$, where d_1 and d_2 are the thicknesses of the InAs and GaSb layers, respectively. Figure 1 gives the potential profile and the calculated subband energies in real space for an InAs-GaSb structure with $d_1 + d_2 = 100$Å. Here E_i corresponds to electrons, and HH_i (LH_i) to heavy (light) holes. The lowest conduction band states are strongly confined in the InAs layers and are shifted in energy, depending upon the well width d_1. Similarly, the valence band states are confined in GaSb, and are also shifted in energy. Figure 3 shows the calculated subband energies and bandwidths of E_i, HH_i and LH_i as a function of d assuming $d_1 = d_2$. For thin layers, the super lattice is a semiconductor and, if E_1 and H_1 are respectively the energies of the ground electron and heavy-hole subbands, the superlattice energy gap, $E_1 - H_1$, is positive. When d is increased, the ground electron subband is lowered and the ground heavy-hole one is raised, so that $E_1 - H_1 = 0$ for d = 170Å, corresponding to a semiconductor-semimetal transition [2,6]. Thus, for d > 170Å, InAs-GaSb superlattices are semimetallic,

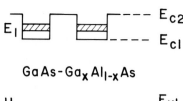

Fig.2 Potential energy profile of a $GaAs-Ga_xAl_{1-x}As$ superlattice

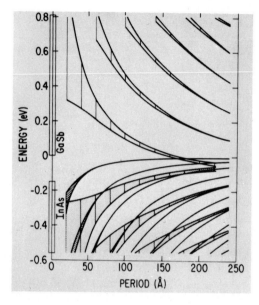

Fig.3 Calculated subband energies and bandwidths for electrons, and light and heavy holes as a function of d, with $d_1=d_2$

and $E_1 - H_1$ becomes negative. In this case, electrons transfer from GaSb to InAs and the Fermi energy E_F lies between E_1 and H_1. The accumulation of electrons and holes in InAs and GaSb, respectively, produces a strong dipole layer, resulting in band bending which pushes E_1 upward and H_1 downward. Exact solutions have not been carried out, which would require extensive formulation and construction of envelop wave functions to account for the spatial distribution of carriers. Instead, the band bending is first calculated in the Thomas-Fermi approximation, assuming a density of electron transfer. Then, the subband energies are determined by the LCAO method [6], and E_F is obtained from the two-dimensional density of states, taking into account that the numbers of electrons and holes should be equal. The self-consistent solution is obtained if the resulting electron density coincides with that assumed initially [7,8]. Experimentally, the semiconductor-semimetal transition was put into evidence from the observation of enhanced carrier densities in Hall measurements as a function of d_1 [7], showing that superlattices become semi-metallic for $d_1 \sim 100$ A, in satisfying agreement with theory [4].

3. The Shubnikov-de Haas Effect

To perform Shubnikov-de Haas experiments, heavily doped n-type InAs-GaSb superlattices were grown on semi-insulating GaAs by MBE, the total thickness of the superlattice being $\sim 2\mu m$ [2,9]. The parameters of four samples (S1 to S4) are given in Table 1. Note that throughout this paper, E_i measures upward from the InAs conduction bandedge [10] and E_F from the bandedge of E_1.

Table 1 Parameters of InAs-GaSb superlattices. ΔE_i is the calculated width of the E_i subband

	S-1	S-2	S-3	S-4
$d_1 = d_2$ (Å)	55	90	150	200
$n_e(10^{18} cm^{-3})$	0.5	1.2	3.4	1.0
$\mu_e(cm^2/V \cdot sec)$	2700	5300	14000	6100
$E_1(\Delta E_1)$ (meV)	215(26)	126(9)	69(11)	37(11)
$E_2(\Delta E_2)$ (meV)	536(26)	331(3)	178(1)	120(1)
$E_3(\Delta E_3)$ (meV)	853(59)	558(3)	313(0)	216(0)
E_F (meV)	40	122	286	168

Assuming that the subband width is sufficiently narrow so that the electron system is essentially two-dimensional, the condition of magnetic quantization is given by $(2\pi eB_z/\hbar)(n+\gamma)=A_t$, where n is the Landau level index, γ is a phase factor between 0 and 1, $B_z = B \sin\phi$ is the field component perpendicular to the plane of the superlattice layers, and $A_t = \pi k_t^2$ is the area of the constant energy surface in k-space parallel to this plane [9,11]. Taking into account the relation between k_t and the number of electrons per unit area, n_s, the oscillatory magnetoresistance is found to exhibit extrema whenever $(eB_z/\pi\hbar)(n+\gamma)=n_s$, and the oscillation period in B_z^{-1} is $\Delta B_z^{-1}/\Delta n = e/\pi\hbar n_s$. If only the ground subband is occupied, n_s can be compared with the measured electron concentration n_e in Table 1 with $n_s = n_e(d_1+d_2)$. To determine the number of occupied subbands, information about E_F is required. Using the simplified Kane model [12] to account for the InAs conduction band nonparabolicity, the electron cyclotron mass is $m^* = m^*_e(1+2E/E_g)$ where $E=E_i+E_t$, and E_g and m^*_e are the energy gap and bandedge mass of InAs, respectively [10]. The density of states is here $\rho = m^*(E)/\pi\hbar^2$, and E_F and the number of electrons in each subband can be obtained by integrating over all the subbands, knowing n_e. Values of E_F thus obtained are given in Table 1 showing, for instance, that only E_1 is occupied for S2.

The magnetoresistance measurements were done with the current flowing in the layer plane and the orientation of the magnetic field varying from perpendicular ($\phi=90°$) to parallel ($\phi=0$) to this plane. Figure 4 shows the observed oscillatory behavior of the transverse magnetoresistance as a function of B at 4.2K in sample S2 [2,9]. Pronounced oscillations are observed for high angles, and disappear progressively as ϕ is decreased. The in-

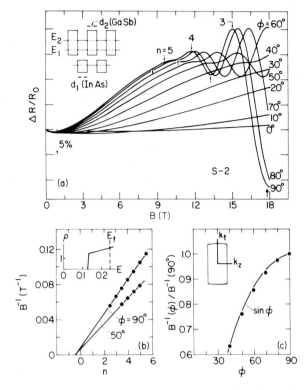

<u>Fig.4</u> Transverse magnetoresistance vs. B at 4.2K in the upper side (a). Lower left: B^{-1} vs. n(b); lower right: orientation effect (c).

sets of Figs. 4(b) and 4(c) show respectively the calculated density of states in units of $10^{10} cm^{-2} meV^{-1}$ and eV and the projection of the Fermi surface on the k_t - k_z plane, both being characteristic of a two-dimensional electron system. Values of B^{-1} at which the extrema occur, as indicated by arrows in Fig.4(a) are plotted versus n in Fig. 4(b) for ϕ=90° and 50°. From the slope of the obtained straight line one finds n_s = $2.45 \times 10^{12} cm^{-2}$ and n_e = $1.36 \times 10^{18} cm^{-3}$, in satisfying agreement with the value given in Table 1. The dependence of the periods on ϕ is given in Fig. 4(c), evidencing the sine-relationship characterizing the two-dimensional nature of the subband [11].

The magneto-oscillation experiments were performed in a temperature range (4.2 - 60K) where n_e and μ_e showed little change. The electron cyclotron masses were determined by fitting the observed amplitudes to the expression $T/sinh$ $(2\pi^2 kTm*/e\hbar B)$ which is valid in the range of sinusoidal oscillations. They lie between 0.048 and 0.068 m_o, and are thus larger than the InAs band-edge mass $m*_e$ (0.023 m_o) [10]. Note finally that Shubnikov-de Haas oscillations were also observed in the other samples, and are analyzed in detail in [2,9].

4. Cyclotron Resonance in Semiconducting Superlattices

Far infrared magneto-transmission experiments were done in an InAs-GaSb superlattice with d_1 = 65Å and d_2 = 80Å. The calculated values of E_1 and E_F are 186 and 39 meV, respectively, and Shubnikov-de Haas measurements gave E_F = 40meV. The absorption maxima, observed for B perpendicular to the InAs and GaSb layers, are attributed to electron cyclotron resonance; namely, to transitions from the last occupied to the first empty Landau level of the E_1 subband. From the slope of the observed linear dependence of the energy position of the absorption maxima versus B, the electron cyclotron mass m* is 0.043 m_0. One can also calculate the Landau levels of E_1, taking into account the effect of the InAs conduction band nonparabolicity in the simplified Kane model [12]. If $\omega_c = eB/m^*_e$ is the electron cyclotron frequency in bulk InAs, one obtains easily:

$$E_{1,n} = -E_g/2 + ((E_g/2)^2 + E_g((n+1/2)\hbar\omega_c + E_1(1+E_1/E_g)))^{1/2} \quad (1)$$

where E_g is the band gap of bulk InAs. The calculated effective mass is then $m^* = m^*_e(1 + (E_{1,n+1} + E_{1,n})/E_g)$. Fits of the data using this model give m^* = 0.048 m_0, in good agreement with the previous value. Thus, it is clear that m^* is larger than m^*_e, as a result of the nonparabolicity of the InAs conduction band.

5. Magnetic Field Effects in Semimetallic Superlattices

The experiments described in this Section were done on two semimetallic InAs-GaSb superlattices grown by MBE on (100)GaSb substrates. The total thickness of the superlattice region was ~2µm, while the thicknesses of the InAs and GaSb layers were respectively 120 and 80Å for sample A and 200 and 100Å for sample B. The samples were not intentionally doped, but residual impurities gave a background electron density of $10^{16} cm^{-3}$.

5.1 Shubnikov-de Haas Measurements

The transverse magnetoresistance was measured in these samples as a function of the magnetic field up to 18 Tesla at 4.2K [7]. For B perpendicular to the layers, pronounced oscillations were observed as in semiconducting structures, but these oscillations start at a much lower field. Indeed, in the case of semimetallic superlattices, the electrons, which result from the flow of carriers between GaSb and InAs layers, are not associated with impurities, and have a much higher mobility. From such experiments, the electron density is $n_s = 7.3 \times 10^{11} cm^{-2}$ in sample B for example, in agreement with the theoretical value of $8 \times 10^{11} cm^{-2}$ calculated as described in Section 2 for semimetallic superlattices. The corresponding bulk concentration is $\sim 3 \times 10^{17} cm^{-3}$, much larger than the residual impurity density. Measurements as a function of the orientation of B with respect to the layer plane give evidence for the quasi-two-dimensional nature of the electron subband. It is also noteworthy that the magnitude of the magnetoresistance in semimetallic structures is much larger than in semiconducting superlattices. This is characteristic of semimetals, and gives further evidence for the semimetallic regime. Finally, we would like to mention that Shubnikov-de Haas experiments have been helpful in studying the process of electron transfer in the semimetallic regime, and also a magnetic field induced semimetal-semiconductor transition [4,8,14]. Indeed, for high magnetic fields, the n=0 Landau level of E_1 can be raised above E_F, while the corresponding hole level is lowered below E_F, resulting in electron transfer from the InAs layers back to the GaSb layers.

5.2 Far Infrared Magneto-Absorption Experiments

Transmission signals at fixed photon energy exhibit as a function of B an oscillatory behavior in both samples when B is perpendicular to the layers [15]. When B is parallel to the layers, no oscillations are detected, as expected in quasi-two-dimensional structures. Figure 5 shows, as a function of B, the infrared energy ($h\nu$) positions of the transmission minima observed in sample A. One can see that the energies at which absorption maxima occur depend approximately linearly on B. Two kinds of behavior are observed: one of the lines extrapolates to $h\nu=0$ at B=0, while all the others converge to - 38 meV. The first one, noted CR in Fig. 5, corresponds to the electron-cyclotron resonance transitions between the last occupied to the first empty Landau level of the E_1 subband. The other curves are attributed to inter-subband transitions from $H_{1,n}$ to $E_{1,n}$ Landau levels, as sketched in Fig. 5. The E_1 Landau levels are given by (1), and the simple quantization relation $H_{1,n} = H_1 - (n + 1/2)\hbar\omega_v$ is used for the heavy-hole Landau levels, where $\omega_v = eB/m^*_h$ is the heavy-hole cyclotron frequency [10]. Electron-cyclotron resonance corresponds, of course, to $h\nu=E_{1,n+1}-E_{1,n}$ with

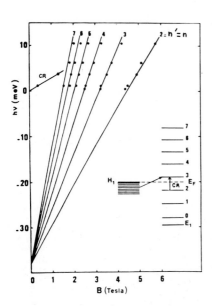

Fig.5 Infrared energy position of the absorption maxima vs. B (Solid lines: theoretical fits; full dots: experimental data)

$E_{1,n+1} > E_1 + E_F > E_{1,n}$, and inter-subband transitions to $h\nu = E_{1,n} - H_{1,n'}$, with $n = n'$ [15]. Due to the high density of states of the heavy-hole subband, the Fermi level is close to H_1, and it is reasonable to take $H_1 = E_1 + E_F$, H_1 being also measured from the InAs conduction bandedge. Fits of the experimental data to this theoretical model are shown in Fig. 5 for sample A. The agreement between experiment and theory is satisfying, and yield for sample A $E_1-H_1 = -(38\pm2)$meV with $E_1 = (115 \pm 15)$meV and $H_1 = (153\pm15)$meV at $B=0$. Similar studies in sample B give $E_1 - H_1 = -(61\mp4)$meV with $E_1 = (60\pm20)$meV and $H_1 = (121\pm20)$meV at zero magnetic field. These results compare favorably with theoretical values calculated as described in Section 2 [15]. Indeed, such calculations give $E_1-H_1 = -42$meV with $E_1=87.5$meV and $H_1 = 129.5$meV for sample A, and $E_1-H_1 = -68$meV with $E_1=64.5$meV and $H_1 = 132.5$meV for sample B. These investigations provide the first experimental determination of the energies of the ground conduction and valence subbands, the semimetallic nature of the superlattices being directly demonstrated by the observation of a negative energy gap.

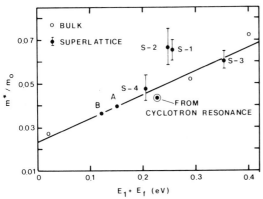

Fig. 6 Cyclotron effective mass versus total energy. The solid line gives the expected dependence.

The cyclotron resonance data provide the superlattice electron cyclotron mass m^*, which is found to be equal to $0.040\ m_0$ in sample A and to $0.036\ m_0$ in sample B. Thus, m^* is enhanced with respect to the InAs bandedge mass m^*_e ($0.023\ m_0$), in accordance with the effect of the conduction band nonparabolicity. This is emphasized in Fig. 6 which shows a plot of the values of m^*/m_0 obtained in several superlattices as a function of E_1+E_F. As already mentioned, taking into account the band nonparabolicity by use of the Kane model [12] gives $m^*=m_e(1+2(E_1+E_F)/E_g)$, which is represented by the straight line in Fig. 6. Also shown in Fig. 6 are data obtained in doped bulk InAs samples where only the E_F term is involved [16,17]. The mass enhancement observed in superlattices can be taken as an indication that electrons remain confined in the InAs layers for semiconducting and semimetallic structures.

6. Conclusion

The MBE technique has made possible the growth of InAs-GaSb superlattices, and thus to synthesize, by a choice of the layer thicknesses, a tailor-made narrow-gap semiconductor or semimetal with an electronic band structure which has little resemblance with that of the host semiconductors. We think that such superlattices, which can be regarded as a novel class of quasi two-dimensional materials with unique characteristics, open a new field of investigations in semiconductor physics.

We have described here some recent investigations performed on InAs-GaSb superlattices under high magnetic fields. It can be concluded that magnetic field experiments have proven fruitful in studying and understanding some of the characteristic properties of these superlattices in both the semiconducting and semi-metallic regimes.

The work presented here is the result of a group effort and is sponsored in part under US Army Research Office Contract.

REFERENCES

*Permanent address: Groupe de Physique des Solides de l'ENS, Université Paris 7, 75005 Paris, France

1. L. Esaki and R. Tsu: IBM J. Res. Dev. 14, 61-65 (1970)
2. See, for example, L. Esaki and L. L. Chang: J. Magn. and Magn. Mat. 11, 208-215 (1979)
3. C.-A. Chang, R. Ludeke, L.L.Chang and L. Esaki: Appl. Phys. Lett. 31, 759-761 (1977)
4. For a recent review, see L. L. Chang: Proc. 15th Int. Conf. on Phys. of Semicon., Kyoto 1980, to be published.
5. G. A. Sai-Halasz, R. Tsu and L. Esaki: Appl. Phys. Lett. 30, 651-653 (1977)
6. G. A. Sai-Halasz, L. Esaki and W. Harrison: Phys. Rev. B18, 2812-2818 (1978)
7. L. L. Chang, N. Kawai, G. A. Sai-Halasz, R. Ludeke and L. Esaki: Appl. Phys. Lett. 35, 939-942 (1980)
8. L. L. Chang, N. J. Kawai, E. E. Mendez, C.-A. Chang and L. Esaki: submitted for publication
9. H. Sakaki, L. L. Chang, G. A. Sai-Halasz, C.-A. Chang and L. Esaki: Solid State Commun. 25, 589-592 (1978)
10. The InAs and GaSb band parameters can be found in: Handbook of Electronic Materials, M. Neuberger ed. (Plenum, N.Y., 1971), Vol. 2
11. L. L. Chang, H. Sakaki, C.-A. Chang and L. Esaki: Phys. Rev. Lett. 38, 1489-1493 (1977).
12. E. O. Kane: J. Phys. Chem. Solids 1, 249-261 (1957)
13. H. Bluyssen, J. C. Maan, P. Wyder, L. L. Chang and L. Esaki: Solid State Commun. 31, 35-38 (1979)
14. N. J. Kawai, L. L. Chang, G. A. Sai-Halasz, C.-A. Chang and L. Esaki: Appl. Phys. Lett. 36, 363-371 (1980)
15. Y. Guldner, J. P. Vieren, P. Voisin, M. Voos, L. L. Chang and L. Esaki: submitted for publication
16. M. Cardona: Phys. Rev. 121, 752-758 (1961)
17. M. B. Thomas and J. C. Wooley: Can. J. Phys. 49, 2052-2060 (1971)

Magnetic Quantization and Transport in a Semiconductor Superlattice

T. Ando

Institute of Applied Physics, University of Tsukuba, Sakura
Ibaraki 305, Japan

>The structure of magnetic energy levels in GaAs-$Al_xGa_{1-x}As$ superlattice has been calculated when a magnetic field is applied perpendicular to the superlattice direction. Transport coefficients have also been calculated and been shown to exhibit Shubnikov-de Haas oscillations in some cases in agreement with recent experiments.

1. Introduction

The evolution of molecular-beam epitaxy has allowed access to man-made semiconductor superlattices, consisting of periodic heterostructures of alternating, ultrathin layers of two semiconductors. It has been known in GaAs-$Al_xGa_{1-x}As$ systems that a simple Kronig-Penney model explains the subband structure if combined with an appropriate band bending effect due to electron transfer [1]. The structure of magnetic energy levels is interesting especially when a field is applied perpendicular to the superlattice direction. There have been some theoretical investigations on this problem [2,3], but they are all inappropriate for realistic systems because of insufficient semiclassical or single-band approximations. In this paper we calculate the magnetic level structure and the transport coefficients in actual GaAs-$Al_xGa_{1-x}As$ superlattices. The calculation predicts that Shubnikov-de Haas type oscillation can appear in some cases, explaining recent experimental results [4].

2. Magnetic Level Structure

We choose the z axis in the superlattice direction and assume that a magnetic field of strength H is applied in the y direction. If we choose a gauge A=(Hz,0,0), we have the Hamiltonian:

$$H = \frac{\hbar^2}{2m} [\,(k_x+\frac{z}{l^2})^2 + k_y^2 + k_z^2\,] + V(z), \qquad (2.1)$$

where $l^2 = c\hbar/eH$ and $V(z)$ is the periodic superlattice potential. The eigen-state is specified by the quantum numbers k_y, which is the momentum in the direction of the magnetic field, and the center coordinate Z, which is related to k_x through $k_x = -Z/l^2$. The energy is periodic in Z with the period of the superlattice d. The level structure can be calculated numerically by diagonalizing a large matrix for each Z. The matrix is expressed in terms of the Landau level wave function in the absence of $V(z)$. This procedure has turned out to give a convergent result except in case that either the barrier height or the period d are large and each layer is well separated from others. In the following we show an example of the results for the case that V=100 meV, $d_1=d_2=45$ Å, and $N_D=1.9\times 10^{18}$ cm^{-3}, which corresponds to the experiments [4]. Here, V is the barrier height for the $Al_xGa_{1-x}As$ layer, d_1 and d_2 are the thickness of GaAs and $Al_xGa_{1-x}As$ layers, respectively, and N_D is the electron concentration in a unit volume. We use $m=0.068m_0$ where m_0 is the free electron mass. In this case electrons are populated in two subbands, the ground and first excited subbands, in the absence of a magnetic field [1]. The ground subband is rather localized and its dispersion in the z direction is small, whereas the first excited subband is extended and has a three-dimensional character.

Figure 1 contains an example of the level structure for $k_y=0$ in H=50 kOe together with the energy-versus-k_x relation in the absence of H. The dashed lines represent the curve $\hbar^2 k_x^2/2m$ which corresponds to the complete two-dimensional case. Below 90 meV the level is very close to this curve reflecting the fact that the lowest subband is rather localized. At the zone boundaries the levels are slightly split because of the small dispersion of the band along z direction. When the energy is larger than the bottom of the second subband the level approaches that of the Landau level with some corrections due

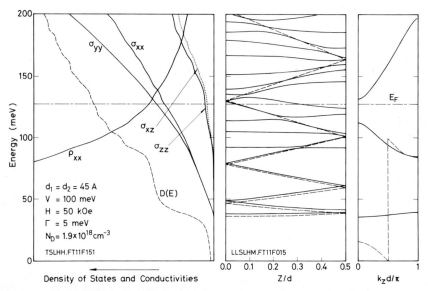

Fig. 1 An example of calculated results in H=50 kOe. The middle part gives the energy-versus-Z relation. The dashed curves correspond to the two-dimensional limit. The right part shows the energy-versus-k_z relation in the absence of H. The thin dashed line represents the potential energy as a function of z. The left figure gives the density of states (dashed line), the conductivities, and the resistivity in the x direction.

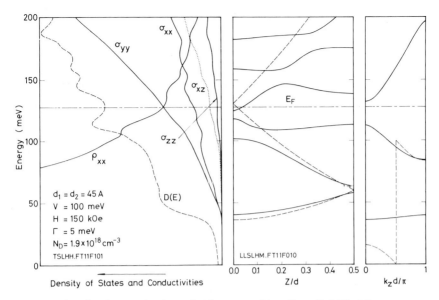

Fig.2 An example of the results for H=150 kOe.

to magnetic breakthroughs. However, the broadening of levels is comparable to level separations. Figure 2 gives an example for H=150 kOe. The levels associated with the lowest subband still have two dimensional characteristics, but strongly couple with Landau levels with higher extended subbands at high energies. Consequently a single-band or semiclassical treatment of the problem cannot give any reasonable answers in such high magnetic fields.

3. Transport Coefficients

Conductivity tensors have been calculated with the use of the energy levels and wavefunctions obtained in the way discussed above. We introduce a constant imaginary part of the self-energy Γ in Green's functions. In the following the level broadening Γ is chosen as 5 meV, which gives a mobility in the absence of H slightly larger than experimentally observed. This constant-Γ approximation is expected to be sufficient in discussing qualitative behaviors of the transport as are considered here.

Figures 1 and 2 contain also the density of states, various components of the conductivity tensor, and the resistivity in the x direction as a function of the energy. At lower energy the density of states is essentially that of two dimension, and at higher energies it has structures associated with Landau levels. One immediately notices that positions of the peak in σ_{xx} can be completely different from those of the density of states. One sees, further, that the resistivity does not behave in the same way as σ_{xx}^{-1} because of the nonvanishing σ_{xz} and σ_{zz}. Details will be discussed elsewhere.

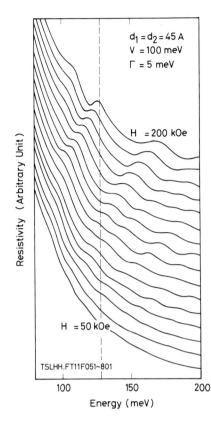

Fig.3 The resistivity ρ_{xx} as a function of the energy for H=50, 60, ..., and 200 kOe. The Fermi level is denoted by the dashed line. The quantum oscillation essentially corresponds to crossing of Landau levels associated with extended excited subbands and is modified by their coupling with levels associated with the localized lowest subband.

Figure 3 gives the resistivity as a function of the energy for different values of the magnetic field. The position of the Fermi level in the absence of H is denoted by a dashed line. It predicts that one can observe Shubnikov-de Haas oscillations if one sweeps the magnetic field. Although the oscillation period is of the same order of magnitude as observed experimentally, the theoretical positions of peaks and dips do not agree with those of the experiments [4]. Since the positions are quite sensitive to the Fermi energy, this disagreement might originate from experimental uncertainties in determining the electron concentration [1,5].

References

1. See, for example, T. Ando and S. Mori, J. Phys. Soc. Jpn. **47**, 1518 (1979) and references therein.
2. R. Tsu and J.F. Janak, Phys. Rev. B**9**, 404 (1974).
3. K. Nakao, J. Phys. Soc. Jpn. **46**, 1669 (1979).
4. L.L. Chang, H. Sakaki, C.A. Chang, and L. Esaki, Phys. Rev. Lett. **38**, 1489 (1977).
5. S. Mori and T. Ando, J. Phys. Soc. Jpn. **48**, 865 (1980).

Part X

Layered Materials and Intercalation

Magnetoreflection and Shubnikov-de Haas Experiments on Graphite Intercalation Compounds

M.S. Dresselhaus[1], G. Dresselhaus[2], M. Shayegan[1], and T.C. Chieu[1]

Massachusetts Institute of Technology
Cambridge, MA 02139, USA

1. Introduction

Graphite intercalation compounds represent a class of compounds with many similarities to the superlattice semiconductors prepared by molecular beam epitaxy. Intercalated graphite has recently attracted considerable attention because the addition of an intercalant can dramatically change the electronic properties of the host graphite material. For example, with the addition of AsF_5, an intercalation compound with room temperature conductivity comparable to that of copper can be achieved [1], while the addition of an alkali metal can result in a superconductor [2]. Similar to the behavior in modulation-doped superlattices, high conductivity in intercalated graphite is achieved by a charge transfer from the intercalate layer, where the mobility is low, to the high mobility graphite layers.

High magnetic fields allow use of the magnetoreflection technique to obtain detailed information on the electronic dispersion relations within a few hundred meV of the Fermi level [3,4], and use of the Shubnikov-de Haas effect and related quantum oscillatory phenomena to provide information on the Fermi surface [5-8]. Initially it seemed difficult to reconcile the results obtained by these two high field techniques because the magnetoreflection experiment provided strong evidence that the electronic structure near the Fermi level remained highly graphitic upon intercalation, while the Shubnikov-de Haas experiment indicated that large changes in the Fermi surface occur through intercalation. Using a recently developed phenomenological model for the electronic structure of intercalated graphite [9], consistency between the magnetoreflection and Shubnikov-de Haas results is demonstrated. This model also provides a basis for the calculation of the magnetic energy level structure which is needed to explain the magnetoreflection spectra in detail.

Graphite intercalation compounds are prepared by the insertion of atomic or molecular layers of a different chemical species between layers of the graphite host material [10]. In this insertion or intercalation process, the layer structure of graphite is almost completely preserved because of the strong intraplanar binding in graphite, and likewise the intercalate layer is closely related to a layer in the parent solid. Structurally, graphite intercalation compounds exhibit the remarkable phenomenon called _staging_, whereby the intercalate layers are periodically arranged in a matrix of graphite layers and are characterized by a stage index n, denoting the number of graphite layers between consecutive intercalate layers. This staging symmetry is the dominant symmetry of these intercalation compounds and results in a

---

1 Center for Materials Science and Engineering and Department of Electrical Engineering and Computer Science

2 Francis Bitter National Magnet Laboratory, supported by the National Science Foundation

superlattice structure similar to that prepared in semiconductors by the molecular beam epitaxy technique. Staging is widespread among graphite intercalation compounds and well-staged samples can be prepared up to high stage (n∼10). The stage determination and the c-axis repeat distance I_c are obtained from (00ℓ) x-ray diffractograms. Because of the strong stage dependence found for the Shubnikov-de Haas (SdH) frequencies in the donor intercalation compounds in this work, it is necessary to carry out the measurements on single-staged samples. By careful preparation of well-staged compounds it is possible to achieve the resonance condition $\omega_c\tau > 1$ required for the observation of Landau level phenomena. For both the magnetoreflection and Shubnikov-de Haas studies, the samples were prepared by the two-zone growth technique [11] and the stage index was determined from (00ℓ) x-ray diffractograms.

2. Magnetoreflection Experiments

The magnetoreflection measurements were made at nearly normal incidence to the c-face of the intercalated graphite samples using the Faraday geometry with magnetic fields in the range $0 < H \leq 15$ Tesla [4]. Magnetoreflection traces were taken at constant photon energy in the range 0.1 eV < $\hbar\omega$ < 0.5 eV, using circularly polarized light obtained with a gold-wire grid and a CsI Fresnel rhomb. Measurements were made using a cold finger dewar operating at 4.2K.

The observed magnetoreflection spectra are in most cases qualitatively similar to those of pristine graphite, but show differences with regard to resonant magnetic fields and in some cases also differences in resonant lineshapes. Illustrative spectra are shown in Fig. 1 for an acceptor compound

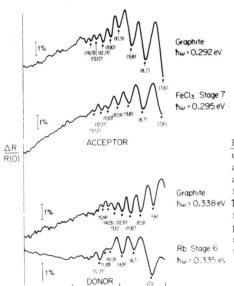

Fig. 1 Magnetoreflection spectrum using (+) circular polarization for an acceptor compound (FeCl$_3$ stage 7) at a photon energy $\hbar\omega$ = 0.295 eV and for a donor compound (Rb stage 6) at $\hbar\omega$ = 0.335 eV. For comparison, traces for graphite are shown at comparable photon energies. The resonances are specified by the quantum numbers for the initial and final states.

(FeCl$_3$ stage 7) and a donor compound (Rb stage 6). For comparison, spectra are also included for pure graphite at similar photon energies. Each structure in Fig. 1 is identified with a K-point Landau level interband transition, specified by the quantum numbers for the initial and final states. For all compounds that have been studied, the resonant magnetic field for a given transition shifts to increasingly lower fields with increasing intercalate concentration for both (+) and (-) senses of circular polarization. The shifts of the resonant magnetic fields for a given transition are much larger for donor compounds than for acceptors of comparable stage. The intensity of the resonances (magnitude of the percent reflectivity change) is reduced with respect to that in graphite because of a reduction in relaxation time and an increase in the magnitude of the zero field reflectivity R(0). For example, at a photon energy of $\hbar\omega \sim 0.2$ eV, R(0) is found to be ~ 0.7, 0.8, 0.9 respectively for pristine graphite, graphite-FeCl$_3$ stage 7, and graphite-Rb stage 6.

Resonant Landau level transitions are observed and are analyzed by taking spectra over a wide range of photon energies. A summary of such results for a typical intercalation compound is shown in Fig. 2, in which a comparison with results for pristine graphite is included. From this analysis it is

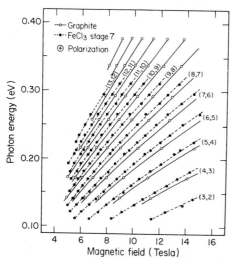

Fig. 2 Summary of resonant magnetic fields "fan chart" for photon energies in the range $0.110 < \hbar\omega < 0.320$ eV for a FeCl$_3$ stage 7 sample using (+) circularly polarized radiation. To emphasize the similarities but measurable differences, a comparison of the results for this compound and for graphite is presented in the same photon energy range.

shown that for graphite intercalation compounds with stages $n \geq 4$, the electronic structure within a few hundred millivolts of the Fermi level is well described by the graphite π-bands, with modifications to the band parameters that can be measured quantitatively.

At constant $\hbar\omega$, the data in Fig. 2 and in "fan charts" constructed for other samples, are well explained by a simple two-band model approximation. Nevertheless, the mass parameters determined from such a model exhibit some dependence on $\hbar\omega$, since the Landau level transitions in Fig. 2 show departures from a linear dependence between $\hbar\omega$ and resonant magnetic field. Therefore the comparison between masses for the intercalated and pure graphite samples is made at a common reference point, taken at the K-point band edge, and found by taking $\hbar\omega \to 0$. The results for the valence and conduction band masses m_c^* and m_v^* thus obtained are given in Table 1 for a number of different samples. We note that in all cases, the mass parameters in the inter-

Table 1   Results of analysis of magnetoreflection experiments

Intercalant	Stage	m_c^*	m_v^*	(γ_0^2/γ_1)
(HOPG)	∞	0.056	0.084	25.1
AlCl$_3$	8	0.056	0.076	26.0
	6	0.054	0.076	26.6
FeCl$_3$	7	0.055	0.079	26.1
	5	0.054	0.075	26.6
Rb	6	0.045	0.065	31.7

calation compounds differ from the masses in graphite by only a few percent, with significantly larger changes in m_c^* and m_v^* occurring in the donor Rb compound than in acceptors of comparable stage.

Analysis of the magnetoreflection spectra in Fig. 2, and in "fan charts" obtained for other samples with $n \geq 4$ show that these spectra can be explained by dispersion relations which have the same form as the three-dimensional Slonczewski-Weiss-McClure (SWMcC) band model for graphite, but with modified values of the band parameters. Included in Table 1 are values for the SWMcC band parameter combination γ_0^2/γ_1, which is directly related to the reduced effective mass m^* [4].

Magnetoreflection resonances such as in Fig. 1 are not observed below a certain photon energy, the cutoff energy E_x, which depends on both stage and intercalate species. The cutoff of K-point interband transitions below E_x is interpreted as due to the introduction of carriers by the intercalant, resulting in a shift in Fermi level E_F. Because of the Pauli exclusion principle, interband Landau level transitions are made from occupied valence to unoccupied conduction states. The lowering of E_F caused by the introduction of holes in acceptor-type compounds leads to a cutoff of Landau level transitions as E_F drops below the extremum of the magnetic subbands for the initial state. With increasing intercalate concentration, E_F moves to lower energies, therefore increasing the magnitude of E_x. In donor-type compounds the introduction of electrons causes E_F to rise, leading to the cutoff of interband transitions as E_F rises above the extremum of the magnetic subbands for the final state. Analysis of the measured cutoff photon energies yields the shift in the Fermi level relative to the K-point band edge $|E_F - E_{3,K}^0|$. Results for several intercalants and stages are presented in Fig. 3, where the + sign in $E_F - E_{3,K}^0$ is for donors and the − sign for acceptors. Of significance is the much larger shift in the Fermi level in donor compounds relative to acceptor compounds of similar stage, indicating a significantly larger charge transfer in the case of donor compounds. To carry out a complete analysis of the magnetoreflection spectra, it is necessary to use Landau levels for intercalated graphite, and this is discussed below in Section 5.

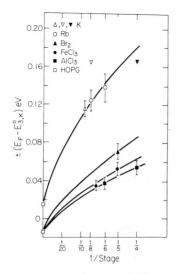

Fig. 3 Dependence on reciprocal stage (1/n) of the Fermi level measured relative to the K-point band edge for several intercalants. The + sign for $(E_F - E^0_{3,K})$ applies to the donor compounds and the − sign to the acceptors. For all intercalants, the graphite value is obtained in the limit $(1/n) \to 0$ and is denoted by open squares. The open triangle ∇ is a Shubnikov-de Haas result from the present work and the closed triangle ▼ is from magnetic susceptibility results of DISALVO ET AL [12].

3. Shubnikov-de Haas Experiments

Having established that the electronic levels near the Fermi level are closely identified with the graphite π-bands, it is of interest to explore the effect of intercalation on the Fermi surface. In this connection Shubnikov-de Haas (SdH) and other quantum transport experiments have been reported on a wide variety of graphite intercalation compounds [5-8]. The multitude of experimental results suggest that with careful sample preparation and characterization, a complete set of frequencies associated with each stage of a given intercalation system can be obtained. A number of careful experiments have already been reported. Perhaps the most complete set of data has been obtained for the graphite-K system where SUEMATSU ET AL [6] give results for stage n = 1,3,4 and we for n = 4,5,8. Also, results for graphite-Rb compounds have been previously reported, showing that the Fermi surfaces are not only stage dependent, but also vary from one alkali metal intercalant to another.

To obtain quantitative information on the stage dependence of these Fermi surfaces, it is important to work with well-staged and characterized samples. Since the Fermi surfaces for the alkali metal compounds with the intercalants K and Rb are stage dependent, staging fidelity is essential for obtaining reproducible experimental results for different samples of the same stage index. Because of the instability of alkali-metal samples in the presence of air and moisture, the samples are encapsulated in ampoules and sample handling is done in an Argon-filled dry box (∿1 ppm oxygen content). The stage of the samples was determined using (00ℓ) x-ray diffraction profiles both before and after the SdH experiments, confirming that the samples were single staged and that no desorption had occurred during the measurements.

The four-point method was used to study the in-plane transverse magnetoresistence in the temperature range 1.4≲T≲4.2K and in magnetic fields up to 15 Telsa. The leads were attached to the sample using conducting epoxy. The sample was then inserted in a helium-filled ampoule and stycast was used to seal the ampoule. The angular dependence of the SdH oscillations could thus be measured by rotating the sample around the direction of the current $\vec{I}$

such that $\vec{I} \perp \vec{H}$ for all angles, and hence transverse magnetoresistence was always measured. Data acquisition was by computer, and the data were manipulated to obtain a Fourier power spectrum of resistance vs. 1/H, thereby yielding the frequencies of SdH oscillations. These frequencies are related to the extremal cross sections of the Fermi surface. The Fourier power spectra for the SdH frequencies for stages 4,5,8 graphite-potassium are presented in Fig. 4.

During the course of these studies, a great deal of care was given to ensure the fidelity and reproducibility of the data. For this reason, the experiment was performed on five potassium stage 5 samples and on one of these, the measurements were repeated several weeks later after attaching new leads. While x-ray profiles after each experimental step confirmed that the samples retained their single stage identity, the magnetoresistance oscillations and the relevant Fourier power spectra revealed nearly identical traces for these stage 5 samples. Measurements done on potassium samples with different stages or on rubidium samples, however, showed distinctly different SdH frequencies. In all cases the SdH frequencies were very different from those found in pristine graphite. Hence our conclusion is that there is a unique set of SdH frequencies and Fermi surfaces associated with each stage and donor intercalant (K,Rb). SUEMATSU ET AL [6] have also shown stage-dependent de Haas-van Alphen frequencies in stages 3 and 4 graphite-potassium compounds.

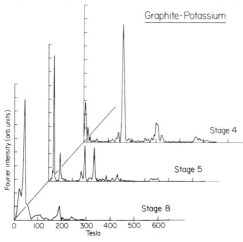

Fig. 4 Shubnikov-de Haas Fourier Transform Power Spectra for stages 4, 5 and 8 graphite-K. These power spectra were obtained by a Fourier transform of an experimental resistivity vs 1/H trace for magnetic fields $\overline{H}$ < 15 Tesla. The peaks in the power spectra correspond to SdH frequencies, which are given in Tesla and the same scale is used for each stage.

Poorly staged samples also give distinct SdH frequencies. In fact more frequencies are found in mixed than in single-staged samples. Samples showing the greatest stage fidelity also show the simplest SdH spectra. The effect of mixed staging introduces additional frequencies associated with the minority phase and shifts the Fermi level to some intermediate value between those corresponding to the pure stage values. This shift in Fermi level is most sensitively observed in the very low frequency cross sections where a small change in Fermi level can result in large shifts in frequency. There is also some indication that different in-plane intercalate densities will yield shifts in Fermi level. In fact, a detailed interpretation of the SdH spectra is expected to provide a critical test for any model of the electronic structure for these materials.

Measurements of the angular dependence of the Shubnikov de Haas frequencies (shown in Fig.5) suggest that the Fermi surface is segmented into

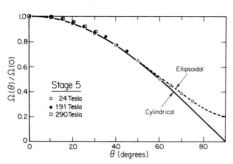

Fig. 5 Angular dependence $\Omega(\theta)/\Omega(0)$ of several dominant SdH frequencies where $\theta=0$ corresponds to $H \parallel$ c-axis and the indicated frequency values are for $\theta=0$. The 290, 191 and 24 T frequencies are not observed above $\sim 30°$, $\sim 45°$ and $\sim 75°$, respectively. The angular dependence of the 290 T and 191 T frequencies is within experimental error given by a $\cos\theta$ dependence (solid line), characteristic of a cylindrical Fermi surface, while the 24 T frequency exhibits departures from the $\cos\theta$ dependence at large angles.

cylindrical rings. Deviations from two-dimensional behavior is measured by small departures from the cylindrical angular dependence and by the observation of doublet SdH frequencies (Fig.4) which indicate different cross-sectional areas at the K and H points in the Brillouin zone.

4. Application of Energy Band Model

To interpret these Shubnikov-de Haas results a band model appropriate to a range of intercalate stages is necessary. Such a model for the electronic dispersion relations $E(\vec{k})$ for graphite intercalation compounds is already available in terms of a Hamiltonian based on the SWMcC three-dimensional model for the graphite π-bands [9]. In this model the c-axis superlattice periodicity is included through a k_z-axis zone-folding of the energy levels. A unitary transformation then transforms the matrix Hamiltonian into a layer representation. For a stage n compound, n graphite layers are retained and the (n+1)st layer is replaced as a first approximation by an empty intercalate layer and for more quantitative results by an intercalate layer which interacts with the adjacent graphite bounding layers. In the "empty intercalate layer" model, the matrix Hamiltonian depends only on the graphite band parameters, and these are already known from previous experiments on pristine graphite. The magnetoreflection experiment provides revised values for some of these band parameters, as modified by the intercalation process. Intercalant-specific interactions between the intercalant and the graphite bounding layer can then be introduced to obtain the final dispersion relations.

This model for the electronic dispersion relations has been applied to the interpretation of the observed SdH frenquencies. Since the most complete set of experimental data is available for stage 5 graphite-K, the application of the energy band model is made for this case. Thus in Fig. 6 are shown the electronic dispersion relations for the five valence and five conduction π-bands appropriate to a stage 5 intercalation compound. The empirical position of the Fermi level E_F shown in Fig. 6 is adjusted to yield the best fit to the experimental SdH ffequencies, in this case corresponding to four partially occupied conduction bands. The Fermi surface parameters obtained from this model are listed in Table 2, including calculated and observed SdH frequencies $\Omega(0)$ for $H \parallel$ c-axis, the cyclotron mass m^*/m_0 at E_F, the trigonal warping anisotropy listed as k_1/k_2 [8], and the electron density n_i for each carrier pocket. The values for k_1/k_2 in Table 2 suggest that trigonal warping is

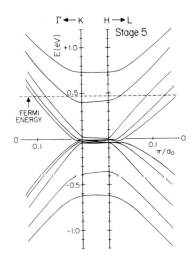

Fig. 6 Electronic band structure in the vicinity of the HK axis near the Fermi level for a stage 5 "empty intercalate layer" model for a graphite intercalation compound. The Fermi level is determined to fit the observed SdH frequencies. For the indicated Fermi level four of the conduction bands are partially occupied and give rise to Fermi surfaces and carrier pockets.

Table 2 Fermi surface parameters associated with stage 5 graphite-K

Fermi Surface Parameters	K-point Band Designations			
	K_1	K_2	K_3	K_4
m^*/m_0	0.143	0.115	0.0828	0.0511
$n_i (\times 10^{20} cm^{-3})$	2.36	1.76	0.975	0.138
Calculated SdH Frequencies $\Omega(0)$	401	300	163	26.7
Observed $\Omega(0)$ (Tesla)	453 430	290 267 243	191 152 135	24 18
Trigonal Warping Anisotrophy k_1/k_2	0.68	0.77	0.91	1.00

important for the larger cross-sectional areas with heavier masses, whereas the smaller light mass cross sections are circular. The generally good agreement of the observed SdH frequencies with the empty intercalate layer model of Fig.6 suggests that the effect of intercalant-graphite bounding layer interactions can be treated as a perturbation and evaluated by fitting the model quantitatively to the observed SdH frequencies. It should be noted that the volumes of the carrier pockets for the four occupied bands in Fig. 6 correspond to a charge transfer of ∼0.3 electrons per intercalant into these carrier pockets, suggesting that other carrier pockets could be present elsewhere in the Brillouin zone.

5. Landau Level Calculation

The empty intercalate layer model used to calculate the Fermi surface can also be applied to compute the Landau levels. The results of such a calculation are shown in Fig. 7 for a stage 3 intercalation compound. For a

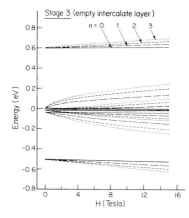

Fig. 7 Magnetic field dependence of the four lowest quantum number Landau levels at $k_z=0$ for a stage 3 graphite intercalation compound, calculated on the basis of an empty intercalate layer model. The middle set of levels correspond to the E_3 levels of graphite while the upper and lower sets correspond respectively to the E_1 and E_2 levels.

stage 3 compound, model calculations yield three conduction bands and three valence bands, such that two of the three conduction and valence bands lie close to each other in the vicinity of the HK axis, leaving one higher lying conduction band and one lower lying valence band. (See Fig.6 where the levels for a stage 5 compound are shown.) The near degeneracy of the four bands for the case of stage 3 results in magnetic level crossings as shown in Fig.7. In this calculation the trigonal warping is also neglected ($\gamma_3=0$) and only linear $k_\perp$ terms in the off-diagonal magnetic Hamiltonian are retained. Under these assumptions the Landau levels can be calculated exactly and the usual optical selection rules apply ($\Delta n=\pm 1$). The central set of levels in Fig. 7 correspond to the K and H point transitions between the pristine graphite E_3 bands, including levels arising from k_z-axis zone folding. On the other hand, the upper and lower Landau levels of Fig. 7 are associated with the E_1 and E_2 graphite bands. Landau level transitions involving these bands are analogous to the H-point transitions in pristine graphite.

Assuming that the Fermi level in a stage 3 donor compound lies between the central and upper group of magnetic energy levels (see Fig.7), additional sets of Landau level transitions are predicted, corresponding to the excitation of an electron from the central set of levels to the upper set. Such excitations require a photon energy of ~0.6 eV which is in excess of the photon energies employed to date in the magnetoreflection experiments on intercalated graphite. The Landau level calculation itself is vital for the detailed analysis of the magnetoreflection experiment with regard to identification of the observed resonant structures, selection rules for Landau level transitions and lineshape analysis for determination of the resonant point within the experimental linewidth. To carry out such a detailed analysis it is necessary to include explicitly the interaction between the intercalant and the graphite bounding layers. In turn, such an analysis provides a sensitive method for the experimental determination of the magnitude of these interaction parameters.

The Shubnikov-de Haas oscillations in the conductivity are also related to the Landau levels shown in Fig. 7. However, for those cross sections with fast frequencies where trigonal warping (see Table 2) is important, it is necessary to also include explicitly the trigonal warping parameter γ_3 and higher terms in $k_\perp$ in order to make a quantitative fit to be experimental SdH frequencies. A systematic fit of both Shubnikov-de Haas and magnetoreflection data for a single series of compounds is now underway and is expected to result in precise determination of the interaction between the intercalant and the graphite bounding layers.

6. Acknowledgements

We gratefully acknowledge S. Y. Leung for valuable discussions and Mr. L. Rubin and Dr. R. L. Aggarwal of the Francis Bitter National Magnet Laboratory for technical assistance and AFOSR Grant #77-3391 for support of this research.

1. G.M.T. Foley, C. Zeller, E.R. Falardeau and F.L. Vogel, Solid State Commun. 24, 371 (1977).
2. Y. Koike, H. Suematsu, K. Higuchi and S. Tanuma, Physica 99B, 503 (1980).
3. D.D.L. Chung and M.S. Dresselhaus, Solid State Commun. 19, 227 (1976); Physica 89B, 131 (1977).
4. E. Mendez, T.C. Chieu, N. Kambe and M.S. Dresselhaus, Solid State Commun. 33, 837 (1980).
5. A.S. Bender and D.A. Young, J. Phys. C. Solid State Phys. 5, 2163 (1972).
6. H. Suematsu, K. Higuchi and S. Tanuma, J. Phys. Soc. Japan 48, 1541 (1980).
7. I. Rosenman, F. Batallan and G. Furdin, Phys. Rev. B20, 2373 (1979).
8. G. Dresselhaus, S.Y. Leung, M. Shayegan and T.C. Chieu, Synthetic Metals (in press).
9. G. Dresselhaus and S.Y. Leung, Solid State Commun. (in press).
10. A. Hérold, Physics and Chemistry of Materials with Layered Structures (ed. F. Lévy), Reidel, Dordrecht, Holland 6, 323 (1979).
11. D.E. Nixon and G.S. Parry, J. Appl. Phys. 1, 291 (1968).
12. F.J. DiSalvo, S.A. Safran, R.C. Haddon, J.V. Waszczak, and J.E. Fischer, Phys. Rev. B20, 4883 (1979).

Electrical Properties of Layered Materials at High Magnetic Fields

S. Tanuma, R. Inada, A. Furukawa, O. Takahashi, and Y. Iye
Institute for Solid State Phyics, University of Tokyo, Roppongi, Minato-ku, Tokyo, 106, Japan

Y. Onuki
Saitama Institute of Technology, Fusaiji, Okabe, Saitama, 369-02, Japan

We present extended results of former work reported in the Oxford MFSP Conference[1]. Part 1 deals with the $1T\text{-}TaS_{2-x}Se_x$ system and part 2, with graphite and its intercalation compounds.

Part 1 $1T\text{-}TaS_{2-x}Se_x$

$1T\text{-}TaS_2$ is known to have the largest amplitude of charge density wave among the family of transition metal dichalcogenides[2] and shows a remarkable electrical property. Namely, DISALVO and GRAEBNER have found that the low temperature conduction of this compound exhibits a variable range hopping type which is characteristic of two dimensional Anderson localization[3]. We have reported an anomalously large positive magnetoresistance of $1T\text{-}TaS_2$ in the temperature range from 4 to 0.5K[1]. An even larger negative magnetoresistance has been found by KOBAYASHI and MUTO at lower temperatures from 0.4 to 0.05K[4]. This negative effect has been explained by FUKUYAMA and YOSIDA[5] in such a way that the highest occupied state has an upward linear Zeeman shift to make the energy difference between the mobility edge E_c and the Fermi energy E_F smaller under magnetic fields and to make the hopping conduction larger. The positive effect has been left unexplained. The present experiment reveals that the positive magnetoresistance is enhanced and the negative one depressed by substituting the isoelectronic selenium for sulfur, as shown in Fig.1. The magnetoresistance $\Delta\rho/\rho=\{\rho(H)-\rho(0)\}/\rho(0)$ of ca.0.8 for x=0.001 at 0.59 K is remarkably large compared to the value of 0.26 for $1T\text{-}TaS_2$ as shown by the dashed line in Fig.1. We furthermore show the magnetoresistance for several values of composition x in Fig.2. The

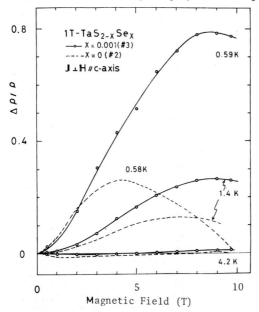

Fig.1 Transverse magnetoresistance of $1T\text{-}TaS_{2-x}Se_x$ (x=0, 0.001)

magnetoresistances $\Delta\rho/\rho$ at 1.4K for x of 0.001, 0.1 and 0.4 are 0.26, 0.32 and 0.47, respectively with the maximum values appearing around 9 to 12T.

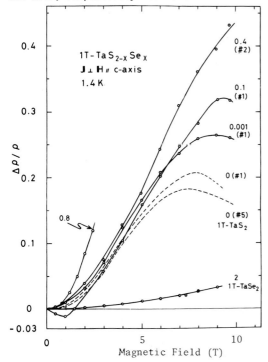

Fig.2 Transverse magnetoresistance of $1T\text{-}TaS_{2-x}Se_x$ (x=0, 0.001, 0.1, 0.4, 0.8, 2)

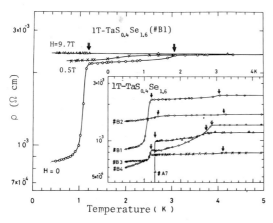

Fig.3 Temperature dependence of electrical resistivity of $1T\text{-}TaS_{0.6}Se_{1.4}$

Therefore, the positive magnetoresistance increases with an increase of disorder in the potential energy by alloying. Recently YOSIDA and FUKUYAMA[6] have improved the theory by considering the change in E_c due to magnetic field as $E_c-E_c(0) \propto H^2$, and obtained the reasonable account of positive magnetoresistance of $1T\text{-}Ta_{1-x}Ti_xS_2$ as well as of the present system.

In the corresponding localized states, the resistivity is described as $\rho^{-1} = A\exp(-T_0/T)^{1/n}$ where T_0 is a characteristic temperature representing the degree of disorder and the n-value is 3 in a two dimensional system. The values of T_0 are also remarkably enhanced as 110K and 2200K for x=0.001 and 0.6 respectively to be compared to ~50K for $1T\text{-}TaS_2$.

The nonmetallic and localized character at low temperatures, being enhanced by Se alloying up to x=0.8, is followed by a poorly metallic region 0.8<x<1.6 where the low temperature resistivity does not change much throughout the temperature range of 300K to 4K. For 1.6<x<2.0, the materials show a metallic behavior, i.e., the low temperature resistivity falls down to the residual values. In the poorly metallic range, the resistivity is found to make a rather steep decrease having two steps at ca. 3K and 1.2K, as shown in Fig.3. The magnetoresistance measurement up to 10T reveals that the resistivity recovers up to the temperature independent resistivity as the extention of residual resistance above ca.3K. This fact means that the system becomes superconducting, at least in parts of the material. Although these decreasing behaviors in resistivity with

317

decreasing the temperature are rather sample dependent for a fixed x value as shown in the insert of Fig.3, two steps of resistivity change always exist in this range of composition. The reason why superconductivity occurs only in the poorly metallic region is interesting but presently unexplained.

Part 2   Graphite and its Intercalation Compounds

Graphite is known to have such an anomalous magnetoresistance that the $\Delta\rho/\rho$ versus H relation is linear at liquid helium temperature ignoring the de Haas-Shubnikov oscillatory part[7]. The linearity holds above about 2T to the highest measured field (15T), and this behavior was explained by the change of the scattering range of ionized impurities for degenerate electrons due to magnetic field in the quantum limit region[7]. We re-examined this behavior on highly oriented pyrolytic graphite (HOPG) and single crystal graphite (SCG) at the temperatures of liquid nitrogen and liquid helium up to the impulsive magnetic field of ca.30T. As is shown in Fig.4, HOPG behaves as follows; for H//c-axis and T=77K, $\rho(H)$ changes as H^2 until about 4T, as H^1 up to about 9T and tends to saturate at higher fields. For lower temperatures the H-linear dependence is more extended to lower fields except the oscillatory part where the dip at 7T is due to the quantum limit. For H⊥c-axis and T=4.2-2.1K, $\rho(H)$ is near to $\propto H^1$ throughout the field range. The saturation at higher fields for H//c-axis has been found by WOOLAM et al.[8] on HOPG under the fields up to 21T and this tendensy has been explained by SUGIHARA and WOOLAM[9] assuming a magnetic freeze out effect of impurity levels. Fig.5 shows the case of SCG. A remarkable behavior is the sudden increase of $\Delta\rho/\rho$ in the highest field range; the field strength at which the increase occurs becomes smaller as the temperature decreases. We have not been able to identify the origin of this new phenomenon.

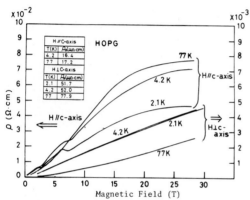

Fig.4   Transverse magnetoresistance of HOPG in the field configurations of H//c-axis and H⊥c-Axis.

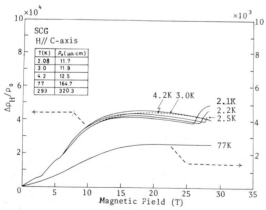

Fig.5   Transverse magnetoresistance of SCG at 77K and liquid helium temperatures

The de Haas-van Alphen effect of graphite intercalation compounds (GIC) of various intercalant species are under investigation in order to look for the Fermi surface of GIC's. Major studies are done by using a 5T superconducting magnet, but some measurements are undertaken by the use of a 15T superconducting

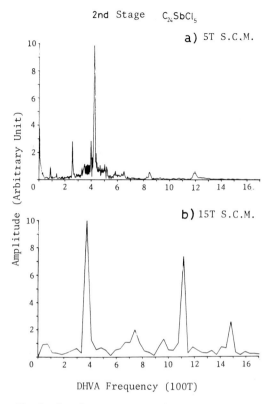

Fig.6 Fourier spectra of de Haas-van Alphen frequencies for 2nd stage $SbCl_5$-GIC obtained by the use of (a) 5T and (b) 15T SCM.

magnet in order to search for the higher frequencies which are not detected by the 5T magnet if they exist. We report in this paper this high field test only, and the major results are reported elsewhere[10]. It is likely that existing but missing higher frequencies in such low stage GIC's as 2nd or 1st are detectable by the use of the 15T magnet. Fig.6 shows a comparison of Fourier spectra of de Haas-van Alphen frequencies for the 2nd stage $SbCl_5$-GIC by the use of the 5T and 15T magnets. Although oscillations under analysis for the latter case are fewer in number and the field calibration is less precise, it is clear that the new frequency of ca.1.5×10^3T is found in the latter case.

The experiment by the use of a pulsive magnet has been done in cooperation with Prof. N. Miura and Dr. G. Kido.

[1] S.Tanuma, H.Suematsu, K.Higuchi, R.Inada and Y.Ōnuki, Booklet of the Conference of the Application of High Magnetic Fields in Semiconductor Physics, Oxford (1978) p.85.
[2] J.A.Wilson, F.J.DiSalvo and A.Mahajan, Adv. in Phys. 24, 117 (1975).
[3] F.J.DiSalvo and J.E. Graebner, Solid State Commun. 23, 825 (1977).
[4] N.Kobayashi and Y.Muto, Solid State Commun. 30, 337 (1979).
[5] H.Fukuyama and K.Yosida, J. Phys. Soc. Jpn. 46, 1522 (1979).
[6] K.Yosida and H.Fukuyama, J. Phys. Soc. Jpn. 48, 1879 (1980).
[7] J.M.McClure and W.J.Spry, Phys. Rev. 165, 809 (1967).
[8] J.A.Woolam, D.J.Sellmyer, R.O.Dillon and I.L.Spain, Proceedings of the 13th International Conference on Low Temperature Physics vol.4, 358 (1974).
[9] K.Sugihara and J.A. Woollam, J. Phys. Soc. Jpn. 45, 1891 (1978).
[10] e.g., S.Tanuma, Yamada Conference IV, Physics and Chemistry of Layered Materials, (Sendai, 1980), to be published in Physica B (1981).

Possibility of the Magnetic Breakdown for 2H-MX$_2$ in High Field

M. Naito, and S. Tanaka

Department of Applied Physics, Faculty of Engineering, University of Tokyo Bunkyo-ku, Tokyo 113, Japan, and

N. Miura

Institute for Solid State Physics, University of Tokyo, Roppongi, Minato-ku Tokyo, Japan

1. Introduction

Layered group VB transition-metal dichalcogenides (MX$_2$) show a great variety of charge-density-wave (CDW) transitions. In 2H-polytypes (2H-NbS$_2$, 2H-NbSe$_2$, 2H-TaS$_2$ and 2H-TaSe$_2$), the observed CDW transitions are moderate and second-order-like, and their metallic natures are retained even at low temperatures. These four compounds are grouped into one family (2H-MX$_2$ family) [1] because in the normal state their crystal structures including lattice parameters and electronic (band) structures are similar to each other. But they have different CDW transition temperatures. NbS$_2$ has no CDW transition, while the other three compounds, NbSe$_2$, TaS$_2$ and TaSe$_2$, exhibit transitions from the normal (N) to the incommensurate (I) phase at 35K, 80K and 122K, respectively. In 2H-NbSe$_2$ the CDW remains incommensurate down to the lowest measured temperature. 2H-TaSe$_2$, on the other hand, shows the further lock-in transition — the incommensurate-commensurate (I-C) transition at 91K. Thus those compounds have different CDW strengths from the weakest NbS$_2$ with no CDW transition to the strongest TaSe$_2$. Therefore the comparative measurements of the physical properties give much information on CDW in this family [1].

We have measured systematically the magnetoresistance $\Delta\rho/\rho$ and the Hall coefficient R_H of these compounds at low temperatures in fields up to 150kOe, using samples of high quality. The most important effect of CDW on the galvanomagnetic properties is to introduce new small energy gaps in the one electron spectrum [2]. In the CDW state a superlattice, the periodicity of which is determined by the wave vectors $\vec{q}$ of the CDW, is formed. This new periodicity induces energy gaps at the values of $\vec{k}$ which satisfy $\varepsilon(\vec{k}) = \varepsilon(\vec{k}+n\vec{q}+\vec{G})$, where $\vec{G}$ is the reciprocal lattice vector. For the commensurate CDW (CCDW), the new periodicity defines the new Brillouin zones in reciprocal space. The energy gaps appear at the boundaries of the new Brillouin zones. These energy gaps modify the shape of the Fermi surface, and change the galvanomagnetic properties drastically. But the magnitude of the energy gaps, which depends on the CDW strength, is generally small. Therefore the magnetic breakdown through these energy gaps is expected to occur easily in relatively low fields. The influence of the magnetic breakdown would be reflected in the galvanomagnetic properties [3]. WILSON [1] has investigated the reconstruction of the Fermi surface of 2H-TaSe$_2$ under the CCDW state by remapping, on the basis of the normal-state band calculations and the dHvA results [4], and suggested the possibility of the magnetic breakdown.

2. Experiments and Results

For 2H-NbSe$_2$ and 2H-TaSe$_2$, single crystals were prepared by the usual iodine vapor transport method using very pure starting materials. On the other hand,

Table 1 Values of resistivity and R.R.R.

	$\rho(300K)[\Omega cm]$	$\rho_0[\Omega cm]$	R.R.R.
NbS_2	0.96×10^{-4}	10.0×10^{-7}	96
$NbSe_2$	1.00×10^{-4}	6.3×10^{-7}	160
$TaSe_2$	1.32×10^{-4}	5.3×10^{-7}	250

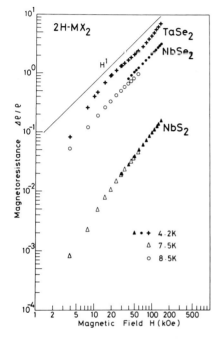

Fig.1 The field dependence of the transverse magnetoresistance (H // c-axis ⊥ current J)

2H-NbS$_2$ single crystals were grown by the direct vapor transport method without iodine. The values of resistivity at room temperature and residual resistivity ρ_0 of NbS$_2$, NbSe$_2$ and TaSe$_2$ are summarized in Table 1, together with the residual resistivity ratio (R.R.R.). In superconducting NbS$_2$ and NbSe$_2$, ρ_0 was estimated by assuming the power-law temperature dependence for the ideal resistivity. The field dependence of the magnetoresistance and the Hall coefficient of these compounds is shown in Fig.1 and Fig.2. The following features have been observed in each compound.

2H-NbS$_2$: The magnetoresistance is small, ~0.16 at 145kOe. Its field dependence varies smoothly from $H^{1.8}$ at low fields to $H^{1.3}$ at high fields. The Hall coefficient is +5.1~5.6 cm^3/C and nearly field independent within 2%.

2H-NbSe$_2$: $\Delta\rho/\rho$ is relatively large, ~3.3 at 145kOe. The field dependence is H^1 over a wide range from 10 to 145 kOe and $H^{1.2}$ below 10kOe. R_H is negative and has a strong field dependence with a sharp minimum at ~11kOe.

2H-TaSe$_2$: $\Delta\rho/\rho$ is relatively large, ~7.5 at 145kOe. Its field dependence is roughly H^1 from 20 to 145kOe and $H^{1.7}$ below 10kOe. There is an inflection upward around 70kOe. R_H has a drastic field dependence. It decreases rapidly from ~0 at the lowest field and flattens near 50kOe. Then above 90kOe it increases gradually. The broad minimum around 70kOe of R_H coincides with the inflection of $\Delta\rho/\rho$.

3. Discussion

In this section we discuss the possibility of the magnetic breakdown in 2H-NbSe$_2$ and 2H-TaSe$_2$. 2H-NbS$_2$ has no CDW transition, and is considered to be roughly an anisotropic normal metal. The smallness of the magnetoresistance may be expected from the theoretical predictions that the carriers originate from d-bands of the transition metal with a heavy mass and that the Fermi surface is nearly cylindrical. The carrier density +1.2x10^{22} cm^{-3} calculated from R_H is comparable with the value 1.78x10^{22} cm^{-3} estimated from one electron/Nb-atom. On the other hand, in NbSe$_2$ and TaSe$_2$ the galvanomagnetic effects are anomalous and exhibit the following properties in common which are absent in NbS$_2$.
1) The magnetoresistance is relatively large (one or two orders of magnitudes larger than NbS$_2$) and its field dependence is expressed by H^n with $n \simeq 1$.

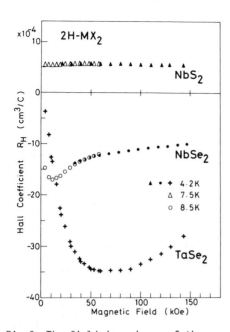
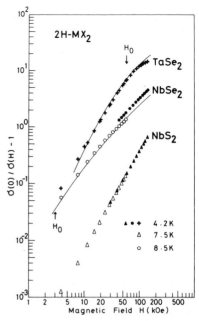

Fig.2 The field dependence of the Hall coefficient (H // c-axis ⊥ J)

Fig.3 The field dependence of $\sigma(0)/\sigma(H)-1$. The solid lines are the fitting curves based on (2)

2) The Hall coefficient shows a strong magnetic field dependence. It has a minimum, followed by a gradual increase.
These features are partly understandable if one takes account of the fact that the Fermi surface gets complicated in the CDW state. But the H^1 dependence of the magnetoresistance and the behavior including a minimum of the Hall coefficient can not be explained by a simple two-band model. Thus the possibility of the magnetic breakdown in these compounds is considered. The qualitative analysis based on this possibility has been performed as follows.

Magnetoresistance
Generally when there are the points, that give rise to magnetic breakdown, in the path of an electron, the electron undergoes an additional scattering at these points [3]. By including this scattering process, the effective relaxation time τ_{eff} of the electron is given as

$$\frac{1}{\tau_{eff}} = \frac{1}{\tau_0} + \frac{1}{\tau_m} \quad , \quad \frac{1}{\tau_m} = \omega_c f(P) \tag{1}$$

where τ_0 is the intrinsic relaxation time and τ_m is due to magnetic breakdown, which is proportional to ω_c^{-1}, because the electron passes through the breakdown points as many times as proportional to ω_c per unit time. P is the breakdown probability, i.e. $P = \exp(-H_0/H)$ and $H_0 = K\Delta^2 mc/\varepsilon_F e\hbar$. H_0 is the characteristic field of breakdown, ε_F the Fermi energy, Δ the gap between two bands involved in the breakdown and K a numerical factor of order 1. The function f depends on the individual case. Using this effective relaxation time, the conductivity in a field is expressed by

$$\sigma(H) = \frac{ne^2}{m}\tau_{eff}\frac{1}{1+(\omega_c\tau_{eff})^2} \quad , \quad \frac{\sigma(0)}{\sigma(H)} = \frac{\tau_0}{\tau_{eff}} + \frac{\tau_{eff}}{\tau_0}(\omega_c\tau_0)^2 . \quad (2)$$

An H-linear field region appears in $\Delta\rho/\rho$ calculated from (2). In order to compare the experiments with this semiquantitative formula, $\sigma(0)/\sigma(H)-1$ for each compound was calculated from $\Delta\rho/\rho$ and R_H by tensor inversion. The results are shown in Fig.3 on a log-log scale. For NbS_2 the slope is constant, and $\sigma(0)/\sigma(H)-1 = H^{1.8}$, which is considered to be nearly normal. In $NbSe_2$, however, $\sigma(0)/\sigma(H)-1$ is almost H-linear except below $\sim$15kOe. In $TaSe_2$ $\sigma(0)/\sigma(H)-1$ is $H^{1.7}$ below 50kOe, then deviates downward in higher fields. By comparing these results with (2) assuming $f = C \cdot \exp(-H_0/H) \times (1 - \exp(-H_1/H))$ (this formula would be applied to the case that the breakdown field has a certain width ΔH; H_0 and $H_1 = H_0 + \Delta H$ are the onset and termination fields of breakdown), the following values were obtained. $H_0 \approx 2 \sim 3$kOe, $H_1 \gtrsim 150$kOe for $NbSe_2$ and $H_0 \approx 55 \sim 65$kOe for $TaSe_2$.

Hall coefficient

First we concentrate on $TaSe_2$. On the basis of theoretical energy-band calculations it is considered that in the normal state the carriers are p-type and that their density n_h^0 is 1.54×10^{22} cm^{-3} ($R_H = 4.1 \times 10^{-4}$ cm^3/C). On the other hand it is supposed that in the CCDW state the n-type carriers become major and that the density $n_e - n_h$ is 1.71×10^{21} cm^{-3} ($= n_h^0/9$, $R_H = -36 \times 10^{-4}$ cm^3/C) unless the real Fermi surface is much different from that obtained by WILSON's remapping [1]. This value is close to the experimental value in 50$\sim$90kOe, so it is considered that in this region R_H takes its high-field value. According to this interpretation the drastic variation of R_H below 50kOe is thought to be due to the change of galvanomagnetic behavior from the low- to the high-field condition. On the other hand, the fact that R_H increases gradually above 90kOe may be understood by the magnetic breakdown, which changes the carrier trajectories from electron-like to hole-like ones, as expected from the theoretical Fermi surface. H_0 is estimated to be 80$\sim$90kOe. In the case of $NbSe_2$ the situation is more complicated. The behavior of R_H might be comprehended by regarding that before the high-field condition is attained, breakdown starts, and $H_0 \lesssim 10$kOe is obtained.

Breakdown Field and Energy Gap

By considering both the results of $\Delta\rho/\rho$ and R_H, the onset field of the magnetic breakdown is thought to be $\sim$4kOe for $NbSe_2$ and 70$\sim$80kOe for $TaSe_2$. For $TaSe_2$ this estimated value agrees with that conjectured from dHvA measurements [4]. The corresponding energy gaps Δ are 4.3 meV for $NbSe_2$ and 18 meV for $TaSe_2$ (if $m=m_0$, $\varepsilon_F=0.4$eV, K=1). It is noted that $\Delta(NbSe_2)/\Delta(TaSe_2)$ is qualitatively equal to the ratio of To. The energy gaps by CDW in the one-electron spectrum are not uniform in k-space. Besides the main first-order gap Δ_1, there are a number of higher-order gaps Δ_2, Δ_3, etc. [3]. The complete breakdown through Δ_1 would change R_H to its normal-state value. So the gaps estimated from our experiments probably correspond to Δ_2 or the lowest limit of Δ_1. Finally at least for $NbSe_2$ the breakdown field is thought to have a considerable width. This fact might be due to the dependence of the gaps on the wave vector k_z parallel to the hexad axis and/or the coexistence of the first- and higher-order gaps. The incommensurate nature of CDW, however, may have to be taken into account in $NbSe_2$.

References

1 J.A.Wilson, Phys.Rev. B15, 5748 (1977)
2 A.W.Overhauser, Phys.Rev. B3, 3173 (1971)
3 L.M.Falicov and P.R.Sievert, Phys.Rev. 138, A88 (1965)
  L.M.Falicov and M.J.Zuckermann, Phys.Rev. 160, 372 (1967)
4 J.E.Graebner, Solid State Commun. 21, 353 (1977)

Part XI

Magnetism and Magnetic Semiconductors

High Field Magnetism

R. Pauthenet

Service National des Champs Intenses - CNRS
Institut National Polytechnique, BP 166 X
38042 Grenoble-Cedex, France

The magnetic phenomena are ruled by two main energy terms: the exchange energy and the magnetocrystalline anisotropy energy. According to the amplitude of the exchange interaction, with respect to the thermal energy; according to the relative importance of the terms of exchange energy, magnetocrystalline energy and of coupling energy of the moments with the applied field, the magnetic moments can adopt different relative orientations. High magnetic fields produce on some substances transitions from an antiferromagnetic to a ferromagnetic state, which we call metamagnetism [1]; numerous examples exist [2], particularly with the rare earth elements and compounds [3][4][5]. On substances with high anisotropy, high magnetic fields allow to precise some specific properties, such as the anisotropy of the magnetic moment [6], the quenching of the moment due to the crystalline field [7]. On the other hand, the analysis of the isotherms of the magnetization with high magnetic fields allow to interpret interesting properties of itinerant magnetism and enhanced magnetism [8][9]. In the present article, I discuss two original results that we have obtained recently at the SNCI Grenoble: the first one deals with the saturation of paramagnetism and the effect of the crystalline field, the second one concerns the properties of the spin waves.

1. The Magnetization of the Rare Earth Ultraphosphates GdP_5O_{14} and YbP_5O_{14}

This study has been performed in collaboration with M. BAGIEU and J.C.PICOCHE. The earliest determination of the magnetic moments of the rare earth ions have been carried out by B. CABRERA [10] from measurements of the paramagnetic susceptibility of some hydrated rare earth salts. W. HENRY [11] studied the saturation of Gd^{3+} ion in fields up to 52,000 œ, on $(SO_4)_3Gd_2 \cdot 8H_2O$. The advantage of highly hydrated salts is that the magnetic ion is highly diluted and behaves as a free paramagnetic ion, without exchange interactions; however, we have to face an experimental problem of sample preparation, due to the fact that the number of the water molecules is not exactly known, when experiment is performed. The rare earth ultraphosphates [12], with the general formula TP_5O_{14}, where T is a rare earth ion, present the advantage of being well defined compounds as well as having highly diluted magnetic ions, 1 magnetic ion among 19 non-magnetic, compared to 1 to 20 in hydrated salts.

The magnetization of GdP_5O_{14} and YbP_5O_{14} has been measured at temperature T, between 1.75 and 4.21 K, and in a magnetic field H up to 198,000 œ. The moment M, in Bohr magneton μ_B, has been represented versus the ratio H/T (refer to Fig.1).

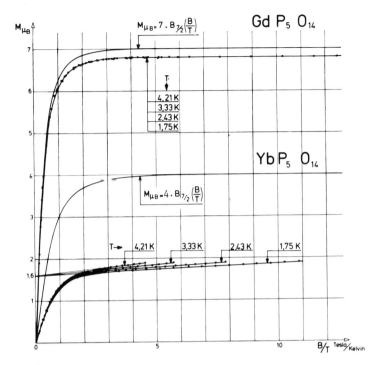

Fig.1 The variation of the magnetic moment M of GdP$_5$O$_{14}$ and YbP$_5$O$_{14}$ versus the ratio B/T of the magnetic field on the temperature

For GdP$_5$O$_{14}$ all experimental points, for the different temperatures, are located on the same continuous curve, which, however, lies slightly below the calculated 7/2 Brillouin law; this discrepancy is due to a defect of stoichiometry of 3.1%; it is not perfect but it is better than for hydrated rare earth salts, for which we have not yet succeeded in obtaining a discrepancy less than 8%. Concerning YbP$_5$O$_{14}$, as it could be expected, the experimental points lie much below the calculated 7/2 Brillouin law for the free ion. The same result has already been obtained on the garnet compound 5Ga$_2$O$_3 \cdot$3Yb$_2$O$_3$ [13]. The moment of the Yb^{3+} ion is quenched by the electric crystalline field of the neighbouring O^{2-} ions. However, during the measurements, I noticed that above 150,000 œ, for the same magnetic field, but at different temperatures, we measure exactly the same moment with an accuracy of 10^{-3} and the (M,H) variation is a straight line, indicating that the corresponding susceptibility is temperature-independent; moreover, in the representation (M,H/T), below H/T < 0.5, which corresponds roughly to H < 20,000 œ, the experimental points are located on a unique curve, indicating a Curie paramagnetism. The quantitative interpretation of the crystalline field effect confirms this point of view. YbP$_5$O$_{14}$ crystallises in the monoclinic system [12]; the Yb^{3+} ion is surrounded with 8 O^{2-} ions, situated at the apexes of a hendecadron; it is a cube with large distortions. Calculations of the magnetic properties of the Yb^{3+} ion in a crystalline field have been accomplished by Y. AYANT and J. THOMAS [14], in the case of a cube without distortions; the eight-fold degeneracy of Yb^{3+} is splitted into two doublets and one quadruplet; the system of lowest energy is a doublet; at about 900 K above lies the quadruplet. The energy levels of the doublet with the field are

$$W_{1,2} = (-18-12s) A \mp \frac{3}{2} g\mu_B H - \frac{3g^2\mu_B^2 H^2}{(20+28s)A} \tag{1}$$

where g is the Landé factor, $A = -\frac{A_4}{77}$ and $s = -\frac{35}{39} \cdot \frac{A_6}{A_4}$, with A_4 and A_6 the coefficients of the spherical harmonics terms of the 4th and 6th order in the hamiltonian of the crystalline field. At temperature T, the magnetization is given by the expression

$$M = \frac{3g\mu_B}{2} \cdot \left[th(\frac{3g\mu_B}{2} \cdot \frac{H}{k_B T}) + \frac{g\mu_B}{(5+7s)A} \cdot H \right] \tag{2}$$

where k_B is the Boltzmann factor.

As previously said, the magnetization is the resultant of two contributions: the first one saturates at 1.71 μ_B, the second one varies linearly with the field and is temperature independent. To render an account of the experimental results, the only adjustable parameter $(5+7s)A$ has to be taken equal to 89 k_B; the temperature-independent susceptibility is $15 \cdot 10^{-6}$ e.m.u/g. However, there is a defect of stoichiometry of 7% in the composition of the compound. It is the first time that the temperature-independent susceptibility term has been observed on the Yb^{3+} ion; for that purpose it is necessary to perform accurate measurements above 150,000 œ. Incidentally, the quality of the agreement between theory and experiment means that the magnetic properties of the Yb^{3+} ion are not very sensitive to the distortions of the surrounding cube of the oxygen ions.

2.1 The Magnetization of Nickel and the Spin Waves

This study has been performed in collaboration with J.C. PICOCHE and P. RUB. The magnetization of a single crystal of nickel has been measured, along its easy axis, between 4.21 K and room temperature and under magnetic fields up to 190,000 œ. The sample has a spherical shape of 6 mm diameter. We calibrate the absolute value of the moments on the very accurate measurements of H. DANAN [15]. We estimate the accuracy at 0.1 e.m.u/g. The magnetic field is known with an error less than 100 œ and the temperature is stabilized at 0.1K.

The relative accuracy on the determination of the moments is at least 10^{-4}; this point has to be emphasized for the following of the discussion; it even happened that we reproduced experiment after several days, having studied other substances in the mean time, with a relative accuracy better than 10^{-4}. The isotherms (σ, He) are shown on Fig.2; σ is the specific moment in e.m.u/g.

It is commonly thought that the interesting part of the magnetization curve is that in high fields, which we consider linear with the field and we define the tangent to the curve in high fields to determine a term of linear susceptibility which is superimposed to the spontaneous magnetization; this spontaneous magnetization is determined with the interception of the tangent with the demagnetizing line. The experimental points below the tangent, which ap-

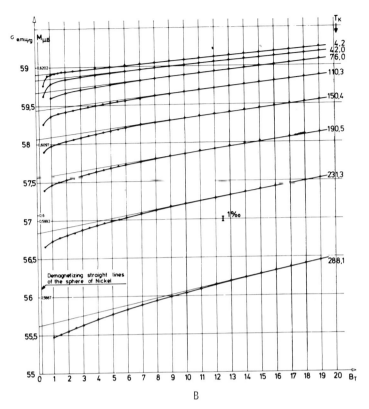

Fig. 2 The isothermal variation of the specific moment versus applied magnetic field He at different temperatures

proach the tangent when the field is increased, are not generally taken into account; they are considered to belong to some not well-defined law of approach to the saturation; it is however surprising to observe a decreasing of the incurved part of this law of approach to saturation, when the temperature decreases (refer to Fig.2).

2.2 The $H^{1/2}$ Holstein-Primakoff Test

F. BLOCH created in 1930 the concept of spin waves [16] and established that the spontaneous magnetization σ_s of a ferromagnet decreases with temperature according to the law

$$\sigma_s = \sigma_o (1 - a_{3/2} T^{3/2}) \tag{3}$$

σ_o is the absolute saturation moment. T. HOLSTEIN and H. PRIMAKOFF studied the "field dependence of the intrinsic domain magnetization of a ferromagnet" [17], within the framework of the spin waves; they establish that, with a good approximation, the magnetization $\sigma(T,H)$ could follow the law

$$\sigma(T,H) = \sigma_s + A(T) \cdot H^{1/2} \tag{4}$$

the coefficient A is a linear function of the temperature and of $a_{3/2}$ in the Bloch law. Moreover, the experiment shows a superimposed linear term with the field of susceptibility χ, that we observe mainly at low temperature, when the coefficient A is small. I attributed the curvature of the experimental isotherm (σ, H)(Fig.2) to the spin wave contribution in $H^{1/2}$ and I tried to represent the experimental $\sigma(T,H)$ curve with the equation

$$\sigma(T,H) = \sigma' + A(T) \cdot H^{1/2} + \chi H; \tag{5}$$

as we shall see later on, σ' is slightly different form the spontaneous magnetization σ_s.

Which field H has to be considered to represent the spin-waves mechanism? According to R.E. ARGYLE and al. [18] and F. KEFFER [19], H is the sum of the applied field $\vec{H}_e$, dipolar field $\vec{H}_{dip}$, demagnetizing field $\vec{H}_{dem}$, and anisotropy field $\vec{H}_A$, along the easy axis of magnetization. Why this latter field? To take the reinforcement of the moment by the anisotropy energy into account. As the sample is spherical, $\vec{H}_{dip} + \vec{H}_{dem} = 0$, hence

$$H = H_e + H_A \tag{6}$$

The anisotropy field is calculated from the first anisotropy constant K_1, the thermal variation of which is known from several authors[20]. In nickel, H_A is not very large; its value does not exceed 2,900 œ. Under these conditions, the spontaneous magnetization σ_s, which by definition is the moment of the atoms in the absence of external applied field, is

$$\sigma_s = \sigma' + A(T) \cdot H_A^{1/2} + \chi H_A; \tag{7}$$

σ' is almost equal to σ_s when H_A is small.

To determine the three unknown quantities σ', $A(T)$ and χ in (7), we estimate that the best method is the following: we fix a value of the applied field H_{eo}, at which corresponds the magnetization $\sigma(T, H_{eo} + H_A)$; we establish the linear relation

$$\frac{\sigma(T, H_e + H_A) - \sigma(T, H_{eo} + H_A)}{(H_e + H_A)^{1/2} - (H_{eo} + H_A)^{1/2}} = A(T) + \chi(T)\left[(H_e + H_A)^{1/2} + (H_{eo} + H_A)^{1/2}\right]$$

between the quantities $y = \dfrac{\sigma(T, H_e + H_A) - \sigma(T, H_{eo} + H_A)}{(H_e + H_A)^{1/2} - (H_{eo} + H_A)^{1/2}}$

and $x = (H_e + H_A)^{1/2} + (H_{eo} + H_A)^{1/2}$, that we can calculate for the different values of H_e; we carry out a linear regression using a least squares method to determine $A(T)$ and $\chi(T)$ and then we calculate σ_s. We try the tests for different couples $\left[H_{eo}, \sigma(T, H_{eo} + H_A)\right]$, without finding any important differences. The correlation coefficient, or "degree of fit" of the given points to the least squares straight line is between 0.988 and 0.998 for the different isotherms. The results are collected in table 1; although the precision on A is ± 2% and that on χ is ± 4%, we keep two decimals in A and three decimals in χ, in order to find calculated values of σ (T, H_e + H_A) very near the experimental values; the values of A and χ are correlated; with different couples (A, χ), σ_s has been determined with an accuracy of 1°/$_{oo}$.

Table 1

T_K	σ' e.m.u/g	$A \cdot 10^6$ e.m.u/g	$\chi \cdot 10^6$ e.m.u/g	σ_s e.m.u/g
4.21	58.892$_1$	103.91	1.529	58.90$_0$
42	58.740	506.16	1.109	58.76$_8$
76	58.519$_4$	847.60	0.956	58.56$_3$
110.3	58.229$_2$	1,144.26	0.710	58.28$_0$
150.4	57.838$_7$	1,241.33	0.868	57.88$_8$
190.5	57.297$_6$	1,427.93	1.064	57.34$_3$
231.3	56.544$_5$	1,614.70	1.440	56.58$_4$
288.1	55.238$_8$	1,877.42	2.083	55.26$_8$

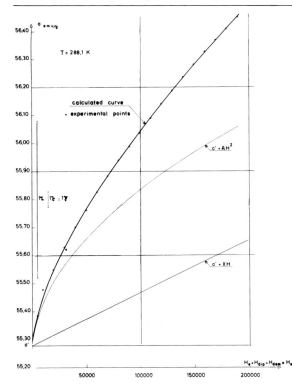

Fig.3 The calculated curve σ (T, H_e) = σ' + A $(H_e + H_A)^{1/2}$ + χ $(H_e + H_A)$ at T = 288.1 K and the experimental points, with the contribution of the A $H^{1/2}$ term and the χ H term

The figure 3 shows an example of the calculations for the isotherm T = 288.1 K; note the precision of the results, the accuracy between the calculated isotherm and the experimental points, the continuous curvature of the σ (T, H_e) curve and the marked bend in weak fields caused by the A (T). $H^{1/2}$ contribution. At this high degree of accuracy, the $H^{1/2}$ Holstein-Primakoff test is well verified; it is the first time as far as we know. We do not detect any other term in H^n.

2.3 The Bloch Law of the Spontaneous Magnetization Variation with Temperature

The most elaborated development of the Bloch law is that established by F. DYSON [21]

$$\sigma_s = \sigma_0 (1 - a_{3/2} T^{3/2} - a_{5/2} T^{5/2} - a_{7/2} T^{7/2} - a_4 T^4). \tag{9}$$

The first experimental test of the $T^{3/2}$ law was by M. FALLOT [22]; since then different physicists [18][23] have tried to test the law in its more elaborated form.

Although it has been established theoretically that the a_i coefficients depend upon temperature - I shall give an experimental proof of this statement later on - I found it of some interest to determine the values of these coefficients, assuming that they are constant; we then obtain useful orders of magnitude. On the variation of the spontaneous magnetization with temperature, we determine by extrapolation the absolute saturation moment σ_0 = 58.903 e.m.u/g, which corresponds to 0.619 μ_B per atom; this value has to be compared with the value 0.616 μ_B determined by H. DANAN and al. [24].

We calculate the difference $\sigma_0 - \sigma_s$ (T) at different temperatures, and plot the curve $\sigma_0 - \sigma_s$ (T) versus $T^{3/2}$; we take only into account the points at 4.2 K, on which the error is 4%, and the points above. The slope of the tangent to the curve passing through the origin gives σ_0 $a_{3/2}$ from which we determine $a_{3/2}$ = (8 ± 0.4) · 10^{-6} deg$^{-3/2}$. With the same method, we determine $a_{5/2}$ = (1 ± 0.4) · 10^{-6} deg$^{-5/2}$ and $a_{7/2}$ is of the order of (0.2 ± 0.2)· 10^{-10} deg$^{-7/2}$; the a_4 coefficient cannot be estimated. These values have to be compared with those determined by B.E. ARGYLE and al. [18], according to another method: $a_{3/2}$ = (7.5 ± 0.2) · 10^{-6}, $a_{5/2}$ = (1 ± 1) · 10^{-8}.

The coefficient $a_{3/2}$ is expressed by the following relation [19]:

$$a_{3/2} = \zeta \left(\frac{3}{2}\right) \cdot \frac{g\mu_B}{\rho\sigma_0} \cdot \left(\frac{k_B}{4\pi D}\right)^{3/2}, \tag{10}$$

in which $\zeta \left(\frac{3}{2}\right)$ is the RIEMANN zeta function, g the Landé factor equal to 2.20 for nickel [25], ρ the density; D is the coefficient of the dispersion law of the spin waves, ε_k = D k^2, with ε_k the energy and k the wave vector; in a cubic substance D = 2 S J a^2 where S is the spin number, J the exchange integral and a the lattice parameter. We determine D equal to 59.4 · 10^{-30} ergs - cm^2 or 371 meV. $\overset{\circ}{A}^2$, to be compared to 555 meV. $\overset{\circ}{A}^2$ as determined by neutron scattering of the spin-wave spectrum [26]; the value of $\frac{J}{k_B}$ is 347 K.

2.4 The Coefficients $a_{3/2}$, $a_{5/2}$, $a_{7/2}$ Vary with Temperature

In the spin-wave theory, the general expression of the variation of magnetization with temperature and field is written as follows [19]

$$\sigma(T,H) = \sigma_0 \left\{ 1 - a_{3/2} \left[F(3/2, t_H)/\zeta(\tfrac{3}{2}) \right] \cdot T^{3/2} \right.$$
$$\left. - a_{5/2} \left[F(5/2, t_H)/\zeta(\tfrac{5}{2}) \right] T^{5/2} \ldots \right\} \quad (11)$$

in which $F(s, t_H)$ is the Bose-Einstein integral function and $t_H = \dfrac{k_B T}{g\mu_B H}$. Following S.H. CHARAP [27], we use a development of $F(s, t_H)$ to write

$$\sigma(H,T) = \sigma_0 \left(1 - a_{3/2} T^{3/2} - a_{5/2} T^{5/2} - a_{7/2} T^{7/2} \right)$$
$$+ 1.355 \cdot \left(\dfrac{g\mu_B}{k_B} \right)^{1/2} \cdot (\sigma_0 a_{3/2}) \cdot T \cdot H^{1/2}$$
$$+ \left(\dfrac{g\mu_B}{k_B} \right) \cdot \left[\sum_{n=0,1,2} \dfrac{\zeta(n+1/2)}{\zeta(n+3/2)} \cdot (\sigma_0 a_{n+3/2}) \cdot T^{n+1/2} \right] \cdot H$$
$$- \sigma_0 a_{5/2} \cdot \dfrac{2.36}{1.341} \cdot \left(\dfrac{g\mu_B}{k_B} \right)^{3/2} \cdot T \cdot H^{3/2} + \ldots \quad (12)$$

This development is convergent for $\dfrac{g\mu_B H}{k_B T} \leq 2\pi$, which is proved to be true for nickel in a wide range of temperature; at 4.21 K, the field limit is 178,000 œ.

The A(T) term, determined in section 2.2, has to be identified with

$$A(T) = 1.355 \left(\dfrac{g\mu_B}{k_B} \right)^{1/2} \cdot (\sigma_0 a_{3/2}) \cdot T \quad (13)$$

From the experimental values of A(T), we determine the values of $a_{3/2}$ at different temperatures (table 2) and establish for the first time the temperature variation of $a_{3/2}$.

Table 2

T_K	4.21	42	76	101.3	150.4	190.5	231.3	288.1
$a_{3/2} \cdot 10^6$ deg$^{-3/2}$	12.3	12.4	11.5	10.7	8.5	7.7	7.2	6.7
$D \cdot 10^{30}$ ergs-cm^2	44.66	44.88	47.44	48.91	57.42	61.05	64.08	67.12

The error on $a_{3/2}$ is estimated at $\pm$ 3%. The value of $a_{3/2}$ determined with this procedure, $12.3 \cdot 10^{-6}$, is 50% higher than the value $8 \cdot 10^{-6}$ determined in section 2.3. This result has a peculiar consequence: the calculated values of $\sigma_s + \sigma_0 a_{3/2} \cdot T^{3/2}$, for T below 110.3 K exceed the absolute saturation moment σ_0 (Fig.4); however, the excess is small, 0.08 e.m.u/g or 1.3 °/$_{\circ\circ}$ of the moment. It is tempting to attribute this result to some specific properties of nickel, such as antiferromagnetism of a weak fraction of nickel atoms; however, this difference is in the limits of error.

The value of D at 0 K is $44.66 \cdot 10^{-30}$ ergs-cm^2 or 279 meV $\cdot$ Å^2, from which we determine $\dfrac{J}{k_B} = 261$ K. The variation of $a_{3/2}$ with temperature is associated to the variation of coefficient D with temperature (table 2)[28][29].

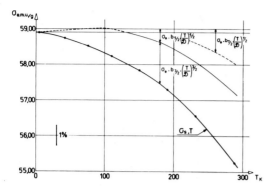

Fig.4 Spontaneous magnetization versus T/D and the different contributions of the $(T/D)^{3/2}$, $(T/D)^{5/2}$ and $(T/D)^{7/2}$ terms

To take this temperature variation of D into account, the variation of the spontaneous magnetization with temperature has to be written

$$\sigma_s = \sigma_0 \left[1 - b_{3/2} \left(\frac{T}{D}\right)^{3/2} - b_{5/2} \left(\frac{T}{D}\right)^{5/2} - b_{7/2} \left(\frac{T}{D}\right)^{7/2} \cdots \right] \quad (15)$$

With the same method as explained in section 2.3, we determine $b_{3/2} = (366 \pm 10) \cdot 10^{-50}$; $b_{5/2} = (35.9 \pm 3) \cdot 10^{-80}$; $b_{7/2} = (5.8 \pm 5) \cdot 10^{-110}$, which gives at 0 K, $a_{3/2} = (12.3 \pm 0.4) \cdot 10^{-6}$, $a_{5/2} = (2.6 \pm 0.25) \cdot 10^{-8}$, $a_{7/2} = (0.27 \pm 0.2) \cdot 10^{-10}$.

2.5 The Equation of the Magnetic Moment in Nickel

The magnetic moment of nickel at different temperatures T and magnetic fields H can be expressed by the general equation

$$\sigma(T, H) = 58.903 \left[1 - (366 \pm 10) \cdot 10^{-50} \left(\frac{T}{D}\right)^{3/2} - (35.9 \pm 3) \cdot 10^{-80} \left(\frac{T}{D}\right)^{5/2} \right.$$
$$\left. - (5.8 \pm 5) \cdot 10^{-110} \left(\frac{T}{D}\right)^{7/2} \right] \quad (16)$$
$$+ \frac{(0.319 \pm 0.005)}{\rho(T)} 10^{-46} \cdot \frac{T}{D^{3/2}} \cdot \left[(H_e + H_A)^{1/2} - H_A^{1/2} \right] + \chi H_e$$

where χ (T) is given in table 1 and D (T) in table 2. This equation represents the experimental points with an accuracy of 0.6%.

2.6 The Superimposed Susceptibility

Table 1 gives the different values of the superimposed susceptibility χ with temperature. At the absolute zero, $\chi = 1.53 \cdot 10^{-6}$ e.m.u/g; a minimum is observed near 115 K. Several determinations of χ had already been made at 4.2 K [23][30]; they led to the values between 1.87 and 2.2 · 10^{-6}; in these experiments, the $H^{1/2}$ spin-wave contribution of the magnetization had not been separated. The resulting susceptibility is the sum of several contributions, which are at 4.2 K [30][31]

diamagnetism	: $\sim 0.054 \cdot 10^{-6}$
orbital susceptibility	: between 0.46 and 1.36 · 10^{-6}
spin susceptibility	: between 0.68 and 0.94 · 10^{-6}
4s electron susceptibility	: $\sim 0.06 \cdot 10^{-6}$
H term in the spin wave contribution	: ~ 0 (at 4.2 K)

so that the resulting susceptibility is between $1.146 \cdot 10^{-6}$ and $2.306 \cdot 10^{-6}$ e.m.u/g.

As concerns the temperature variation of χ, it seems that we have the superimposition of two contributions, one increasing slowly and the other one decreasing when the temperature increases.

References

1 - L. Néel. Comptes-rendus du 10ème Conseil Solvay . Bruxelles, 251 (1955)
2 - E. Stryjewski and N. Giordano. Advances in Physics, 26, 487 (1977)
3 - R. Lemaire. Cobalt 33, 201 (1966)
4 - J.L. Féron, G. Hug and R. Pauthenet. Colloque CNRS N° 180, 19 (1970)
5 - H. Bartholin and O. Vogt. Phys. Stat. Sol. 52, 315 (1979)
6 - R. Aléonard, P. Morin and J. Pierre. Colloque CNRS N° 242, 39 (1975)
7 - B. Barbara, M.F. Rossignol, H.G. Purwins and E. Walker. Colloque CNRS N° 242, 51 (1975)
8 - D.M. Edwards and E.P. Wohlfarth. Proc. Roy. Soc. London, 303, 12 (1968)
9 - J. Beille, D. Bloch and J. Voiron. Colloque CNRS N° 242, 237 (1975)
10 - B. Cabrera. C.R. Acad. Sci. 180, 668 (1925)
11 - W. Henry. Phys. Rev. 88, 555 (1952)
12 - M. Bagieu-Beucher and Duc Tranqui. Bull. Soc. Fr. Mineral. Cristallogr. 93, 505 (1970)
13 - M. Guillot and R. Pauthenet. J. Appl. Phys. 36, 1 003 (1965)
14 - Y. Ayant and J. Thomas. C.R. Acad. Sci. 248, 387 (1960)
15 - H. Danan. Thèse Strasbourg 1958; J. Phy. Rad. 20, 203 (1959)
16 - F. Bloch. Z. Phys. 61, 206 (1930)
17 - T. Holstein and H. Primakoff. Phys. Rev. 58, 1 098 (1940)
18 - R.E. Argyle, S.H. Charap and E.W. Pugh. Phys. Rev. 132, 2 051 (1963)
19 - F. Keffer, Handbuch des Physik, 18-2, 1 (1966)
20 - R.M. Bozorth, Ferromagnetism, 569 (1951)
21 - F. Dyson, Phys. Rev. 102, 1 217, 1 230 (1956)
22 - M. Fallot, Ann.Phys. 6, 305 (1936)
23 - J. Foner, A.J. Freeman, N.A. Blum, R.B. Frankel, E.J. Mc Niff and H.C. Praddande. Phys. Rev. 181, 863 (1969)
24 - H. Danan, A. Herr and A.J.P. Meyer. J. Appl. Phys. 39, 669 (1968)
25 - A.J.P. Meyer and G. Asch. J. Appl. Phys. 32, 3 305 (1961)
26 - H.A. Mook, J.W. Lynn and R.M. Nicklow. Phys. Rev. Lett. 30, 556 (1973)
27 - S.H. Charap. Phys. Rev. 119, 1 538 (1960)
28 - T. Izuyama and R. Kubo. J. Appl. Phys. 35, 1 074 (1964)
29 - J. Mathon and E.P. Wohlfarth. Proc. Roy. Soc. A 302, 409 (1968)
30 - J.P. Rebouillat. Thèse Grenoble 1972
31 - M. Yasui, H. Yamada and M. Shimizu. J. Phys. F. Metal Physics (to be published).

Spectroscopy of Magnetic Insulating Transition Metal Dihalides in High Magnetic Fields*

J. Tuchendler

Laboratoire PMTM, Université Paris Nord, Avenue Jean-Baptiste Clément
93430 Villetaneuse, France, and
Laboratoire de Physique des Solides de l'Ecole Normale Superieure
24 rue Lhomond
75231 Paris 05, France

1. Introduction

The transition metal dihalides have attracted considerable interest for many decades. Originally, attention was drawn to compounds such as iron dichloride because they represent archetypes of a variety of materials known as metamagnets. These exhibit, at low temperatures, an antiferromagnetic state which can be ferromagnetically saturated in relatively weak external magnetic fields. This behaviour, first explained by LANDAU [1], originates in the fact that the crystals have a layered structure in which the intralayer exchange coupling is strong and ferromagnetic, while that between layers is weak and antiferromagnetic. In the last decade, it has been the quasi-two-dimensional behaviour due to the lamellar structure which renewed the interest in these compounds in connection with new theoretical and experimental developments on phase transitions and critical phenomena. A large set of references can be found in the review paper on magnetic model systems of DE JONGH and MIEDEMA [2].

The magnetic properties have been studied in great detail by specific heat and susceptibility measurements, elastic and inelastic neutron scattering experiments, magnetic resonance, Fourier transform spectroscopy in the far infrared, Raman light scattering experiments, etc... Within the series of transition metal dihalides one can find examples of both easy-axis ($FeCl_2$) and easy-plane systems ($CoCl_2$, $CoBr_2$, $NiCl_2$, $NiBr_2$) and examples of large ($FeCl_2$, $CoCl_2$, $CoBr_2$) and small ($NiCl_2$, $NiBr_2$) single ion anisotropies.

More recently, it was the appearance of new concepts in the problem of disorder which attracted attention to the alloys of the transition metal dihalides which are rather simple and well-known systems. It turns out that mixed antiferromagnetic systems are ideal for testing theories of disordered materials. Dilute magnetic systems are now largely well understood [3] but concentrated systems, for which much less data are available, pose interesting theoretical questions [4].

In general, the magnetic excitations associated with impurities are closely connected with the magnitude of the magnetic interactions between the impurity and the host spins. In the most usual case, when the magnetic properties of the impurity and host are similar, the presence of the impurities disturbs the host excitation spectrum and it is difficult to get experimental evidence of physical features associated with the impurities. On the other hand, if the magnetic interactions between the impurity spin and the host are very different compared with those between the host spins themselves, it is possible to get well-defined localized impurity modes. Two cases may be observed.

The energy of the localized mode can be large or small compared with that of the host spin waves. The first situation has been widely investigated using optical techniques, for instance Ni^{2+} impurities in MnF_2 [5], while there are only a few studies of the second case, most probably because the energies fall in the very far infrared, a range much more difficult to reach.

It seemed to me that the spectrometer which had been developed for many years at E.N.S. could be used with much profit for magnetic resonance experiments in this kind of study. We will describe in this talk the experiments on the $FeCl_2$-$MnCl_2$ mixtures in which we have studied both the manganese and iron spin excitations in both the antiferromagnetic and ferromagnetic phases.

2. Experimental Details : Samples and Spectrometer

We have chosen to study the layered-structure mixed antiferromagnets $Fe_{1-x}Mn_xCl_2$ for several reasons. $FeCl_2$ and $MnCl_2$ both have the $CdCl_2$ type structure. Large single crystals are easily prepared for the entire concentration range because the pure compounds are isomorphous and the constituent Fe and Mn ions have similar masses and radii. The resulting mixed crystals are expected to be randomly disordered and have lattice constants which vary linearly with x between those of $FeCl_2$ and $MnCl_2$. The magnetic properties of $FeCl_2$ and $MnCl_2$ are very different. In $FeCl_2$ (T_N = 23.5K), the spins align along the crystal c axis, whereas in $MnCl_2$ (T_N = 1.96K) they lie in the basal plane. $FeCl_2$ simulates a 2D antiferromagnet, whereas the magnetic structure of $MnCl_2$ is more complex and not yet fully understood, which is the reason why our experiments were performed mainly above 4.2K.

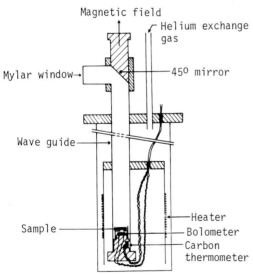

Fig.1 Schematic diagram of the sample holder

There are antiferromagnetic resonance data on pure $FeCl_2$ [6] and impurity spin resonance data of Fe^{3+} and Mn^{2+} ions in antiferromagnetic $FeCl_2$ [7] and of Mn^{2+} ions in $FeBr_2$ [8]. There are also ferromagnetic resonance data on pure $FeCl_2$ [9]. In addition, there have been antiferromagnetic resonance studies of the lowest magnon-like excitations in $Fe_{1-x}Mg_xCl_2$ [10] and in $Fe_{1-x}Cd_xCl_2$ [11].

The samples used in the course of our study were prepared from single crystals grown by S. LEGRAND of CEN Saclay using a Bridgman method. Two samples were given to us by D. BILLEREY of the University of Nancy. Each crystal was analysed by atomic absorption spectroscopy in order to measure the manganese concentrations. Eleven concentrations were used. The spectrometer used has been described in detail elsewhere [12]. Carcinotrons made by Thomson-CSF are used as sources, while the external magnetic field is provided by a superconducting magnet which can be swept up to 60kG. The frequency range is from 70 to 600GHz. Variable temperature inserts and variable temperature controllers are available. A schematic diagram of the sample holder is shown on Fig.1.

3. Experimental Results   A. Mn^{2+} Excitations

For the lowest manganese concentration (x ~ 0.04) in both the AF and F phases, five distinct well-defined resonance lines corresponding to $\Delta S_z = 1$ transition between different states of the S = 5/2 manganese multiplet are observed. The frequency field dependences of the resonance lines are shown on Fig.2. The experimental spectra are well interpreted by the effective spin Hamiltonian [8]

$$\mathcal{H}_{eff} = g'_{//}\mu_B(H_z+\varepsilon H'_{int})S'_z+D'[S'^2_z - \frac{1}{3}S'(S'+1)]+\varepsilon a_3 S'^2_z .$$

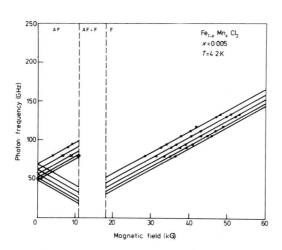

Fig.2 Frequency field dependence of the resonance lines observed for x = 0.005 in $Fe_{1-x}Mn_xCl_2$ at 4.2K

It yields a set of values for the parameters H_{int}, D' and a_3. The main term in the effective Hamiltonian is a result of the important effective field H_{int} which describes the magnetic exchange and dipolar interactions between the impurity and the host. One has $H_{int}=H'_{exch}+H'_{dip}$. H_{dip} can be calculated and one obtains H^{AF}_{exch} = 29.48kG and H'^F_{exch} = 5.08kG, from which we can separate the contributions to the exchange field at an Mn impurity from the two α and β sublattices of the $FeCl_2$ host.

$$H'_{exch}(AF)=H'^{\alpha\alpha}_{exch}+H'^{\alpha\beta}_{exch}$$

$$H'_{exch}(F) =H'^{\alpha\alpha}_{exch}-H'^{\alpha\beta}_{exch}$$

and we get $H'^{\alpha\alpha}_{exch}$ = 17.27kG and $H'^{\alpha\beta}_{exch}$ = 12.19kG for the intralayer exchange field equivalent to $2J'_1$ = + 7.8 GHz and $2J'_2$ = - 5.4 GHz when the Hamiltonian which characterises the exchange between an impurity and the host is $\mathcal{H} = -\sum_i 2J'_i \vec{S}_i.\vec{S}'$.

Figure 3 compares the shape of the resonance spectra obtained for several concentrations. The fine structure has disappeared, the main line shifts to higher fields and a second resonance line is noticeable. The shift of the main line and the intensity of the second line increase with x, while the distance between the two lines remains of the order of 9kG, Fig.4. The frequency field of the two lines is the same, corresponding to a spectroscopic factor of 1.94. We attribute the first resonance to an excitation on a single Mn defect surrounded by six Fe ions, and the second resonance to an excitation on an Mn pair. The probability of the occurrence of clusters with 0,1,2... manganese is given by a binomial distribution if the Mn are randomly distributed.

If one treats the exchange interactions between the Mn impurities and

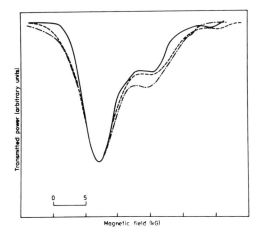

Fig.3 Absorption spectra are compared for Mn concentrations x = 0.105 (full curve), 0.141 (broken curve) and 0.251 (chain curve) for temperature T = 4.2K and for photon frequency ν=76GHz. Spectra are shifted in field and affinity is done on the transmitted power, in order to superpose the single-defect resonance line.

their Fe nearest neighbours in the Ising approximation, the Hamiltonian for a single Mn defect surrounded by six Fe ions can be written:

$$\mathcal{H}_s = g'_{//}\mu_B(H'_{dip}+H_{ext})S'_z + 12J'_1 S'_z$$

which gives the energy of an excitation between $-5/2$ and $-3/2$ states:

$$\hbar\omega_s = g'_{//}\mu_B(H'_{dip}+H_{ext}) + 12J'_1$$

where $2J'_1$ is the intralayer Fe-Mn exchange parameter. For a cluster with two Mn defects and five Fe ions, we have the following Hamiltonian:

$$\mathcal{H}_p = g'_{//}\mu_B(H'_{dip}+H_{ext})(S'_{1z}+S'_{2z}) + \\ +10J'_1(S'_{1z}+S'_{2z}) - 2J''_x S_{1x}S_{2x} - \\ -2J''_y S_{1y}S_{2y} - 2J''_z S_{1z}S_{2z}.$$

The energy of an excitation between the ground state of the pair $|-\frac{5}{2},-\frac{5}{2}\rangle$ and the first excited state $1/\sqrt{2}(|-\frac{5}{2},-\frac{3}{2}\rangle+|-\frac{3}{2},-\frac{5}{2}\rangle)$ is given by

$$\hbar\omega_p = g'_{//}\mu_B(H'_{dip}+H_{ext}) + 10J'_1 + 5J''_z - \\ -\frac{5}{2}(J''_x+J''_y).$$

The field interval between the single Mn mode and pair Mn mode

$$g'_{//}\mu_B \Delta H = 2J'_1 - [5J''_z - \frac{5}{2}(J''_x+J''_y)]$$

is a function of the Fe-Mn exchange energy but also a function of the

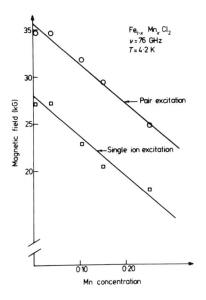

Fig.4 Magnetic field position of the two resonance lines (single Mn mode, pair Mn mode) in relation to the Mn concentration for T=4.2K and ν=76 GHz

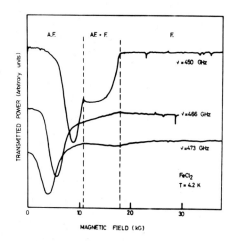

Fig.5 Typical recording traces for $FeCl_2$ at 4.2K and at different frequencies. The center of the resonance line is identified as the minimum of the transmitted power. Broken lines indicate the magnetic field range where both AF and F phases are simultaneously present

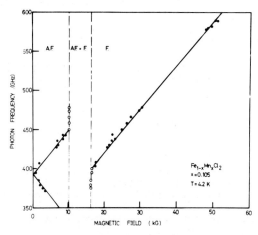

Fig.6 Field dependence of the resonance line observed at T = 4.2K in $Fe_{1-x}Mn_xCl_2$ (here x = 0.105) in both AF and F phases

anisotropy of the exchange between the two Mn spins. We obtain

$$J_z'' - \frac{1}{2}(J_x'' + J_y'') = -0.112 \text{ cm}^{-1}.$$

It is much more difficult to explain the important shift observed and we have no model so far. We may suggest a strong dependence of the Fe-Mn exchange coupling with the lattice parameter.

B. Fe^{2+} Excitations

Figure 5 shows recorder traces obtained at 4.2K for several frequencies in $FeCl_2$. The frequency-field diagram for a manganese concentration x = 0.105 is shown on Fig.6. The AF and F magnons gap energies are obtained by extrapolating linearly to zero magnetic field the magnons energies in the presence of field. Fig.7 shows the dependence on concentration of the magnons gap energies. For $x \leqslant 0.151$, the magnon energies slope linearly with a slope $d(h\nu(x)/h\nu(o))dx = -1.80$.

The linewidths of the AF and F resonance lines at half maximum are about 20 GHz, they increase with x as expected. A broadening and a shift in field are observed for instance in $FeCl_2$ when the temperature is raised above 4.2K, but no structure of the line appears. The magnetic field dependence of the magnons yields a g value close to 4 as expected.

For larger x, the absorption broadens rapidly and disappears for x > 0.757 and the experimental uncertainty increases drastically. One can obtain an estimate of the value of the single ion anisotropy of Fe^{2+} in $MnCl_2$ from the energy dependence of the magnons. $D \sim 150$ GHz, which is also expected.

The appropriate Hamiltonian for $FeCl_2$ has been extensively

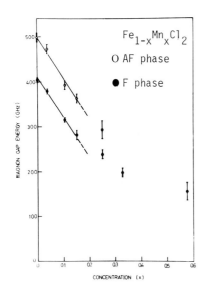

Fig.7 Concentration dependence of of the zero field magnon gap energy in AF and F phases. The full lines give the theoretical concentration dependence using the cluster model

developed by ONO et al. [13] and ALBEN [14]. Within the lowest S = 1 triplet, the effective spin Hamiltonian including the residual trigonal field, the Zeeman term with an external field along the c axis and the exchange interactions can be written :

$$\mathcal{H} = - \sum_i D(S_i^{z^2} - \frac{2}{3}) - \sum_{<ij>} 2\alpha_{//}^2 J_{ij} S_i^z S_j^z -$$

$$- \sum_{<ij>} 2\alpha_\perp^2 J_{ij}(S_i^x S_j^x + S_i^y S_j^y) -$$

$$g_{//}\mu_B H \sum_i S_i^z.$$

Using a spin wave treatment [15], we obtain the energy of the AF magnon at $\vec{k} = 0$.

$$h\nu_{AF} = g_{//}\mu_B(H_A \pm H_0)\left[1 - \frac{2\eta \times 12 J_2}{g_{//}\mu_B(H_A \pm H_0)}\right]^{1/2}$$

with $\eta = (\frac{\alpha_\perp}{\alpha_{//}})^2$.

$J_1 = \alpha_{//}^2 J_{ij}$, i and j being nearest neighbours in the plane,

$J_2 = \alpha_{//}^2 J_{ij}$, i and j being nearest neighbours in adjacent planes and

$g_{//}\mu_B H_A = D + (1-\eta)(12J_1 - 12J_2)$, D and η describe, respectively, the single ion and magnetic anisotropies.

A good estimate of J_2 is obtained from the field of the metamagnetic transition. It gives $12J_2 = -61$ GHz.

We have $g_{//}\mu_B H_A \gg 24\eta J_2$ since D is of the order of 200 to 300 GHz and $\eta \sim 0.78$. The AF magnon gap energy becomes :

$$h\nu_{AF} = D + (1 - \eta) 12J_1 - 12J_2 \pm g_{//}\mu_B H_0.$$

In the same way, one can obtain the ferromagnetic energy at $\vec{k} = 0$:

$$h\nu_F = D + (1 - \eta)(12J_1 + 12J_2) + g_{//}\mu_B(H_0 - H_d)$$

where H_d is the demagnetizing field which can readily be obtained from the extension in magnetic field of the metamagnetic transition $H_d = 7.3$ kG. Taking D = 219 GHz and the observed value $H_{AF} = 498$ GHz, we obtain the in-plane exchange parameter $12J_1 = 991$ GHz. The experimental value $h\nu_{AF}(0) - h\nu_F(0) = 94$ GHz is in good agreement with the theoretical one

$$h\nu_{AF}(0) - h\nu_F(0) = (\eta-2)12J_2 + g_{//}\mu_B H_d = 115 \text{ GHz}.$$

To explain the linear dependence with x of the gap energies, one can use a mean field model neglecting the Fe-Mn exchange interactions which (see above) are much weaker than the Fe-Fe exchange interactions. Using the expression given above for the AF gap energy, with the restriction that the exchange term varies linearly with x, the AF magnons gap energy becomes :

$$h\nu_{AF}^x = h\nu_{AF}^0 - x(1-\eta)12J_1 - 12J_2 \ .$$

Unfortunately, using $12J_2 = -61$ GHz and any of the values found in the literature, the dependence on concentration of the gap energy is always too small, the highest value being -0.65.

One way to overcome this difficulty is to use a cluster model solved in the Ising and single ion approximations. Here we consider the cluster mode of a central spin surrounded by six nearest neighbours in the plane, r Mn spins and (6-r) Fe spins. Out-of-plane nearest neighbours are neglected. For a given x the probability of the occurrence of such a cluster is given by a binomial distribution, and we suppose that the energy of the magnetic excitation is the mean value of the energies corresponding to various clusters. The parameters J_1 (Fe-Fe) and J_1' (Fe-Mn) are supposed independent of x, and the magnetic order is believed to be the same as that of pure $FeCl_2$ with mean values of the Fe and Mn spin components $<S_z> = -1$ and $<S_z'> = -5/2$. In this simple model, the difference between the energy of the cluster with zero Mn^{2+} spin and that with r Mn^{2+} spins is given by

$$\Delta(h\nu_r) = -2r\ J_1 S + 2r\ J_1' S' \ .$$

The probability of the occurrence of each type of cluster is P(r,x) which gives an observed variation of the excitation energy

$$\Delta(h\nu^x) = \sum_r P(r,x)\Delta(h\nu_r)$$
$$= -(12J_1 S - 12J_1' S')x \ .$$

The $12J_1$ exchange coupling parameter between two nearest neighbour Fe^{2+} spins can be determined by fitting the experimental concentration dependence, and we obtain $12J_1 = 1010$ GHz in surprisingly good agreement with the value obtained above.

For x value larger than $x = 0.15$, deviations from linear dependences are observed. This may be due to a concentration dependence of the exchange interactions.

4. Conclusions

I would like to draw three main conclusions. Firstly, it is obvious that far infrared spectrometers like those which many of us have developed for semiconductor and semimetal studies are perfectly adequate to study magnetic insulators, and in particular disordered alloys of antiferromagnets with ordering temperatures of the order of a few Kelvins. They are very well fitted to the range of energies involved and can prove very useful.

Secondly, I would like to say that we have tackled the problem of disordered alloys with simple-minded theories. An arsenal of complicated theories exists : renormalization group theory [16], coherent potential approximation [17], concentration expansion techniques [18], etc... It is my own belief that in the complicated problem of disordered alloys one must stay simple, retain only essentials and look only at model systems made of well-known constituents.

Thirdly, I would like to say that new concepts such as frustration cannot yet be applied directly to even simple systems such as $Fe_{1-x}Mn_xCl_2$. More theoretical work has to be done before they can be used by experimentalists. Conversely, more data on simple systems such as those with competing anisotropies can be useful to test new ideas.

Finally, I would like to acknowledge the collaboration with S. LEGRAND and thank him for the quality of his samples and the extreme generosity with which he provided them.

References

1. L. Landau : Z. Physik. Sowjetunion 4, 675 (1933)
2. L.J. De Jongh and A.R. Miedema : Advances in Physics 23, 1 (1974)
3. See for example R.A. Cowley and W.J.L. Buyers : Rev. Mod. Phys. 44, 406 (1972) and
R.J. Elliott, J.A. Krumhansl and P.L. Leath:Rev. Mod. Phys. 46,465 (1974)
4. R.A. Cowley : AIP Conf. Proc. 29, 243 (1976)
5. L.F. Johnson, R.E. Dietz and H.J. Guggenheim : Phys. Rev. Lett. 17, 13 (1966)
6. I.S. Jacobs, S. Roberts and P.E. Lawrence : J. Appl. Phys. 36,1197 (1965) and D.Petitgrand and P. Meyer : Journal de Physique 37, 1417 (1976)
7. See for example M. Motokawa and M. Date : J. Phys. Soc. Japan 23, 1216 (1967) and M. Date and M. Motokawa : J. Appl. Phys. 39, 820 (1968)
8. G. Mischler, P. Carrara and Y. Merle d'Aubigné : Phys. Rev. B 15, 1568 (1977)
9. A.R. Fert, J. Léotin, J.C. Ousset, D. Bertrand, P. Carrara and S.ASKENAZY Solid State Commun. 18, 327 (1976)
10. W. Hayes, P.J. Walker and M.C.K. Wiltshire : J. Phys. C Letters Solid State 9, L255 (1976) and M.C.K. Wiltshire and W. Hayes : J. Phys. C Solid State 11, 3701 (1978)
11. P. Meyer and A. Brun : J. Appl. Phys. 49, 2198 (1978)
12. See for example J. Tuchendler, J. Magariño and J.P. Renard : Phys. Rev.B 20, 2637 (1979)
13. K. Ono, A. Ito and T. Fujita : J. Phys. Soc. Japan 19, 2119 (1964)
14. R. Alben : J. Phys. Soc. Japan 26, 261 (1969)
15. See for example F. Keffer : Handbuch der Physik XVIII/2
16. A. Aharony : Phys. Rev. B 13, 2092 (1976) and Phys. Rev. B 18,3318 (1978)
17. A.R. McGurn and R.A. Tahir Kheli : J. Phys. C 11, 2845 (1978) and R.A. Tahir Kheli and A.R. McGurn : J. Phys. C 11, 1413 (1978)
18. T. Idogaki and N. Uryu : J. Phys. Soc. Japan 35, 1627 (1973) ; J. Phys. Soc. Japan 43, 845 (1977) and J. Phys. Soc. Japan 45, 1498 (1978).

* not presented at the Seminar

Semimagnetic Semiconductors in Magnetic Fields

T. Dietl

Institute of Physics, Polish Academy of Sciences, Al.Lotników 32/46
02-668 Warsaw, Poland

1. Introduction

In this paper we will review recent work on II-VI compounds containing Mn: $Hg_{1-x}Mn_xTe$, $Hg_{1-x}Mn_xSe$, $Cd_{1-x}Mn_xTe$, $Cd_{1-x}Mn_xSe$. From many points of view these materials, called semimagnetic semiconductors (SMSC), are very similar to the well-known systems such as $Hg_{1-x}Cd_xTe$ and $Cd_{1-x}Zn_xTe$. In particular, over a wide range of compositions they crystallize in zinc-blende (HgMnTe, HgMnSe, CdMnTe) or wurzite (CdMnSe) structure, they exhibit relatively low deviation from stoichiometry, low concentration of intrinsic defects, and they can be made either n-type or p-type by intentional doping or by appropriate anealing. At first sight, their band structure is the same as that of other II-VI compounds. In particular there is a wide conduction band of s-type symmetry (of p-type in the zero-gap case) and there is a wide p-like valence band. Briefly, they exhibit typical semiconducting properties and, consequently, we have access to their structure through the most powerful techniques that have been developed to study the canonical semiconductors. However, using these techniques, especially those which involve the presence of a magnetic field, we quickly realise that the properties of SMSC are not at all similar to those of other II-VI compounds. If we want to find a reference point, they are, in some aspects, similar to those observed in magnetic semiconductors [1], examples being exchange-induced splitting of bands and the formation of magnetic polarons. Of course, all this is due to the presence of the half-filled $3d^5$ shell of Mn and thus due to the presence of localized spins. In fact, early EPR [2] and magnetic susceptibility [3] studies of Mn in II-VI materials have shown that $g_{Mn}= 2$ and the total spin $|\vec{S}| = 5/2$, just as in the free Mn atom.

Our review, of course, does not cover all aspects of SMSC. In particular, we omit important problems of crystal growth and characterization, possible applications, details of theoretical models. For further information and further references to original papers we refer readers to other review articles already available [4].

2. Band Structure

The starting point for the description of the band structure of SMSC is the Vonsovskii model. According to that model one assumes that electrons can be divided into two groups.

The first consists of band electrons from the bonding and antibonding sp^3 orbitals. The energy structure of extended states in SMSC resembles that of other II-VI compounds, implying in particular the applicability of the virtual crystal approximation to the description of band states in the alloys in question. The energy gap is direct and located in the center of the Brillouin zone. The gap is either zero or narrow in mercury compounds (Fig.1), or wide in cadmium compounds (Fig.2), and smoothly increases with Mn concentration. The latter property is observed in gaps away from the Γ point [16,17]. The gap and Mn concentration can be independently changed in quaternary compounds, eg. CdHgMnTe [18].

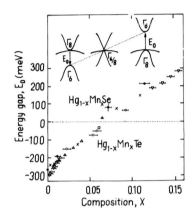

Fig.1 Energy gap in $Hg_{1-x}Mn_xTe$ (x[5], o[6], △[7], □[8], ▽[9]) and in $Hg_{1-x}Mn_xSe$ (▲[10], ●[9]) at 4.2K

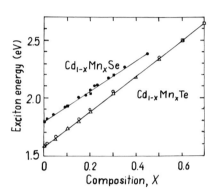

Fig.2 Free exciton energy in $Cd_{1-x}Mn_xTe$ (o[11], □[12] △[13]) and in $Cd_{1-x}Mn_xSe$ (■[14], ●[15]). Straight lines are drawn through experimental points

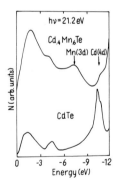

Fig.3 Shape of the photoemission spectra in CdMnTe [19] and in CdTe [20]. Energy is calculated from the top of the valence band. Arrows denote $Mn(3d^5)$ and $Cd(4d^{10})$ bands

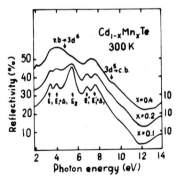

Fig.4 Reflectivity in $Cd_{1-x}Mn_xTe$. Arrows denote structures attributed to transitions from and on $3d^5$ Mn shell (after [17])

345

The second group comprizes localized electrons from the half-filled 3d shell of Mn. These electrons, according to photoemission studies in CdMnTe [19], are about 6eV below the top of the valence band (Fig.3).

This finding is consistent with UV reflectivity measurements (Fig.4), a structure around 10eV being attributed to the transitions from the $3d^5$ level to a point in the conduction band away from the center of the Brillouin zone [17] (with an uncertainty concerning crystal field splitting of the final state $3d^4$; crystal field parameters can be estimated by analysing the intra $3d^5$ absorption band observed near 2.2eV [12]). The same measurements reveal a structure around 4eV. This structure may be attributed to the transitions from the valence band on the $3d^5$ level which, when occupied ($3d^6$), lies above the valence band because of a strong Coulomb repulsion between d-electrons of opposite spins. The above measurements result in a picture for the density of states, which is very schematically presented in Fig.5. According to that picture, the stability of the magnetic moment is due to the localized character of the $3d^5$ Mn shell and a large intra-atomic correlation energy for d-electrons, $U \approx 10eV$. The important feature of that picture is also a large distance of d-levels from the Fermi level. This a posteriori explains why the band structure near the Fermi level and $\vec{k}\cdot\vec{p}$ parameters of the materials in question are so similar to those of Mn-less II-VI compounds.

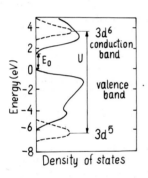

Fig.5 Schematic picture for the density of states in II-VI compounds containing Mn. Arrows mark energy gap E_O and correlation energy U (after [17])

However, it should be noted that there is some information which may indirectly indicate that the mixing of d-levels and band states does not play a negligible role. First, one observes a strong modification and broadening of the UV reflectivity spectra in comparison with other II-VI alloys [16,17]. Second, the p-d exchange integral is relatively large $\beta/\Omega \approx -1eV$ and about two times larger than the s-d integral $\alpha/\Omega \approx 0.5eV$ (for definition of symbols see next section). This is probably hard to interpret without allowing for the p-d hybridization.

3. Exchange Interaction

There is, of course, a Coulomb interaction between the two electron systems described in the previous section. A part of this interaction depends both on the spin of the band electron and the total spin localized on the Mn ion. It is assumed that this spin-dependent part of the interaction, the exchange interaction, can be taken into account by completing a hamiltonian of the problem in question with the Heisenberg term:

$$H' = -J(\vec{r} - \vec{R}_i)\vec{\sigma}\vec{S}, \qquad (1)$$

where J is the short range exchange operator acting on a one-electron orbital part of a band electron wave function; $\vec{\sigma}$ is the spin operator of a band electron; $\vec{S}$ is the total spin operator of a Mn ion located at the point $\vec{R}_i$. As is usual in the study of magnetic materials, this is the basic hamiltonian for further discussion.

4. Exchange-Induced Splitting of the Effective Mass States in a Magnetic Field

4.1 General remarks

The exchange-induced corrections to an effective mass electron energy are calculated within the framework of the molecular field approximation and the first order perturbation theory (if necessary generalized for degenerate or closely lying bands). The former approximation leads to the interaction of the Ising form:

$$H' = -\sigma_z \langle S_z \rangle \sum_i J(\vec{r} - \vec{R}_i), \qquad (2)$$

where the z-axis is chosen along the average value of the spin $\langle \vec{S} \rangle$ and the summation runs over all Mn ions in the crystal. We see that in the absence of a magnetic field or spontaneous magnetization there are no exchange corrections to electron energy. If, however, a magnetic field is applied there appears a nonvanishing spin polarization along the magnetic field, and for electrons in a simple s-like band the corrections, according to the first order perturbation theory, are given by: $E = \pm \frac{1}{2} N_S \alpha \langle S_z \rangle$ for "spin down" and "spin up" electrons, respectively. Here N_S is the concentration of Mn ions and $\alpha = (S|J|S)$ is the exchange integral, where the integration is over the unit cell volume Ω. Therefore, the exchange interaction produces an additional spin-splitting $E_s = \bar{g}\mu_B B$ with $\bar{g}$-factor given by:

$$\bar{g} = \frac{\alpha \chi}{g_{Mn}\mu_B^2} \qquad (3)$$

where we used the standard relations between $\langle S_z \rangle$, magnetization M, and magnetic susceptibility χ : $M = -g_{Mn}\mu_B N_S \langle S_z \rangle$, $\chi = M/H$. Inserting a typical value of the exchange integral $\alpha/\Omega = 0.5$ eV and a typical value of χ at low temperatures $\chi = 4 \cdot 10^{-4}$ emu (i.e. magnetic permeability $\mu = 1.005$) into (3) we find $\bar{g}$ as large as 100.

Of course, a more quantitative theory has to take into account the direct influence of magnetic field on electron states, $\vec{k}\cdot\vec{p}$ interaction between bands, symmetry of bands, spin-orbit interaction [21,22,23,24] and in the case of excitons [25,26] or impurities [27] - appropriate Coulomb interactions. Clearly, all the above factors are very important but they do not affect the main property of the alloys in question: in SMSC there are large exchange-induced spin splittings which reflect the temperature and magnetic-field dependence of the magnetization.

For an S-state ion (such as the Mn-ion in the ground $3d^5$ state) the exchange coupling is determined by two nonvanishing and independent matrix elements: $\alpha = (S|J|S)$ and $\beta = (X|J|X)$ [28]. They are to be determined by experiment, the typical values obtained being $\alpha/\Omega = 0.5$ eV, $\beta/\Omega = -1$ eV [29].

4.2 Spectroscopic Observations

In this section we will discuss the spectroscopic evidence for the exchange-enhanced spin splitting of Landau levels, shallow impurities, and excitons.

In narrow gap SMSC (HgMnTe, HgMnSe), due to large $\omega_c \tau$ and small exciton corrections, the positions of magnetic sublevels (spin-split Landau levels) can be accurately determined by interband [6,9,21,30] and intraband [31,32] magnetoabsorption. Fig.6 shows the sum of cyclotron energies of electrons and light holes, and the spin splitting of the first Landau level in the conduction band, as obtained from interband magnetoabsorption in HgCdTe and HgMnTe [21]. The cyclotron energy for a given value of E_0 is similar in both materials indicating similar value of $\vec{k} \cdot \vec{p}$ parameters. The spin splitting is however much larger in HgMnTe than in HgCdTe because of the exchange interaction in the former material. For the sample $Hg_{0.97}Mn_{0.03}Te$ ($E_0 \simeq -150$meV) the spin splitting in 2T is as large as 20meV, which corresponds to $|g^*|=170$. Furthermore, spin splitting in HgMnTe is temperature dependent, which reflects the temperature dependence of the magnetization.

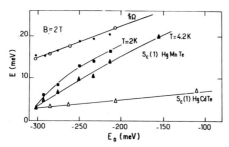

<u>Fig.6</u> Sum of the cyclotron energy of electrons and light holes $\hbar\Omega$ and spin splitting of the first Landau level in the conduction band, $s_c(1)$ as a function of energy gap in HgCdTe (o △) and in HgMnTe (● ▲ ■). Solid lines are drawn through experimental points and can be reproduced by the exchange-modified Pidgeon-Brown model (after [21])

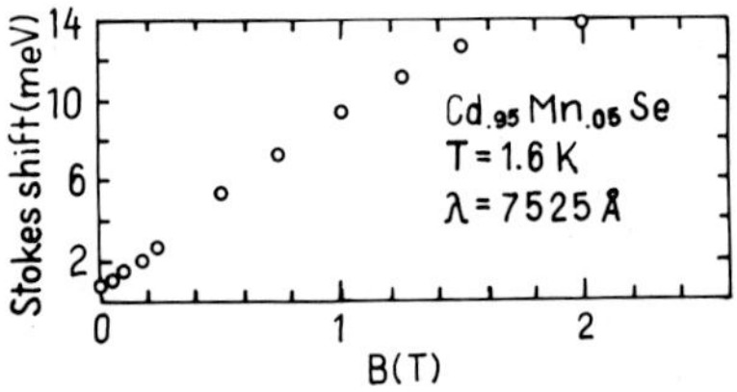

<u>Fig.7</u> Stokes shift in $Cd_{0.95}Mn_{0.05}Se$ at 1.6K. Note the large value of shift, saturation in high fields, and non-zero value in zero field (after [33])

In Fig.7 we see the magnetic-field dependence of the energy difference between incoming and scattered light (Stokes shift) in $Cd_{0.95}Mn_{0.05}Se$ at 1.6K [33]. In CdMnSe at low temperatures electrons are localized on donors and thus the Stokes shift is a direct measure of the Zeeman splitting of a donor level. Again, the splitting is very large and tends to saturation in high fields reflecting the saturation of the Mn spin polarization.

Finally we turn to the exchange-induced splitting of excitons in CdMnTe [23,25,26,34,35]. In Fig.8 the energy difference ΔE between the stronger components of the free exciton observed in magnetoreflection up to 6T in the σ^- and σ^+ configuration is plotted as a function of mean spin value, $N_s\Omega<S_z>$ [35]. The latter value has been deduced from direct magnetization measurements. The strict proportionality of ΔE and $N_s\Omega<S_z>$ implies that the assumptions of the model are basically correct and that the exchange integrals are composition independent.

4.3 Influence on Transport Phenomena

4.3.1 Quantum Transport

The unusual temperature dependence of the positions of magnetic sublevels strongly influences the Shubnikov-de Haas effect. In narrow gap SMSC with temperature-independent electron concentration and mobility one observes the following effects. First, temperature dependence of the positions of oscillation maxima [10,22,36]. Second, in constant magnetic field, oscillations of magnetoresistance as a function of temperature [37]. This is because temperature variations induce transitions of the magnetic sublevels through the Fermi level. Third, nonmonotonic dependence of the SdH-oscillation amplitude on temperature [10,22] (see Fig.9).

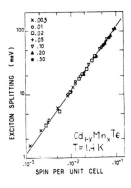
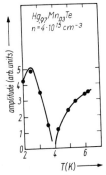

Fig.8 Free exciton splitting in $Cd_{1-x}Mn_xTe$ as a function of mean spin value deduced from magnetization. Solid line is theoretical (after [35])

Fig.9 Amplitude of Shubnikov-de Haas oscilations as a function of temperature in $Hg_{0.97}Mn_{0.03}Te$. Solid line is theoretical (after [22])

The latter effect can be easily explained [38] if we note that the amplitude is determined by the ratio of the distance between magnetic sublevels to their broadening. In particular the maximum in Fig.9 at ~2.5K means that the distance between the magnetic sublevels has reached its maximum value $\hbar\omega_c$ and thus spin-splitting is equal to $n\hbar\omega_c$, n being an integer (n=2 in this particular case). The minimum at ~4K occurs because spin-splitting has been reduced to $3/2\hbar\omega_c$ and thus the distance between the magnetic sublevels has its minimum value $\tfrac{1}{2}\hbar\omega_c$.

4.3.2 Hopping Conductivity

In n-CdMnSe, with $N_D = 10^{16} - 10^{17} cm^{-3}$ and a donor binding energy $E_D \simeq 25 meV$, a magnetic field below 5T does not cause shrinking of donor wave functions in a degree sufficient to affect magnetoresistance in the hopping region. In spite of this, one observes a large and complex magnetoresistance [39] (absent in CdSe, see Fig.10). The negative magnetoresistance in high fields is attributed to the presence of a magnetic polaron [39] (see next section). Several mechanisms can under certain circumstances lead to the positive magnetoresistance [40]. First, increasing magnetic field makes the spin-flip transitions less effective because of large energy distance between spin-down and spin-up states in a magnetic field. The spin-flip transitions in SMSC can be originated in the exchange-induced hopping, spin and energy for the hops being taken from the magnetic subsystem.

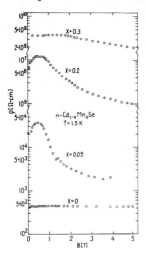

Fig.10 Transverse magnetoresistance in the hopping region in n-$Cd_{1-x}Mn_x$Se at 1.5K. Net donor concentration changes from $2 \cdot 10^{17} cm^{-3}$ for CdSe to $10^{16} cm^{-3}$ for $Cd_{0.7}Mn_{0.3}Se$ (after [39])

Second, magnetic field lifts the spin degeneracy of donor levels, and the form of the electron distribution function in the impurity band changes. Thus, assuming constant number of electrons and density of states, we have a change (increase) of the Fermi energy. Third, the field dependence of the energy of the donor level is sensitive to the distance from other occupied donors when taking into account the exchange interaction between electrons on the neighbouring donors. This may lead to increase of the width of the impurity band in a magnetic field.

The exchange interaction affects in an interesting way the hopping conductivity in narrow-gap p-type SMSC [27]. In $B \gtrsim 3T$ the exchange-enhanced splitting shifts the top valence-band magnetic sublevel $b_v(-1)$ from the other levels to a distance larger than the acceptor binding energy E_I. Therefore, in spite of a small magnetic field ($\gamma = \hbar\omega_c/E_I \ll 1$), we can construct the ground state acceptor wave function from the $b_v(-1)$ wave function. In the wave function thus obtained the longitudinal radius a_B^* is only slightly affected by the magnetic field, but the transverse radius is now determined by the magnetic length λ. Thus the magnetic field extends the acceptor wave function ($\lambda \approx 3 a_B^*$ in 3T) leading to a giant negative magnetoresistance, as indeed observed in p-type $Hg_{1-x}Mn_xTe$, $x \sim 0.15$, in the hopping conductivity region [27,41].

5. Magnetic Polaron

So far we have considered the influence of the exchange on the effective mass states. On the other hand, each Mn ion is in a molecular field produced by effective-mass electrons. The mutual interaction between these subsystems leads, under some conditions, to nonvanishing spin polarization of both subsystems even in the absence of magnetic field, the loss of entropy being outweighed by the lowering of the system energy.

In SMSC the spontaneous magnetization produced by the above mechanism has so far been found to occur within the Bohr radius of occupied impurity states (bound magnetic polaron BMP), as evidenced by observation of zero-field splitting of impurity states in bound exciton luminescence studies in p-type CdMnTe [42] and spin-flip Raman scattering measurements in n-CdMnSe[33] (see Fig.7).

The formation of BMP strongly reduces hopping conductivity through an increase of hopping activation energy and a decrease of the Bohr radius of an occupied donor. The negative magnetoresistance in a large field in n-CdMnSe can thus be attributed to the vanishing of BMP, occuring if all the Mn spins are polarized by an external magnetic field (see Fig.11). A similar effect occurs in p-type HgMnTe [27].

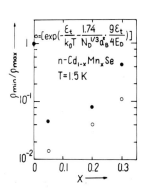

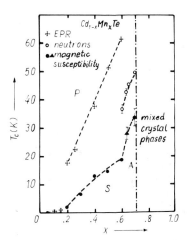

Fig.11 Ratio of min. and max. resistivity in magnetic field in n-CdMnSe at 1.5K ●[39] (see Fig.10). The calculated ratio of resistivity without and with magnetic polaron o[40]. Here polaron binding energy is identified with half of spin splitting in zero magnetic field [33] (see Fig.7); magnetizations from [35] are used

Fig.12 Critical temperature in $Cd_{1-x}Mn_xTe$ as obtained from the broadening of the EPR line ☩ [46], low field magnetic susceptibility ●[45] (B=1.5mT,f=0)▲ [47] (B=4mT,f=38Hz), and elastic neutron scattering o[48]

6. Magnetic Structure

The dominant spin-spin coupling in SMSC is believed [43,44] to originate from an indirect exchange interaction produced by virtual interband excitations induced by the exchange (1), i.e. so-called Bloembergen and Rowland mechanism.

The phase diagram of $Cd_{1-x}Mn_xTe$ [45] is shown in Fig.12. For $x_c \lesssim 0.17$ no phase transition is observed because of the short range and antiferromagnetic sign of the interaction, and the disorder in the magnetic system (some broadening of the EPR line is however seen at low temperatures [46], which may be indicative of the onset of a contribution from the long range dipole-dipole interaction). Above $x_c \gtrsim 0.17$ the spin-glass phase occurs as is shown by the cusps in the low field magnetic susceptibility and the linear temperature dependence of the specific heat [45]. Interestingly enough, the spin-glass phase is due to purely antiferromagnetic interactions and not to a competition between ferro and antiferro interactions as is the case in canonical spin-glasses. Finally, for x>0.6 an antiferromagnetic ordering takes place, as is demonstrated most directly by a maximum in the magnetic susceptibility and in the specific heat [45] as well as by elastic neutron scattering [48].

7. Final Remarks

From the above discussion we see that II-VI compounds containing Mn do not exhibit the key feature of magnetic semiconductors: the influence of narrow bands and critical phenomena (related to the well defined phase transition) on their behaviour. This makes these materials very interesting from the point of view of semiconductor properties. They possess all features of a typical and simple semiconductor with one difference: the exchange interaction. Consequently one can study, say, the amplitude of the SdH effect, excited states of excitons, hopping conductivity,.. by affecting independently spin and orbit. In addition, these materials form an interesting class of disordered magnetic systems. In this review we have limited our considerations to HgMnTe, HgMnSe, CdMnTe, CdMnSe. However, the family of materials with similar properties is quite large and contains other II-VI compounds (eg. ZnMnTe [49]) with other magnetic ions (eg.HgFeTe [50]) as well as, for instance, IV - VI, and II - V compounds with Mn (eg.PbMnTe [51], $(CdMn)_3As_2$ [52]).

The above arguments have lead R.R.Gałązka [4] to propose a separate name for the family of materials exhibiting the properties mentioned above: semimagnetic semiconductors. This name seems to be appropriate not only because it is simple and thus facilitates communication, but also because it correctly reflects the absence of several characteristic features of magnetic semiconductors and simultaneously stresses the excellent semiconducting properties of the family.

Acknowledgements

The extensive study of II-VI compounds with Mn was initiated by Dr.R.R.Gałązka and most of the results presented above have been obtained by teams headed by him and Prof.L.Sosnowski. The author is also indebted to the authors of most of the papers listed below for many valuable discussions. A part of this paper was prepared during the author's stay at Munich Technical University and Osaka University. It is a pleasant duty to thank Prof.F.Koch, Prof.S.Narita,and S.Takeyama for their hospitality.

References

1. see for example: T.Kasuya, Proc.11th Int.Conf.Phys.Semic. ed.by M.Miasek,PWN Warsaw 1972,p.141;E.L.Nagaev, Physics of Magnetic Semiconductors,Nauka, Moscow 1979 (in Russian); Magnetic Semiconductors, J.de Physique C 5 (1980).
2. see for example:K.Leibler, W.Giriat, Z.Wilamowski, R.Iwanowski, phys.stat.sol.(b) 47,405(1971)
3. see for example: M.Savage, J.J.Rhyne, R.Holm, J.R.Cullen, C.R.Carroll, E.R.Wohlfarth, phys.stat.sol.(b) 58,685(1973)
4. R.R.Galazka, Proc.14th Int.Conf.Phys.Semic. Edinburgh 1978, ed.by B.L.H.Wilson, Conf.Series No 43,p.133; R.R.Galazka and J.Kossut,Proc.Int.School Phys.Narrow-Gap Semic. Nimes, 1979, Springer, in press; G.Bastard, J.A.Gaj, R.Planel, C.Rigaux J.de Physique C5,247(1980); J.Mycielski, Proc.Annual Conf.Condensed Mat.Div.,European Phys.Soc.,Antwerpen 1980, in press; J.A.Gaj, Proc.15th.Int.Conf.Phys.Semic.,Kyoto 1980, in press.
5. J.Kaniewski,Ph.D.thesis,Warsaw 1976, unpublished
6. G.Bastard,C.Rigaux,A.Mycielski,phys.stat.sol.(b)79,585(1977)
7. M.Jaczynski,Ph.D.thesis, Warsaw 1978;see M.Jaczynski,J.Kossut R.R.Galazka,phys.stat.sol.(b)88,73(1978)
8. S.W.McKnight,P.M.Amirtharaj,S.Perkowith,Sol.State Comm.25, 357(1978)
9. M.Dobrowolska,W.Dobrowolski,M.Otto,T.Dietl,R.R.Galazka,Proc. 15th Int.Conf.Phys.Semic.,Kyoto 1980, in press
10. S.Takeyama and R.R.Galazka,phys.stat.sol.(b)96,413(1979)
11. J.A.Gaj,R.R.Galazka,M.Nawrocki,Sol.State Comm.25,193(1978)
12. Nguyen The Khoi and J.A.Gaj,phys.stat.sol.(b)83,K133(1977)
13. J.Stankiewicz,N.Bottka,W.Giriat, in Ref.[9]
14. J.Antoszewski and E.Kierzek-Pecold,Sol.State Comm.34,733(1980)
15. W.Giriat and J.Stankiewicz,phys.stat.sol.(a)59,K79(1980)
16. T.Kendelewicz and E.Kierzek-Pecold,Sol.State Comm.25,579(1978)
17. T.Kendelewicz,Ph.D.thesis,Warsaw 1979; to be published
18. M.Dietl,G.Grabecki,E.Janik,E.Kierzek-Pecold,M.Klimkiewicz, U.Debska,Proc.10th Polish Sem.Semic.Compounds,Jaszowiec 1980, in press
19. B.Orlowski,phys.stat.sol.(b)95,K31(1979)
20. N.J.Shevchik,J.Tejda,M.Cardona,D.W.Langer,phys.stat.sol.(b)59, 87(1973)
21. G.Bastard,C.Rigaux,Y.Guldner,J.Mycielski,A.Mycielski,J.de Physique 39,87(1978)
22. M.Jaczynski,J.Kossut,R.R.Galazka,phys.stat.sol.(b)88,73(1978)
23. J.A.Gaj,J.Ginter,R.R.Galazka,phys.stat.sol.(b)89,655(1978)
24. W.Walukiewicz, J.Mag.Mag.Mat.11,157(1979)
25. M.Z.Cieplak and P.Byszewki,Sol.State Comm.29,81(1979)

26. A.Twardowski,M.Nawrocki,J.Ginter,phys.stat.sol(b)$\underline{96}$,497(1979)
27. A.Mycielski and J.Mycielski, in Ref.[9]
28. J.Kossut,phys.stat.sol.(b)$\underline{72}$,359(1975);$\underline{78}$,537(1976)
29. for compilation of exchange integrals see J.A.Gaj, in Ref.[4]
30. C.Rigaux,G.Bastard,Y.Guldner,G.Rebmann,A.Mycielski,J.Furdyna, D.P.Mullin, in Ref. [9]
31. K.Pastor,M.Grynberg,R.R.Galazka,Sol.State Comm.$\underline{29}$,739(1979)
32. K.Pastor,M.Jaczynski,J.K.Furdyna, in Ref. [9]
33. M.Nawrocki,R.Planel,G.Fishman,R.R.Galazka, in Ref.[9]'
34. A.V.Komarov,S.M.Ryabchenko,O.V.Terletskii,I.I.Zheru,R.D.Ivanchuk,Zh.Eksp.Teor.Fiz.$\underline{73}$,608(1977)
35. J.A.Gaj,R.Planel,G.Fishman,Sol.State Comm.$\underline{29}$,435(1979)
36. P.Byszewski,K.Szlenk,J.Kossut,R.R.Galazka,phys.stat.sol.(b) $\underline{95}$,359(1979)
37. M.Dobrowolska,W.Dobrowolski,R.R.Galazka,J.Kossut,Sol.State Comm.$\underline{28}$,25(1979)
38. J.Kossut,Sol.State Comm.$\underline{27}$,1237(1978)
39. J.Antoszewski,U.Debska,M.Dietl,T.Dietl,E.Janik,J.Jaroszynski, M.Sawicki, in Ref. [18]
40. T.Dietl, unpublished
41. R.T.Delves,Proc.Phys.Soc.$\underline{87}$,809(1966)
42. A.Golnik,J.A.Gaj,M.Nawrocki,R.Planel,C.Benoit,in Ref. [9]
43. G.Bastard and C.Lewiner,Phys.Rev.B$\underline{20}$,4256(1979)
44. J.Ginter,J.Kossut,L.Swierkowski,phys.stat.sol.(b)$\underline{96}$,735(1979)
45. R.R.Galazka,Shoichi Nagata,P.H.Keesom,Phys.Rev.B, in press
46. S.Oseroff,R.Calvo,W.Giriat,J.Appl.Phys.$\underline{50}$,7738(1979)
47. M.Otto and T.Dietl, unpublished
48. T.Giebultowicz,B.Buras,K.Clausen,R.R.Galazka,H.Kepa,in press
49. A.V.Komarov,S.M.Ryabchenko,N.I.Vitrikhovskii,Pisma v Zh.Eksp Teor.Fiz.$\underline{27}$,441(1978)
50. Y.Guldner,C.Rigaux,M.Menant,D.P.Mullin,J.K.Furdyna,Sol.State Comm.$\underline{33}$,133(1980)
51. for example:J.Niewodniczanska-Zawadzka,A.Sandauer,Z.Golacki, W.Dobrowolski,J.Kossut, in Ref.[18]
52. F.A.P.Blom, Proc.1st Int.Symposium Phys.Chem.II-V Compounds Mogilany 1980, in press

Participants of the Seminar

Persons on the Photograph

1 Prof. N. Miura
2 Prof. S. Chikazumi
3 Prof. C. Hamaguchi
4 Mr. M. Konno
5 Dr. C.M. Fowler
6 Mrs. J. Fowler
7 Mrs. C. Kubo
8 Prof. M.S.Dresselhaus
9 Mr. Y. Ichikawa
10 Prof. G. Landwehr
11 Mrs. A. Landwehr
12 Prof. R. Kubo
13 Prof. R. Pauthenet
14 Mrs. C. Pauthenet
15 Mrs. P. Stradling
16 Prof. R.A. Stradling
17 Prof. S. Askenazy
18 Mrs. C. Askenazy
19 Miss M. Tsuchiya
20 Mrs. K. Miura
21 Dr. M. Kido
22 Mrs. M. Chikazumi
23 Dr. R.L. Aggarwal
24 Prof. Y. Uemura
25 Prof. D.G. Seiler
26 Dr. T. Dietl
27 Dr. J. Metzdorf
28 Dr. Y. Ono
29 Prof. T. Morimoto
30 Dr. K.K. Bajaj
31 Prof. K. Cho
32 Prof. Y. Nakagawa
33 Prof. S. Tanaka
34 Prof. S. Tanuma
35 Dr. K. Noto
36 Prof. S. Kawaji
37 Dr. L. Eaves
38 Dr. A. Misu
39 Dr. L. Passari
40 Dr. G. Dresselhaus
41 Prof. P. Halevi
42 Dr. H. Störmer
43 Dr. A. Raymond
44 Prof. J.C. Portal
45 Dr. G. Kido
46 Mr. K. Nakamura
47 Dr. H. Miyajima
48 Mr. T. Itakura
49 Mr. N. Onozawa
50 Dr. Y. Iye
51 Dr. Y. Sasaki
52 Prof. H. Fukuyama
53 Prof. S. Mase
54 Prof. S. Narita
55 Dr. D. Yoshioka
56 Prof. V.K. Arora
57 Dr. J. Wlasak
58 Prof. J. Hajdu
59 Prof. W. Zawadzki
60 Mrs. G. Herlach
61 Prof. F. Herlach
62 Prof. M. Date
63 Prof. Y. Nishina
64 Prof. T. Kasuya
65 Mrs. B. Pascher
66 Prof. K. Murase
67 Dr. M.S. Skolnick
68 Dr. T. Ichiguchi
69 Mr. K. Hiruma
70 Dr. K. Nakao
71 Dr. Y. Onuki
72 Mr. Y. Sato
73 Prof. J.P. Kotthaus
74 Prof. H.D. Drew
75 Prof. Y. Inuishi
76 Prof. J.F. Koch
77 Mrs. D. Störmer
78 Prof. H. Köhler
79 Prof. T. Ando
80 Mr. B. Schlicht
81 Mr. M. Roos
82 Prof. G. Bauer
83 Prof. H.W. Myron
84 Prof. M. Motokawa
85 Dr. T. Englert
86 Prof. M. von Ortenberg
87 Prof. M. Voos
88 Dr. H. Pascher
89 Mrs. B. Fantner
90 Mr. M. Naito
91 Dr. E.J. Fantner
92 Dr. K. von Klitzing
93 Dr. D.M. Larsen
94 Prof. U. Rössler
95 Dr. R. Mansfield
96 Dr. T. Fukami
97 Dr. K. Muro
98 Dr. N.O. Lipari
99 Mrs. G. Lipari

Index of Contributors

Number in *italics* refer to the person in the photograph on page 356

Aarts, C.J.M.   24
Aggarwal, R.L.   105*(23)*
Akihiro, M.   72
Altarelli, M.   180
Ando, T.   301*(79)*
Askenazy, S.   161*(17)*
Aulombard, R.L.   253

Baldwin, S.   224
Bauer, G.   244*(82)*

Caird, R.S.   54
Chieu, T.C.   306
Chikazumi, S.   64,72*(2)*
Cho, K.   190*(31)*

Date, M.   44,150,195*(62)*
Davidson, A.M.   84
Dietl, T.   344*(26)*
Dresselhaus, G.   306*(40)*
Dresselhaus, M.S.   306*(8)*
Dreybrodt, W.   190
Drew, H.D.   224*(74)*

Eaves, L.   130*(37)*
Englert, T.   130,274*(85)*
Erickson, D.J.   54
Esaki, L.   292

Fantner, E.J.   244*(91)*
Fert, A.   161
Fowler, C.M.   54*(5)*
Freeman, B.L.   54
Fukuyama, H.   288*(52)*
Furukawa, A.   316

Garn, W.B.   54
Gaus, C.   116
Goodwin, M.W.   112

Halevi, P.   203*(41)*
Hamaguchi, C.   169*(3)*
Herlach, F.   34*(61)*
Hori, H.   44,195

Ichiguchi, T.   249*(68)*
Inada, R.   316
Ishikawa, Y.   150
Iye, Y.   316*(50)*

Kasaya, M.   150
Kasuya, T.   150*(64)*
Katayama, H.   72
Kawaji, S.   284*(36)*
Kido, G.   64,72*(45)*
Klitzing, K.von   139*(92)*
Knowles, P.   84
Koch, F.   262*(76)*
Konczewicz, L.   253
Kotthaus, J.P.   116*(73)*
Kuroda, N.   195

Landwehr, G.   2*(10)*
Larsen, D.M.   120*(93)*
Lipari, N.O.   180*(98)*
Lopez-Otero, A.   244

Makado, P.   84
Metzdorf, J.   199*(27)*
Miura, N.   64,72,320*(1)*
Miyajima, H.   64*(47)*
Motokawa, M.   44*(84)*
Murase, K.   249*(66)*
Muro, K.   216*(97)*
Myron, H.W.   24*(83)*

Naito, M.   320*(90)*
Nakao, K.   64*(70)*
Narita, S.   216*(54)*
Nimtz, G.   257
Nishina, Y.   195*(63)*

Okuda, K.   44
Ono, Y.   174*(28)*
Onuki, Y.   316*(71)*
Ortenberg, M.von   94*(86)*
Ousset, J.C.   161

Pascher, H.   244*(88)*

Pauthenet, R.   326*(13)*
Porowski, S.   84,253

Raymond, A.   253*(43)*
Robert, J.L.   253

Sakakibara, T.   44,150
Sasaki, Y.   195*(51)*
Schlicht, B.   257*(80)*
Seiler, D.G.   112*(25)*
Shayegan, M.   306
Shimomae, K.   169
Shinoda, M.   195
Skolnick, M.S.   208*(67)*
Stallhofer, P.   116
Steigenberger, U.   94
Stradling, R.A.   84*(16)*
Suga, S.   190
Swierkowski, L.   234

Tachikawa, K.   12
Takahashi, H.   150
Takahashi, O.   316
Takayama, J.   169
Takegahara, K.   150
Tanaka, S.   320*(33)*
Tanuma, S.   316*(34)*
Tuchendler, J.   139,336

Uihlein, C.   130

Voos, M.   292*(87)*
Vroomen, A.R.de   24

Wakabayashi, J.   284
Wasilewski, Z.   84
Wlasak, J.   234*(57)*
Wyder, P.   24

Yamanaka, M.   216
Yoshioka, D.   288*(55)*

Zawadzki, W.   112,234*(59)*

358

Electroluminescence

Editor: J. I. Pankove
1977. 127 figures, 16 tables. XI, 212 pages
(Topics in Applied Physics, Volume 17)
ISBN 3-540-08127-5

Contents:
J. I. Pankove: Introduction. – *Y. M. Tairov, Y. A. Vodakov:* Group IV Materials (Mainly SiC). – *P. J. Dean:* III-V Compounds Semiconductors. – *Y. S. Park, B. K. Shin:* Recent Advances in Injection Luminescence in II-VI Compounds. – *S. Wagner:* Chalcopyrites. – *T. Inoguchi, S. Mito:* Phosphor Films.

Excitons

Editor: K. Cho
1979. 118 figures, 8 tables. XI, 274 pages
(Topics in Current Physics, Volume 14)
ISBN 3-540-09567-5

Contents:
K. Cho: Introduction. – *K. Cho:* Internal Structure of Excitons. – *P. J. Dean, D. C. Herbert:* Bound Excitons in Semiconductors. – *B. Fischer, J. Lagois:* Surface Exciton Polaritons. – *P. Y. Yu:* Study of Excitons and Exciton-Phonon Interactions by Resonant Raman and Brillouin Spectroscopies.

Photoemission in Solids II

Case Studies
Editors: L. Ley, M. Cardona
1979. 214 figures, 26 tables. XVIII, 401 pages
(Topics in Applied Physics, Volume 27)
ISBN 3-540-09202-1

Contents:
L. Ley, M. Cardonna: Introduction. – *L. Ley, M. Cardona, R. A. Pollack:* Photoemission in Semiconductors. – *S. Hüfner:* Unfilled Inner Shells: Transition Metals and Compounds. – *M. Campagna, G. K. Wertheim, Y. Baer:* Unfilled Inner Shells: Rare Earths and Their Compounds. – *W. D. Grobman, E. E. Koch:* Photoemission from Organic Molecular Crystals. – *C. Kunz:* Synchrotron Radiation: Overview. – *P. Steiner, H. Höchst, S. Hüfner:* Simple Metals. – Appendix. Table of Corelevel Binding Energies. – Additional References with Titles. – Subject Index.

H. Raether

Excitation of Plasmons and Interband Transitions by Electrons

1980. 121 figures, 17 tables. VIII, 196 pages
(Springer Tracts in Modern Physics, Volume 88)
ISBN 3-540-09677-9

Contents:
Introduction. – Volume Plasmons. – Tie Dielectric Function and the Loss Function of Bound Electrons. – Excitation of Volume Plasmons. – The Energy Loss Spectrum of Electrons and the Loss Function. – Experimental Results. – The Loss Width. – The Wave Vector Dependency of the Energy of the Volume Plasmon. – Core Excitation Application to Microanalysis. – Energy Losses by Excitation of Cerenkov Radiation and Guided Light Modes. – Surface Excitations. – Different Electron Energy Loss Spectrometers. – Notes Added in Proof. – References. – Subject Index.

Springer-Verlag
Berlin
Heidelberg
New York

Theory of Chemisorption
Editor: J. R. Smith
1980. 116 figures, 8 tables. XI, 240 pages
(Topics in Current Physics, Volume 19)
ISBN 3-540-09891-7

Contents:
J. R. Smith: Introduction. – *S. C. Ying:* Density Functional Theory of Chemisorption of Simple Metals. – *J. A. Appelbaum, D. R. Hamann:* Chemisorption on Semiconductors Surfaces. – *F. J. Arlinghaus, J. G. Gay, J. R. Smith:* Chemisorption and d-Band Metals. – *B. Kunz:* Cluster Chemisorption. – *T. Wolfram, S. Ellialtioğlu:* Concepts of Surface States and Chemisorption on d-Band Perovskites. – *T. L. Einstein, J. A. Hertz, J. R. Schrieffer:* Theoretical Issues in Chemisorption.

Thermally Stimulated Relaxation in Solids
Editor: P. Bräunlich
1979. 142 figures, 1 tables. XII, 331 pages
(Topics in Applied Physics, Volume 37)
ISBN 3-540-09595-0

Contents:
P. Bräunlich: Introduction and Basic Principles. – *P. Bräunlich, P. Kelly, J.-P. Fillard:* Thermally Stimulated Luminescence and Conductivity. – *D. V. Lang:* Space-Charge Spectroscopy in Semiconductors. – *J. Vanderschueren, J. Gasiot:* Field-Induced Thermally Stimulated Currents. – *H. Glaefeke:* Exoemission. – *L. A. LeWerd:* Applications of Thermally Stimulated Luminescence.

Very Large Scale Integration (VLSI)
Fundamentals and Applications
Editor: D. F. Barbe
1980. 130 figures, 37 tables. XI, 279 pages
(Springer Series in Electrophysics, Volume 5)
ISBN 3-540-10154-3

Contents:
D. F. Barbe: Introduction. – *J. L. Prince:* VLSI Device Fundamentals. – *R. K. Watts:* Advanced Lithography. – *P. Losleben:* Computer Aided Design for VLSI. – *R. C. Eden, B. M. Welch:* GaAs Digital Integrated Circuits for Ultra High Speed LST/VLSI. – *E. E. Swartzlander:* VLSI Architecture. – *B. H. Wahlen:* VLSI Applications and Testing. – *R. I. Seace:* VLSI in Other Countries. – Subject Index.

X-Ray Optics
Applications to Solids
Editor: H.-J. Queisser
1977. 133 figures, 14 tables. XI, 227 pages
(Topics in Applied Physics, Volume 22)
ISBN 3-540-08462-2

Contents:
H.-J. Queisser: Introduction: Structure and Structuring of Solids. – *M. Yoshimatsu, S. Kozaki:* High Brillance X-Ray Sources. – *E. Spiller, R. Feder:* X-Ray Lithography. – *U. Bonse, W. Graeff:* X-Ray and Neutron Interferometry. – *A. Authier:* Section Topography. – *W. Hartmann:* Live Topography.

Springer-Verlag
Berlin
Heidelberg
New York